Spatial Impacts of Climate Change

SCIENCES

Geography and Demography, Field Director – Denise Pumain

Physical Geography, Construction of Environments and Landscapes,
Subject Head – Étienne Cossart

Spatial Impacts of Climate Change

Coordinated by
Denis Mercier

iSTE

WILEY

First published 2021 in Great Britain and the United States by ISTE Ltd and John Wiley & Sons, Inc.

ISTE Ltd
27-37 St George's Road
London SW19 4EU
UK

www.iste.co.uk

John Wiley & Sons, Inc.
111 River Street
Hoboken, NJ 07030
USA

www.wiley.com

Library of Congress Control Number: 2020947184

British Library Cataloguing-in-Publication Data
A CIP record for this book is available from the British Library
ISBN 978-1-78945-009-5

ERC code:
LS8 Ecology, Evolution and Environmental Biology
LS8_1 Ecosystem and community ecology, macroecology
LS8_2 Biodiversity, conservation biology, conservation genetics

LS9 Applied Life Sciences, Biotechnology, and Molecular and Biosystems Engineering
LS9_4 Applied plant sciences (including crop production, plant breeding, agroecology, forestry, soil biology)

MIX
Paper from responsible sources
FSC® C013604

Contents

Introduction

Spatial Impacts of Climate Change: Multi-scale Issues

Denis MERCIER
Sorbonne University, Paris, France

Climate change involves a change in the elements of what is known as the climate machine, often summarized in the press as a change in the temperature variable alone. However, the Earth's climate is a complex system inducing interactions between its different components: atmosphere, hydrosphere, cryosphere, pedosphere, biosphere, lithosphere and noosphere.

I.1. The impact of contemporary climate change on forest fires in Australia in 2019–2020: a systemic approach

There is no shortage of examples in the recent past to illustrate the systemic interactions related to climate change in recent decades. Gigantic fires in Australia had destroyed more than 10 million hectares (ha) in the southeast of the country by early January 2020. This is more than five times the area of the California fires of 2018 (1.8 million ha) and ten times the area of the high-profile fires in the Amazon rainforest in the summer of 2019 (0.9 million ha, see Chapter 12). Global warming increases the likelihood that fires may occur by accentuating the expansion potential of these fires, which are devastating for fauna, flora and sometimes also for human life. Indeed, the fire seasons in Australia now begin earlier in September and last longer. Climate change in Australia is measured by a reduction in annual precipitation, which contributes to the drying out of soils, vegetation and thus the potential for fires. The year 2019 was the driest year on record since 1900 according

Spatial Impacts of Climate Change,
coordinated by Denis MERCIER.

to the Australian Bureau of Meteorology[1]. At the same time, temperatures have been rising, and 2019 was also the warmest year on record in Australia since 1910. These same climate changes can be understood on another spatial scale by combining ocean and atmospheric circulation. In the Indian Ocean, the dipole reflects the temperature differential between the western and eastern parts of this ocean basin. During the positive phases of this dipole, ocean temperatures are higher in the west than in the east, favoring rainfall over East Africa and reducing rainfall over Australia and Indonesia (Kämpf *et al.* 2019). These positive phases have become more frequent and intense since the 1950s. When the Antarctic Oscillation enters a negative phase, the westerly winds surrounding the Antarctic continent, the so-called Roaring Forties and Howling Fifties, move away from the continent and move up towards mid-latitudes, generating strong winds towards land areas such as Australia (Feng *et al.* 2019). When these two phenomena occur in time, dry air masses over the eastern Indian Ocean are propelled towards Australia, which then receives less precipitation and experiences strong winds. These air masses then cross Australia's interior deserts and fall back down over the eastern part of the Australian Cordillera, which reinforces the local drying of these air masses over the southeastern part of the country by the foehn effect. In addition, these Australian fires generate dust in the atmospheric circulation and result in dust fallout on New Zealand glaciers, particularly those of Fox and Franz Josef, whose melting is likely to be accelerated by the lowering of their surface albedo, which is brownish in color due to ash and soot fallout. At its scale, this local melting of the cryosphere contributes to a global rise in sea level.

Beyond the Australian case and on a more global scale, recent research shows that a warmer planet increases the risk of forest fires (Johns *et al.* 2020). Increasing temperatures, decreased precipitation and soil moisture, combined with drying strong winds, contribute to an increase in the frequency and severity of fire-prone periods.

I.2. The impacts of contemporary climate change: a multi-scalar approach

The issues are therefore systemic and multi-scale. Societies, whatever they are and wherever they are found, must and will have to adapt to the following changes for which they are not necessarily individually responsible. This book therefore presents the different impacts of climate change according to the areas and territories under consideration.

1 http://www.bom.gov.au/.

On a global scale, all the fundamental elements of the Earth system and their interactions are mobilized by climate change: water cycle, carbon cycle, atmospheric circulation, thermohaline circulation. Although general mechanisms make it possible to understand and explain climate change, regional and local nuances show that geographical elements such as the distribution of land and oceans, the layout of landforms, coastal dynamics and human activities can minimize or, on the contrary, exacerbate the spatial consequences of general physical laws. The example of contemporary climate warming on a global scale and its amplification in the Arctic illustrate the importance of these changes in spatial scales (see Chapter 1). The melting of the marine and terrestrial cryosphere, discussed in Chapter 2, is not spatially uniform. It contributes to changing geopolitical and economic issues in the Arctic (see Chapter 3). The melting of the terrestrial cryosphere induces a rise in global sea and ocean levels, which will not affect coastlines in the same way according to their own typology (cliff coasts, deltas, etc.) or their own dynamics (subsidence, stability or uplift) and according to the way they are occupied by societies (see Chapter 4).

At the regional scale, the spatial impacts of contemporary climate change are being addressed using a variety of approaches. By modifying the cryosphere in high latitudes and high mountains, changes in climate induce changes in the paraglacial sedimentary cascade (see Chapter 5) and periglacial environments (see Chapter 6). River organisms in cold environments (see Chapter 7) and temperate environments (see Chapter 8) record climate change in different ways. The melting glaciers of Central Asia place the consequences of climate change at the heart of geopolitical issues in this region (see Chapter 9).

At the local level, the impact of rainier season droughts in the western Mediterranean basin on Spanish rain-fed agriculture provides a link between regional climate dynamics and local impacts (see Chapter 10).

Multi-scalar approaches also make it possible to show the stakes of contemporary climate change on viticulture (see Chapter 11), on the scale of the Amazon basin (see Chapter 12), on the distribution of biomes (see Chapter 13) or on the distribution of birds (see Chapter 14).

In all the chapters, the examples analyzed underline the importance of geographical approaches for the study of the impacts of contemporary climate change.

I.3. References

Feng, J., Zhang, Y., Cheng, Q., San Liang, X., Jiang, T. (2019). Analysis of summer Antarctic sea ice anomalies associated with the spring Indian Ocean dipole. *Global and Planetary Change*, 181 [Online]. Available at: https://doi.org/10.1016/j.gloplacha.2019.102982.

Jones, M.W., Smith, A., Bettes, R., Canadell, J.G., Prentice, C., Le Quéré, C. (2020). Climate change increases the risk of wildfires. *ScienceBrief* [Online]. Available at: https://sciencebrief.org/topics/climate-change-science/wildfires.

Kämpf, J. and Kavi, A. (2019). SST variability in the eastern intertopical Indian Ocean – On the search for trigger mechanisms of IOD events. *Deep Sea Research Part II: Tropical Studies in Oceanography*, 166, 64–74.

1

Climate Change at Different Temporal and Spatial Scales

Denis MERCIER
Sorbonne University, Paris, France

1.1. Contemporary global climate change

Contemporary climate change refers to the period from 1850 to the present day and covers the period from the Industrial Revolution to the digital revolution. It also covers a period during which humanity experienced a population explosion, reaching 1 billion people for the first time in 1820. On January 1, 2020, the human population was estimated at 7.7 billion and is expected to reach 11 billion by 2100, according to the UN.

Through the use of fossil fuels (coal, oil, gas) and increased agricultural production to feed the world's growing population, these elements contribute to increasing humanity's role in the climate machine.

Since the mid-19th Century, the average global air temperature has increased by 1.1°C. This increase has not been linear over time and Figure 1.1 illustrates the stages of this evolution. Two warming sequences help to understand this increase: the first from 1910 to 1940 and the second from 1980 to the present day, during which the increase in temperature was 0.18°C per decade. According to the World

Meteorological Organization[1], the year 2019 was the second warmest year recorded since 1850. It comes after the year 2016, which experienced a particularly intense El Niño episode, with abnormally high ocean surface water temperatures in the eastern South Pacific. These two periods of warming are interspersed by temporal sequences of cooling (from 1880 to 1910, then from 1940 to 1980).

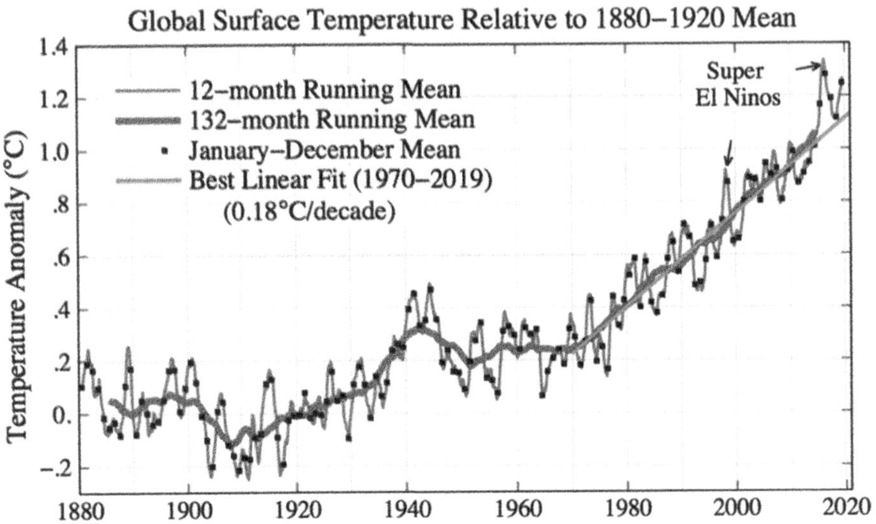

Figure 1.1. *Annual mean surface temperature from 1880 to 2019 compared to the 1880–1920 mean (source: Sato and Hansen, Climate Science, Awareness and Solutions at Columbia University Earth Institute, 2020). For a color version of this figure, see www.iste.co.uk/mercier/climate.zip*

This non-linear temperature evolution over time is not spatially homogeneous (see Figure 1.2). These maps illustrate general trends. Continental land areas record this contemporary global warming better than ocean surfaces; of these continental land surfaces, those with a hypercontinental climate such as Siberia are experiencing the greatest temperature increases.

Although the map projection is not very favorable, Figure 1.2 shows that high latitude regions, especially the Arctic basin and its surroundings, have experienced the greatest increases in temperature.

1 https://public.wmo.int/fr.

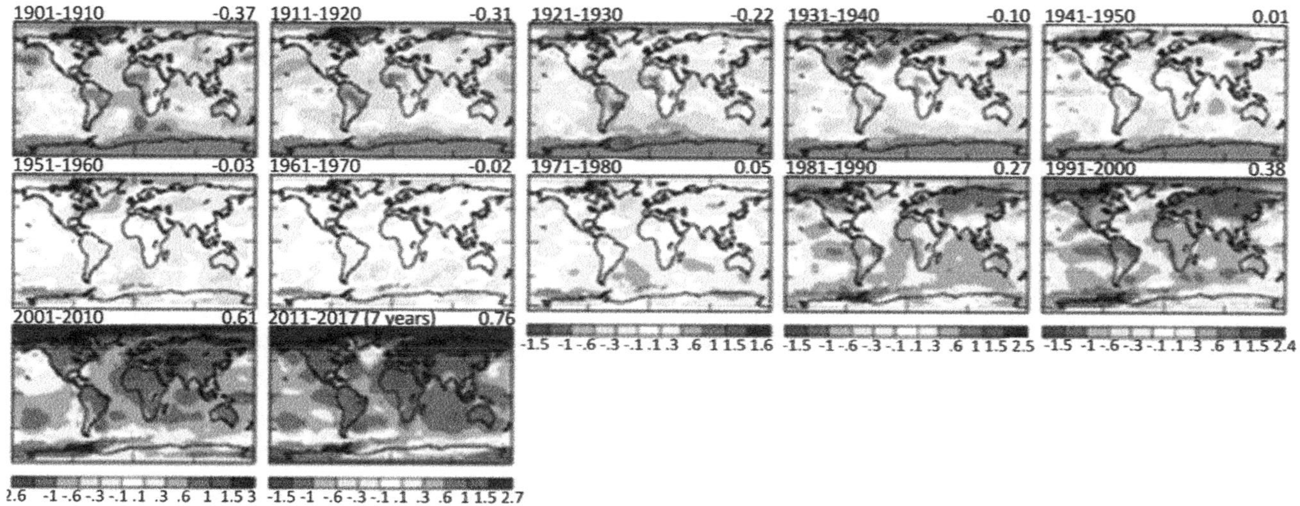

Figure 1.2. *Average surface temperature per decade from 1910 to 2017 compared to the 1951–1980 average (source: 2018 NASA-GISS temperature data, downloaded from https://data.giss.nasa.gov/gistemp/). For a color version of this figure, see www.iste.co.uk/mercier/climate.zip*

Although the oceans are warming less than land areas, they are still warming and store 93% of the excess heat. The last 10 years are the warmest recorded for ocean surface waters since 1955 with a linearly increasing temperature trend since the 1980s (see Figure 1.3) (Cheng *et al.* 2020). For the first period, the warming was relatively constant of approximately 2.1 ± 0.5 Zetta Joules[2] per year. However, the warming in the more recent period is greater than that of the previous warming (9.4 ± 0.2 Zetta Joules per year, or 0.58 watt per m^2 on average on the Earth's surface), hence the significant increase in the rate of global climate change at the ocean scale (Cheng *et al.* 2020).

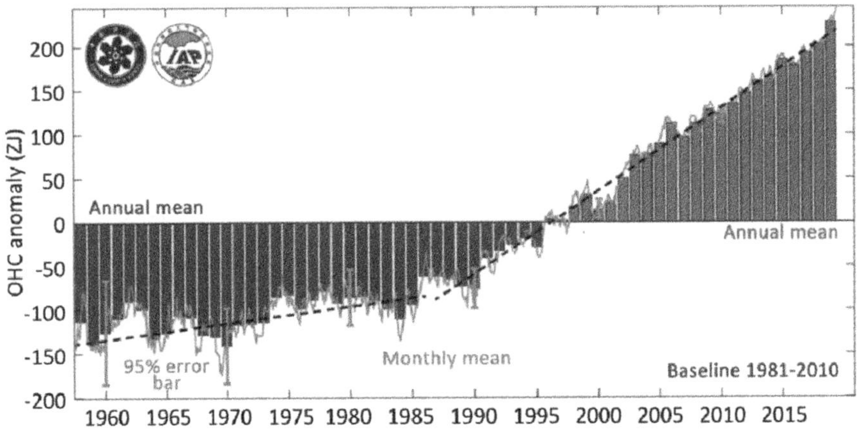

Figure 1.3. *Ocean heat content (OHC) in the upper water section above 2,000 m from 1955 to 2019. For a color version of this figure, see www.iste.co.uk/mercier/climate.zip*

COMMENT ON FIGURE 1.3.– *The histogram represents annual anomalies (ZJ: Zetta Joules, where 1 ZJ = 10^{21} Joules) where positive anomalies relative to a mean calculated between 1981 and 2010 are shown as red bars and negative anomalies are shown in blue. The two dashed black lines represent linear trends for the periods 1955–1986 and 1987–2019 (source: Cheng* et al. *2020).*

The increase in ocean surface temperatures affects all oceans. Although some ocean areas, such as the North Atlantic, experienced a decrease in temperature between 1960 and 2019 (Cheng *et al.* 2020), the penetration of heat into the deep ocean is clear in Figure 1.4, mainly in the Atlantic and Southern Oceans (Cheng

2 Zetta: one trilliard (10^{21}), or one thousand trillion, according to the international system of units.

et al. 2020). These two ocean basins, especially near the Antarctic Circumpolar Current (40° 60° S) show greater warming than most other basins (Cheng *et al.* 2020).

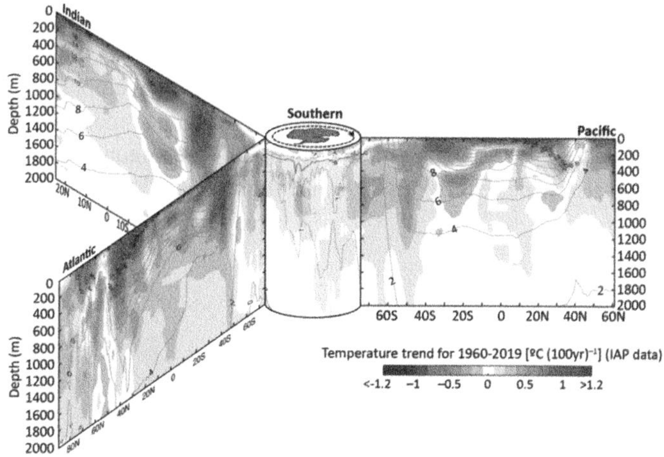

Figure 1.4. *Vertical cross-section of ocean temperature trends from 1960 to 2019 from the sea surface to 2,000 m (60-year ordinary least squares linear trend). For a color version of this figure, see www.iste.co.uk/mercier/climate.zip*

COMMENT ON FIGURE 1.4.– *The zonal mid-sections of each ocean basin are organized around the Southern Ocean (south of 60° S) in the center. The black outlines show the associated mean temperature with 2°C intervals (in the Southern Ocean, 1°C intervals are shown as dashed lines) (source: Cheng* et al. *2020).*

This increase in global ocean surface temperatures leads, through thermal expansion, to a rise in sea level, as an increase in air temperature contributes to the melting of the Earth's cryosphere and thus to the increase in the amount of water in the global ocean (see Chapter 2 on melting of the cryosphere). Similarly, rising ocean temperatures reduce dissolved oxygen in the ocean and significantly affect marine life, especially corals and other organisms sensitive to temperature and water chemistry (IPCC 2019; see Chapter 4 on coasts). Increasing ocean surface water temperature promotes evaporation over the oceans and moisture in the atmosphere, which logically can promote heavy rainfall, and can be associated with more frequent and/or more intense cyclones, and can, depending on the case, lead to flooding (IPCC 2019). The consequence of this change in ocean temperatures is prolonged contemporary warming simply because of the thermal inertia of these gigantic ocean masses.

1.2. Contemporary Arctic-wide climate change

Global warming is not always visible to some, but it is most easily illustrated in the Arctic, particularly with the melting of the cryosphere (see Chapter 2). Indeed, the high latitudes of the boreal regions have recorded an increase in temperature of around 2.5°C since the beginning of the 20th Century, with temperature sequences comparable to those recorded on a global scale. Globally, this temperature increase is mainly due to the last few decades. Seasonally, the winter months (November to April) recorded the greatest temperature increases (see Figure 1.5), although the warmer months also experienced higher temperatures.

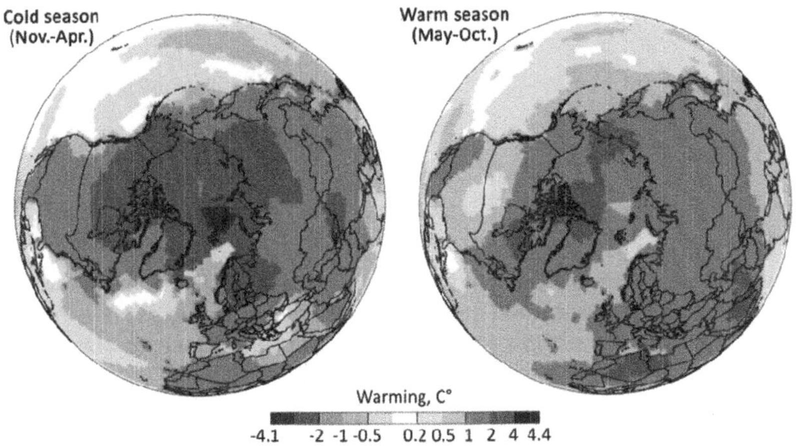

Figure 1.5. *Spatial distribution of Arctic warming for the period 1961–2014 for the cold season (November to April) and the warm season (May to October) (source: AMAP 2017). For a color version of this figure, see www.iste.co.uk/mercier/climate.zip*

In the Arctic Basin, the Svalbard Archipelago is located in the area with the greatest warming. The curves in Figure 1.6, showing the evolution of temperature since the end of the 19th Century, illustrate both this climate warming on all time scales (annual and seasonal, especially winter) and the increase in annual precipitation. The average temperature in Longyearbyen, (Svalbard archipelago) has increased by 4 to 5°C since the beginning of the 20th Century. Like all the meteorological stations of this archipelago, Longyearbyen, being in a coastal position, is all the more sensitive to the spatial retraction of the winter sea ice in recent years, which explains the more significant increase in winter temperatures in recent decades in particular. Temperature trends are not linear, and cycles of different lengths and amplitudes have been obtained by statistical analyses (Fourier and wavelet, see Humlum *et al.*

2011). The similarities between the thermal evolutions at the Longyearbyen station and the North Atlantic Multidecadal Oscillations (AMO) underline the importance of the influence of ocean temperatures on that of the lower layers of the atmosphere (Humlum *et al.* 2011).

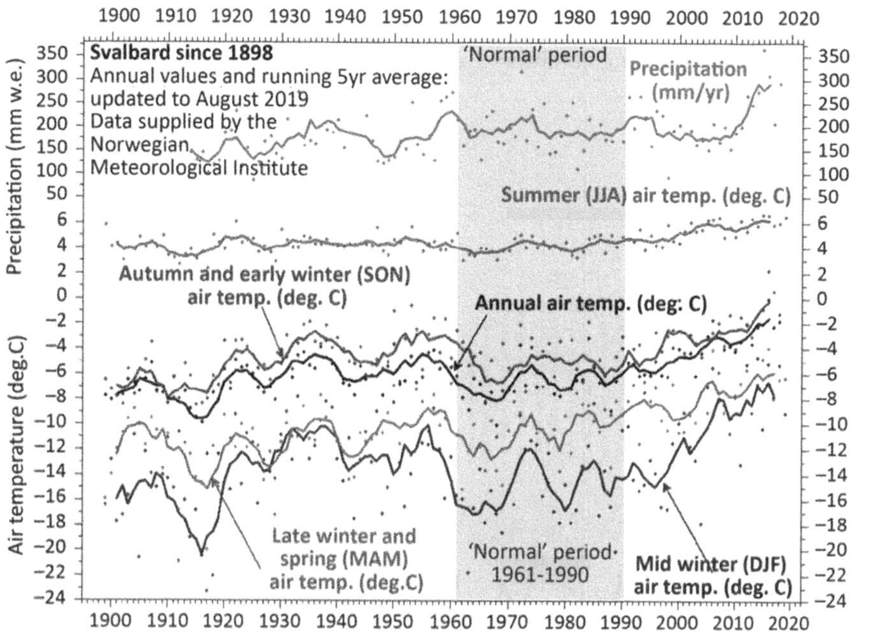

Figure 1.6. *Temporal distribution of air temperature warming at Longyearbyen, 78° 25' N, 15° 47' E, capital of the Svalbard archipelago, for the period 1898–2019 at different time scales, annual in black, summer (June, July, August) in red, autumn (September, October, November) in purple, winter (December, January, February) in blue, spring (March, April, May) in green. For a color version of this figure, see www.iste.co.uk/mercier/climate.zip*

COMMENT ON FIGURE 1.6.– *The baseline average is calculated for the period 1961–1990. Change in mean annual precipitation with a five-year sliding average (in pink) (source: based on data from the Norwegian Meteorological Institute*[3]*).*

For the Ny-Ålesund station, located on the northwestern coast of the Svalbard archipelago (78° 55' N, 11° 55' E), Figure 1.7 shows an increase in mean annual temperatures of 4°C from 1969 to 2016 and the increase in predominantly rainy

3 http://www.climate4you.com/SvalbardTemperatureSince1912.htm.

precipitation from 356 mm per year in 1969 to 546 mm per year in 2016. The increase in temperatures largely explains the increase in precipitation due to an increase in the hygrometric capacity of the air, by the increase in the frequency of oceanic disturbances caused by the North Atlantic drift; in winter, in connection with periods when the North Atlantic Oscillation (NAO) index is positive, and in relation to the reduction of the ice pack in the Arctic basin, which allows open sea water to release heat into the lower layers of the atmosphere (see Chapter 2).

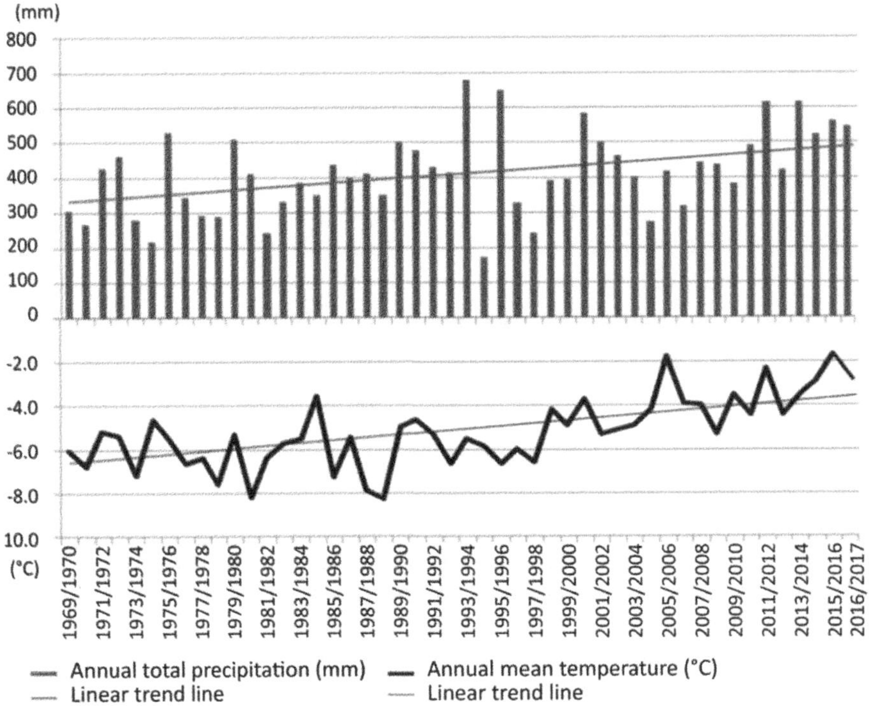

Figure 1.7. *Annual mean precipitation and annual mean temperatures from 1969 to 2016 at the Ny-Ålesund weather station (northwestern Spitsbergen, Svalbard) (source: Bourriquen* et al. *2018, based on data from the Norwegian Meteorological Institute). For a color version of this figure, see www.iste.co.uk/mercier/climate.zip*

Thus, whatever the spatial scales used here, contemporary climate change is illustrated by an increase in temperatures and an associated increase in precipitation.

1.3. Future global climate change

At the current rate of atmospheric warming (0.2°C per decade), a warming of 1.5°C is expected by 2050 (IPCC 2014). Under the worst-case scenario and an increase of 8.5 watts per square meter, the temperature could rise by 4°C. The rise in sea level would be between 50 cm and 1 m by 2100 according to the latest IPCC report on cryosphere and oceans (IPCC 2019 and Chapters 2 and 3).

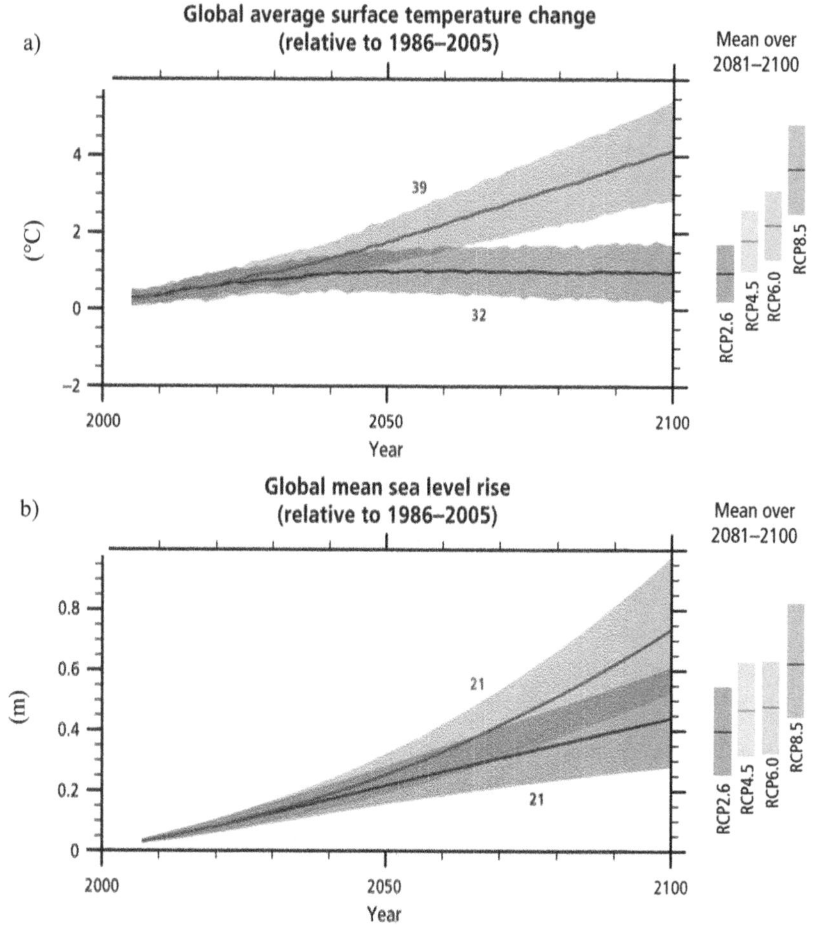

Figure 1.8. *(a) Mean change in surface temperature. (b) Mean sea-level rise from 2006 to 2100 (as determined by multi-model simulations). For a color version of this figure, see www.iste.co.uk/mercier/climate.zip*

COMMENT ON FIGURE 1.8.– *All changes relate to 1986–2005. Time series of projections and a measure of uncertainty (shading) are presented for the RCP2.6 (blue) and RCP8.5 (red) scenarios*[4]. *The average and associated uncertainties averaged over 2081–2100 are given for all RCP scenarios as colored vertical bars on the right side of each panel. The number of models from Phase 5 of the Coupled Model Intercomparison Project (CMIP5) used to calculate the multi-model mean is given (source: IPCC 2014).*

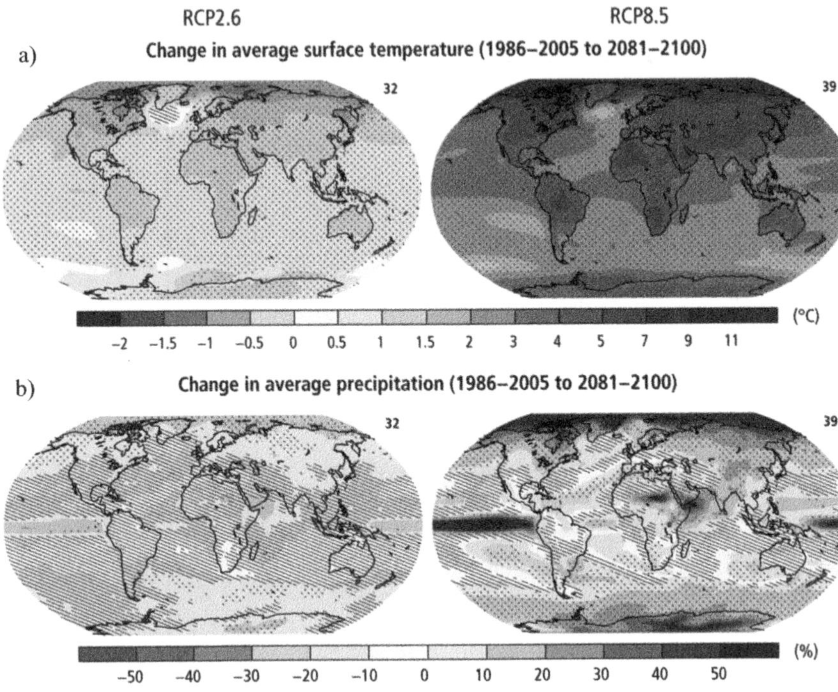

Figure 1.9. *(a) Change in mean surface temperature. (b) Change in mean precipitation based on the multi-model mean projections for 2081–2100 compared to 1986–2005 in the RCP2.6 (left) and RCP8.5 (right) scenarios. For a color version of this figure, see www.iste.co.uk/mercier/climate.zip*

COMMENT ON FIGURE 1.9.– *The number of models used to calculate the multi-model average is shown in the upper right corner of each panel. The dotted lines show*

4 Representative Concentration Pathway (RCP), scenario expressed in watts per square meter.

regions where the projected change is large relative to the natural internal variability, and where at least 90% of the models agree on the sign of the change. Hatching shows regions where the projected change is less than one standard deviation of the natural internal variability (source: IPCC 2014).

Maps of climate change projected to the end of the 21st Century show that the continents and the Arctic basin are most affected by this warming under any scenario with an increase of 2.6 watts per square meter (RCP2.6), accentuated by an increase of 8.5 watts per square meter (RCP8.5). For the amount of average annual precipitation considered by 2100, the maps show that the cold regions of the Arctic and Antarctic should logically receive more water in relation to a higher moisture content of the air in these cold regions associated with the increase in the evaporation potential over the Arctic basin with less ice pack in summer.

On the other hand, regarding land masses, it is mainly the Mediterranean regions and certain regions of South America and South Africa that would experience a decrease in rainfall, which would have a significant impact on agricultural yields (see Chapter 10).

1.4. Future Arctic-wide climate change

The polar projections in Figures 1.10 and 1.11 provide a better understanding of the magnitude of the projected warming in the Arctic Basin by 2050 and 2100 for summer (June to August) and winter (December to February), regardless of the scenario selected, RCP4.5 or RCP8.5. For the summer months, the continental regions surrounding the Arctic Basin would warm up the most.

On the other hand, the magnitude of warming would be exacerbated over the marine areas of the Arctic Basin during the winter months. Indeed, as the melting of the summer ice pack increased in summer during the 21st Century, the thermal inertia of the oceans, which accumulate more heat in summer through solar radiation, also makes it possible to halt the spatial expansion of sea ice in the winter period, even though winter temperatures continue to be favorable to its formation.

This amplifying role of the oceans shows what is at stake in the interactions between the atmosphere and the hydrosphere, with or without the central role of the sea ice filter.

For example, the average winter temperature in the Longyearbyen region of Svalbard at the end of the 21st Century is expected to be about 10°C higher than the current climate (Førland *et al.* 2012).

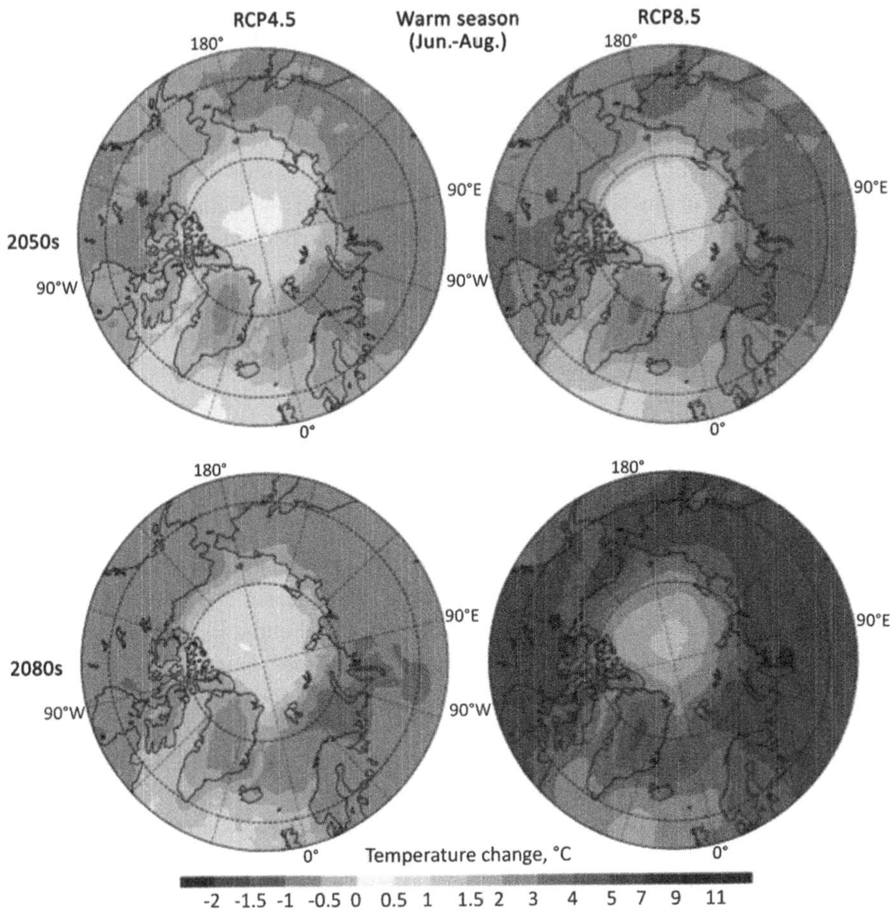

Figure 1.10. *Projected changes in summer surface temperatures (June to August) compared to the 1986–2005 average under scenario 4.5 and 8.5, for the years 2050 and 2080 (source: AMAP 2017). For a color version of this figure, see www.iste.co.uk/mercier/climate.zip*

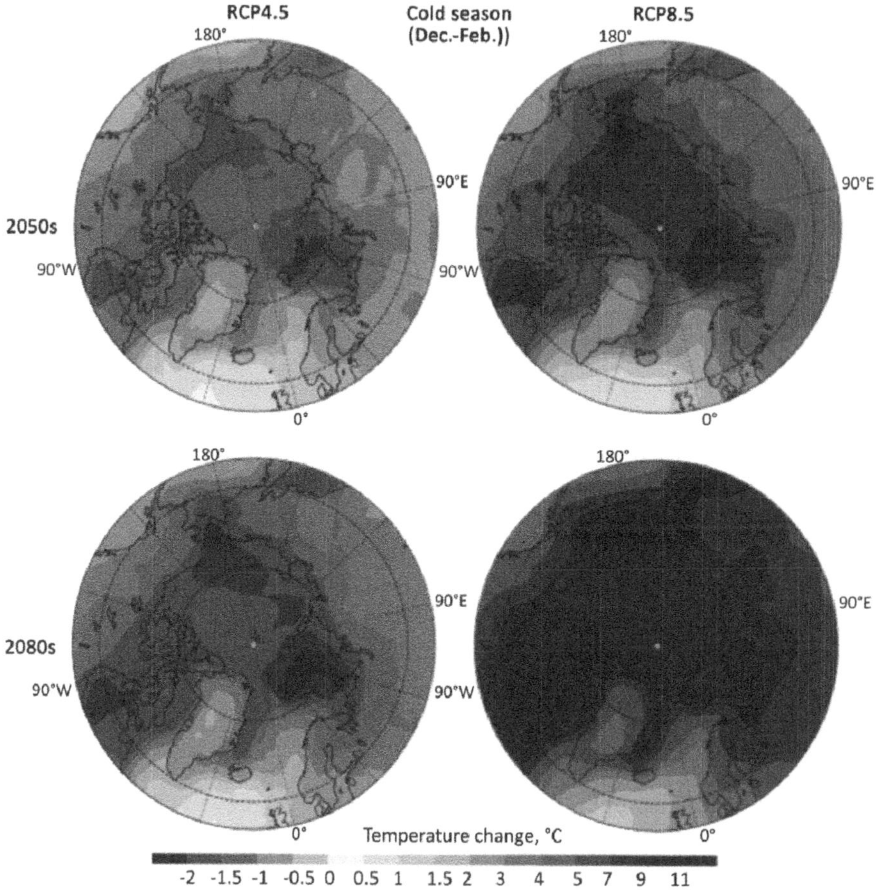

Figure 1.11. *Projected changes in winter (December to February) surface temperatures relative to the 1986–2005 average under scenarios 4.5 and 8.5, for the years 2050 and 2080 (source: AMAP 2017). For a color version of this figure, see www.iste.co.uk/mercier/climate.zip*

1.5. The causes of climate change

1.5.1. *Solar radiation*

Depending on the time scales considered, orbital variations concerning the Earth and the sun play a fundamental role in understanding climate change. Thus, the astronomical cycles (eccentricity, obliquity and precession) highlighted by Milutin

Milanković allow us to understand the large climatic oscillations between long cold sequences separated by shorter interglacial periods. The change in global average surface temperature between the last glacial maximum (21,000 years ago) and the pre-industrial climate is estimated to be about 5°C (between 3 and 8°C). It would have been a few degrees in the tropics and 10°C at the poles (Masson-Delmotte *et al.* 2015). During the Eemian interglacial period (128,000 to 116,000 years ago), estimated temperatures were 3 to 5°C higher than in the pre-industrial period, resulting in the melting of a significant fraction of the cryosphere and a sea level rise of about 6.6 to 9.4 m above the present level (Lageat 2019).

On another time scale, it has been shown that solar radiation varies in intensity with minimums, such as the Maunder minimum in the 17th Century or the Dalton minimum from 1800 to 1830. The latter corresponds to the coldest period of the Little Ice Age. At the century scale, a linear relationship between the air temperature series at the Longyearbyen station in Svalbard and the length of a solar cycle has been demonstrated (Solheim *et al.* 2011). Thus, the contemporary evolution of temperatures, which records sequences of warming (1910–1940 and since 1980), separated by sequences of stagnation or even slight cooling (from 1880 to 1910 and then from 1940 to 1980), could be partly explained by these variations in solar radiation intensity.

1.5.2. *Anthropogenic greenhouse gas emissions*

Since the beginning of the industrial revolution, the quantity of greenhouse gases (GHGs) injected into the atmosphere by human activities (CO_2, CH_4, etc.) has been considerable (see Table 1.1).

These concentrations represent a major disruption in the evolution of the climate compared to pre-industrial natural evolution. These anthropogenic gases reinforce the greenhouse effect and prevent infrared radiation from leaving the lower layers of the atmosphere, which leads to the warming of the air, surface ocean layers and soils. The increase in GHGs is mainly linked to the use of fossil fuels (coal, gas and oil). However, agriculture also contributes to this increase, in particular through deforestation, which is partly responsible for the emission of carbon dioxide (CO_2), rice growing and the breeding of ruminants that release methane (CH_4), and pig farming and the spreading of manure as fertilizer, which is responsible for the increase in nitrous oxides (N_2O). These changes are intimately linked to changes in consumption patterns and to the evolution of the world population explosion since the 19th Century, unprecedented in the history of mankind.

	CO_2	CH_4	N_2O
Average overall abundance in 2018	407.8 ± 0.1 ppm	1,869 ± 2 ppb	331.1 ± 0.1 ppb
Average overall abundance in 1750	278 ppm	722 ppb	270 ppb
Relative abundance in 2018 compared to 1750	+ 147%	+ 259%	+ 123%
Absolute increase between 2017 and 2018	2.3 ppm	10 ppb	1.2 ppb
Relative increase between 2017 and 2018	+ 0.57%	+ 0.54%	+ 0.36%
Annual average of absolute growth over the last 10 years	+ 2.26 ppm per year	+ 7.1 ppb per year	+ 0.95 ppb per year

Table 1.1. *Global annual average surface area abundances and trends of the main greenhouse gases of the Global Atmosphere Watch (GAW) of the World Meteorological Organization (WMO) Global Atmosphere Watch (GAW) global greenhouse gas (GHG) monitoring network*

COMMENT ON TABLE 1.1.– *Units are molecular fractions of dry air and uncertainties are 68% confidence limits. A number of stations are used for the analyses: 129 for CO_2, 127 for CH_4 and 96 for N_2O (source: WMO 2019).*

The most optimistic scenarios of global warming are based on a decrease or stabilization of GHG emissions. However, if we look lucidly at the consumption trajectories of contemporary societies, the most pessimistic scenarios remain the most likely. In order to achieve a neutralization of CO_2 emissions, the necessary changes in energy consumption, transportation, industrial production, agricultural production linked to changes in food consumption, societal choices and therefore political choices are radical and therefore unlikely in the short term, even though many solutions exist.

1.5.3. *Volcanism*

Major volcanic eruptions can inject huge amounts of gases into the stratosphere, including carbon dioxide CO_2, sulfur dioxide SO_2 or hydrogen sulfides H_2S. Sulfur gases (SO_2 and H_2S) lead to the formation of liquid sulfate aerosols in the stratosphere with a lifetime of a few years. These aerosols scatter incoming solar radiation and absorb infrared radiation. They then lead to a clear reduction in the radiation reaching the earth's surface and thus induce cooling.

Major periods of volcanic activity in the Earth's geological past, such as the one that allowed the Deccan traps in India to form, have resulted in climatic changes that have caused major environmental crises, such as the most famous mass extinction of the late Cretaceous period, 65 million years ago, which contributed to the extinction of the dinosaurs. The multi-millennia history of societies also includes a number of volcanic eruptions that have modified the climate, with repercussions at different scalar levels: climatic, agricultural, sanitary, demographic and political (Eldgjá in 939–940, Samalas in 1257, Laki in 1783, Tambora in 1815, Krakatoa in 1883). Two major volcanic eruptions in 536 and 540, the names of the active volcanoes involved are still not known with certainty, are believed to have caused a 2°C drop in temperature in the northern hemisphere during the decade 536–545. A positive feedback loop of spatial extension of the Arctic ice pack would thus have amplified volcanic-induced atmospheric cooling by the combined increase in albedo and reduced ocean-atmosphere interactions (Toohey *et al.* 2016).

Reconstructions of summer temperatures for the Northern Hemisphere can be made for the last 1500 years based on tree ring widths and maximum wood density. For the Samalas eruptions of 1257, the summer cooling is estimated to have been –1.3°C for the extra-tropical regions of the Northern Hemisphere and –0.8°C for the Tambora eruption in 1815. These coolings continued 4–5 years after the Samalas eruption and 2–3 years after the Tambora eruption (Stoffel *et al.* 2015). Analyses based on glacial records from Greenland and Antarctica since 500 BC show that the 20th Century, even though it saw major eruptions such as the Bezymianny in Kamchatka in 1955, the Agung in Bali in 1963 or Mount St. Helens in the USA in 1980, did not experience as much volcanic forcing as in previous centuries (Toohey *et al.* 2017).

Thus, the thermal variations over the last millennium (medieval optimum and Little Ice Age) could be explained by a combination of natural climate forcings, linked in particular to variations in solar activity and volcanic activity (Khodri *et al.* 2015). On the other hand, simulations of future climate change for the 21st Century generally do not take into account likely major volcanic eruptions. One study, however, attempts to model how 60 eruptions could influence climate change by 2100 and concludes that, beyond annual or decadal variability, they are unlikely to be able to mitigate contemporary climate warming on this secular scale (Bethke *et al.* 2017).

1.5.4. *Albedo and the radiation balance*

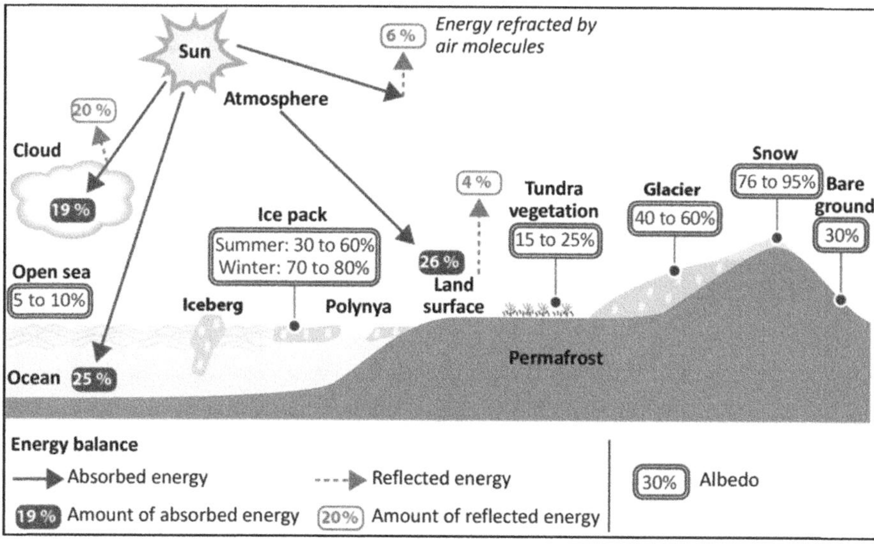

Figure 1.12. *Elements of the global energy balance and albedo of different surfaces in the Arctic (source: design D. Mercier, drawing by F. Bonnaud, Faculty of Arts, Sorbonne University, 2020). For a color version of this figure, see www.iste.co.uk/mercier/climate.zip*

Although albedo occurs everywhere on the planet's surface, its role in today's global warming is best illustrated in the Arctic basin. The ocean (sea ice) and land (glaciers, snow) surfaces of the Arctic cryosphere have a high albedo potential (see Figure 1.12). Increases in air and ocean temperatures are contributing to a decrease in the spatial extent of the cryosphere (see Chapter 2).

Thus, while changes in sea ice and snow cover can be considered as impacts of climate change, the role of these changes in the albedo-temperature feedback also makes them agents of change, through a positive feedback loop (see Figure 1.13). As temperature increases, surfaces with low albedo powers (ice-free ocean and land surface devoid of glaciers and snow) increase and absorb more solar radiation and become warmer.

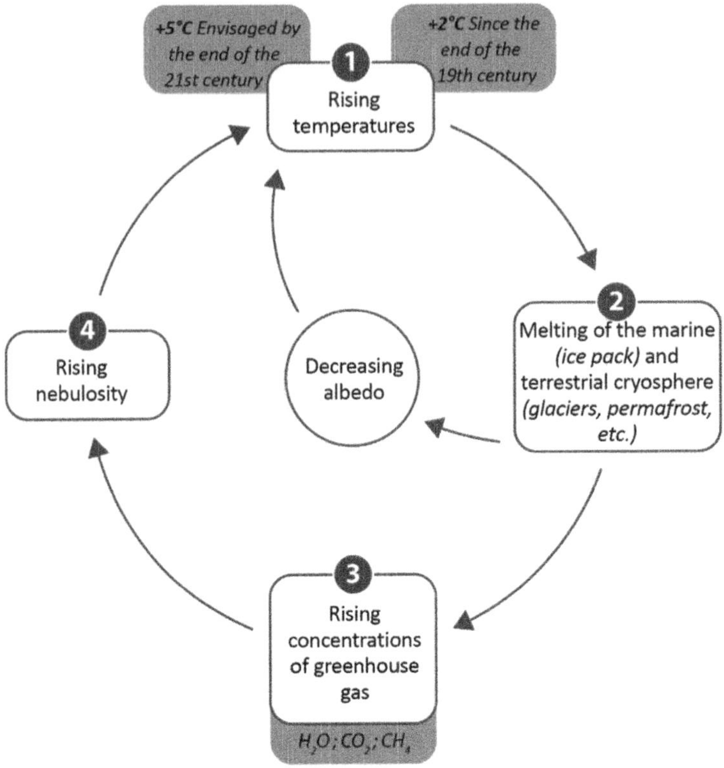

Figure 1.13. *Positive feedback loops explaining the amplification of Arctic climate warming (source: design D. Mercier, drawing by F. Bonnaud, Faculty of Arts, Sorbonne University, 2020). For a color version of this figure, see www.iste.co.uk/mercier/climate.zip*

In addition to albedo, Figure 1.12 illustrates the major components of the Earth's radiation budget, which is simplified by an average solar energy input of 342 watts per square meter to the Earth's surface. The percentages of each component (clouds, ocean, land surface, atmosphere) show that only 47% is absorbed (25% by the oceans and 22% by land surfaces). In the evolution of temperature in the lower layers of the atmosphere, the cloud component plays a fundamental role because it absorbs part of the energy (19%) and reflects 20%. Cloud cover and its temporal evolution therefore appear to be an essential element in understanding the evolution of the radiation balance on the Earth's surface.

1.6. Conclusion

Contemporary climate change is illustrated by recognized and measured thermal and rainfall trends that are neither linear over time nor spatially uniform. Beyond the general logics of the physical laws governing the climate machine on a global scale (radiation balance, importance of astronomical parameters, role of greenhouse gases and volcanism, thermal gradients, air humidity capacity, etc.), regional and local nuances illustrate the importance of geographical factors and interactions between all components, such as the fundamental Arctic-wide interaction between the ocean, sea ice and the atmosphere. Whether during the cold Pleistocene sequences or during contemporary global warming, we are seeing an amplification of changes in high-latitude environments, particularly in the Arctic Basin.

1.7. References

AMAP (2017). *Snow, Water, Ice and Permafrost in the Arctic (SWIPA).* Arctic Monitoring and Assessment Programme (AMAP), Oslo.

Bethke, I., Outten, S., Otterå, O.H., Hawkins, E., Wagner, S., Sigl, M., Thorne, P. (2017). Potential volcanic impacts on future climate variability. *Nature Climate Change*, 7(11), 799–805.

Bourriquen, M., Mercier, D., Baltzer, A., Fournier, J., Costa, S., Roussel, E. (2018). Paraglacial coasts responses to glacier retreat and associated shifts in river floodplains over decadal timescales (1966–2016), Kongsfjorden, Svalbard. *Land Degradation and Development*, 29(11), 4173–4185.

Cheng, L., Abraham, J., Zhu, J., Trenberth, K.E., Fasullo, J., Boyer, T., Locarnini, R., Zhang, B., Yu, F., Wan, L., Chen, X., Song, X., Liu, Y., Mann, M.E. (2020). Record-setting ocean warmth continued in 2019. *Advances in Atmospheric Sciences*, 37, 137–142.

Førland, E.J., Benestad, R., Hanssen-Bauer, I., Haugen, J.E., Skaugen, T.E. (2012). Temperature and precipitation development at Svalbard 1900–2100. *Advances in Meteorology*, 2011(17).

Hanssen-Bauer, I., Førland, E.J., Hisdal, H., Mayer, S., Sandø, A.B., Sorteberg, A. (2019). Climate in Svalbard 2100 – A knowledge base for climate adaptation. Report, 1/2019, NCCS.

Humlum, O., Solheim, J.-E., Stordahl, K. (2011). Spectral analysis of the Svalbard temperature record 1912–2010. *Advances in Meteorology*, 2011.

IPCC (2014). Climate Change 2014: Synthesis report. Contribution of working groups I, II and III to the fifth assessment report of the Intergovernmental Panel on Climate Change. Report, IPCC, Geneva.

IPCC (2019). Special report on the ocean and cryosphere in a changing climate. [Online]. Available at: https://www.ipcc.ch/srocc.

Khodri, M., Swingedouw, D., Mignot, J., Sicre, M.A., Garnier, E., Masson-Delmotte, V., Ribes, A., Terray, L. (2015). Le climat du dernier millénaire. *La Météorologie*, 88, 36–47.

Lageat, Y. (2019). *Les variations du niveau des mers*. Presses Universitaires de Bordeaux, Pessac.

Masson-Delmotte, V., Braconnot, P., Kageyama, M., Sepulchre, P. (2015). Qu'apprend-on des grands changements climatiques passés ? *La Météorologie*, 88, 25–35.

Solheim, J.-E., Stordahl, K., Humlum, O. (2011). Solar activity and Svalbard temperatures. *Advances in Meteorology*, 2012.

Stoffel, M., Khodri, M., Corona, C., Guillet, S., Poulain, V., Bekki, S., Guiot, J., Luckman, B.H., Oppenheimer, C., Lebas, N., Beniston, M., Masson-Delmotte, V. (2015). Estimates of volcanic-induced cooling in the Northern Hemisphere over the past 1,500 years. *Nature Geoscience Letters*, 8(10), 784–788.

Toohey, M. and Sigl, M. (2017). Volcanic stratospheric sulfur injections and aerosol optical depth from 500 BCE to 1900 CE. *Earth System Science Data*, 9, 809–831 [Online]. Available at: https://doi.org/10.5194/essd-9-809-2017.

Toohey, M., Krüger, K., Sigl, M., Stordal, F., Svensen, H. (2016). Climatic and societal impacts of a volcanic double event at the dawn of the Middle Ages. *Climatic Change*, 136, 401–412.

WMO (2019). The state of greenhouse gases in the atmosphere based on global observations through 2018. *Greenhouse Gas Bulletin*, 15.

2

Climate Change and the Melting Cryosphere

Denis MERCIER
Sorbonne University, Paris, France

2.1. Introduction

Contemporary climate change affects the cryosphere; the thermal changes at stake today are limited compared to the great climatic oscillations that affected the Earth, particularly during the past 2.58 million years of the Quaternary Period. Indeed, the areas concerned, and the volumes of ice are undeniably not of the same order of magnitude. During the great cold periods of the Pleistocene (2,580,000 to 11,700 years ago), the terrestrial cryosphere capitalized on the planet's land spaces led to an eustatic decrease of around 120 to 130 m in the global ocean. In retrospect, during previous interglacial periods such as the Eemian (128,000 to 116,000 years ago), the temperature was around 3.5°C higher than today's, which led to a significant melting of the terrestrial cryosphere and a rise in the average sea and ocean level of between 6.6 and 9.4 m above the current level (Lageat 2019). We are now experiencing a few decimeters per century in sea level rise, as a result of the partial melting of what is currently left of the Earth's cryosphere. However, over the next several centuries, the continued melting of the cryosphere could bring the average sea and ocean level back to the Eemian level average, due in part to the melting of Greenland's ice. However, the consequences of this current melting of the cryosphere due to warming air temperatures affect all components of the climatic and hydrological mechanics.

Spatial Impacts of Climate Change,

2.2. The sensitivity of the cryosphere to climate change

The cryosphere is defined as the cold sphere (from the Greek *kruos*, the cold), and occurs on Earth in various forms from water to solid ice. These include the two ice sheets in Greenland and Antarctica, icecap glaciers such as Vatnajökull in Iceland, valley and cirque glaciers in mountains at almost all latitudes, permafrost, snow, lake and river ice, and sea ice and icebergs (see Figure 2.1). The cryosphere is thus seasonal (snow) or multi-millennia (ice sheets, permafrost).

According to sources, the cryosphere on Earth occupies more than 36 million km^2, or 24% of the continents. Depending on the season, the marine cryosphere covers between 19 and 28 million km^2, or 5.3 to 7.8% of the ocean surface. The continental cryosphere mainly represents 30.2 million km^3 of freshwater, which represents a potential theoretical sea level rise of 74 m (65 m for Antarctica, 7 m for Greenland, 1.1 m for permafrost, 24 cm for mountain glaciers, and less than 1 cm for snow; see Table 2.1 and Francou and Vincent 2011). However, since much of the cryosphere is a multi-millennia legacy, some of its components are not subject to contemporary climate change. For example, the Antarctic continent has been frozen for millions of years, since plate tectonics placed it in a polar orbital position, and a huge part of its 27 million km^3 of ice is fortunately unaffected by contemporary warming. However, the sensitivity of the Greenland ice sheet is greater because of its smaller size and its position in the heart of the Arctic, which is particularly affected by current warming (see Chapter 1).

Cryosphere component	Surface area in km^2	Volume in km^3 (water equivalent)	Sea level equivalent (the surface of the oceans represents 361 million km^2)
Antarctica	12.4 million	27 million	65 m
Greenland	1.8 million	2.7 million	7 m
Permafrost	23 million	0.24 million	1.1 m
Mountain glaciers	0.43 million	0.08 million	0.24 m
Snow	4 to 46 million	500 to 5000	0.1 to 1 cm

Table 2.1. *The components of the cryosphere. The marine cryosphere is not included in this table because the melting of the sea ice does not induce sea level rise (source: Francou and Vincent 2011)*

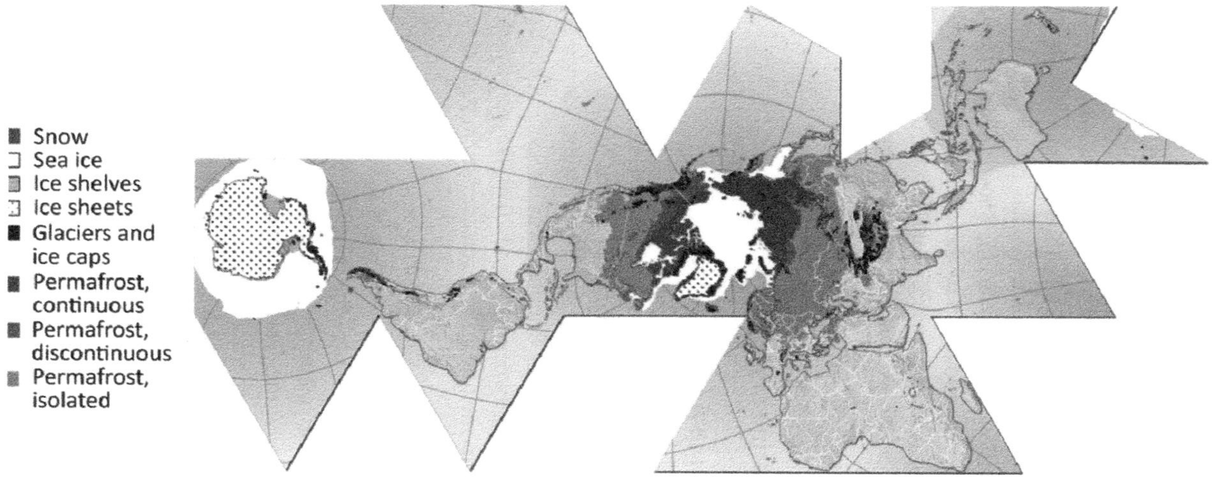

Figure 2.1. *Extension of the cryosphere (source: © Hugo Ahlenius, UNEP/GRID-Arendal). For a color version of this figure, see www.iste.co.uk/mercier/climate.zip*

The cryosphere is highly sensitive to climate change for two physical reasons. The first is the increase in surface air and seawater temperatures, which induces a transition from solid to liquid state. The second reason is correlated with the increase in air temperatures, which induces a decrease in solid (snowy) precipitation, which contributes to a reduction of the mass balance of glaciers of all sizes.

Moreover, the cryosphere is sensitive to the fundamental role played by albedo. The bright surfaces of the marine and terrestrial cryosphere reflect a significant proportion of solar radiation (see Chapter 1). Snow reflects 75–95% of the sun's energy, glaciers 40–60%. The reduction of these areas with a high albedo potential automatically leads to an increase in areas that absorb more solar radiation and, through a positive feedback loop, contribute to the warming of the lower layers of the atmosphere.

2.3. Melting of the marine cryosphere

2.3.1. *The melting of the Arctic sea ice*

Between 1979 and 2019, the Arctic sea ice lost 12.9% of its surface area per decade, representing a loss of half of its surface area by the end of the melt season (see Figure 2.2). As a result, in September 2019, Arctic sea ice occupied just over 4 million km^2, compared to almost 8 million km^2 in 1979. This trend represented a loss of 82,400 km^2 per year between 1979 and 2019. Beyond this four-decade trend, Figure 2.2 shows that the melting of the sea ice has in fact slowed over the last 13 years (2007–2019).

The year 2012 corresponds to the year when the reduction in the ocean surface area covered by Arctic sea ice was the most marked in the last 40 years (see Figures 2.2 and 2.3). It was mainly visible along the Siberian coast, clearing the Northeast Passage. Similarly, the sea ice disappeared from the Beaufort Sea along the coasts of Canada and northern Alaska in the summer of 2012. Sea ice still occupies the oceanic part between the North Pole, northern Greenland and Ellesmere Island, for atmospheric and oceanic reasons related to the Beaufort Gyre. This spatial asymmetry in the extension of the Arctic sea ice is increasing with contemporary global warming.

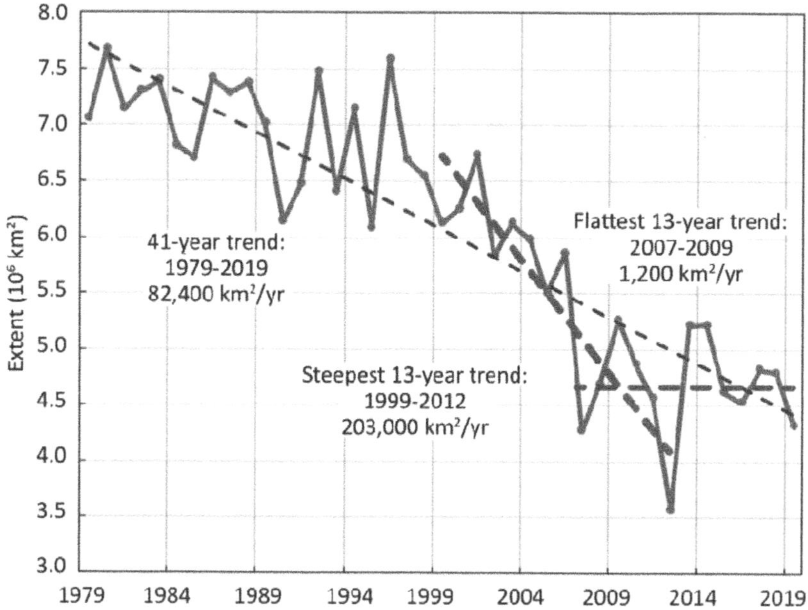

Figure 2.2. *Average Arctic sea ice extent for the month of September between 1979 and 2019. Decade of decline of 12.9%. (source: National Snow and Ice Data Center). For a color version of this figure, see www.iste.co.uk/mercier/climate.zip*

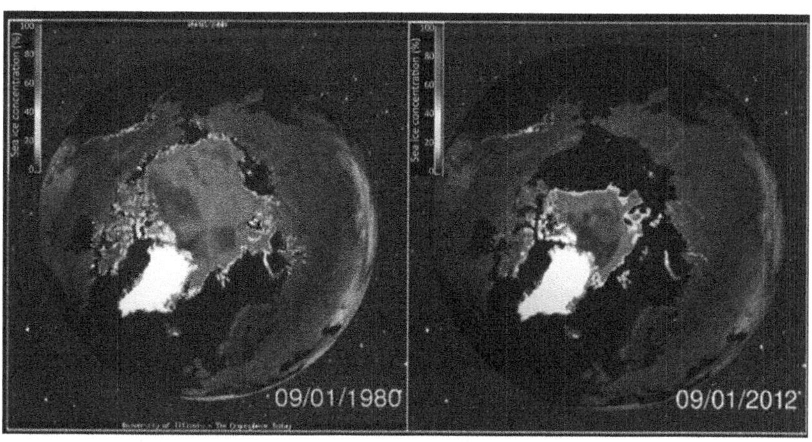

Figure 2.3. *Spatial extent of Arctic sea ice as of September 1, 1980, and on September 1, 2012, the year in which the extension was the least in the last four decades (source: The Cryosphere Today). For a color version of this figure, see www.iste.co.uk/mercier/climate.zip*

In addition to the loss of surface area, the average thickness of the Arctic sea ice has decreased by half, from 3 m to 1.5 m. This is due to the spatial reduction of multi-year sea ice in favor of the young annual sea ice (AMAP 2017). Thus, the younger, thinner Arctic sea ice is also more fragile, more brittle, and more easily displaced from the Arctic Basin via Fram Strait by transpolar drift (see Figure 2.4). This positive feedback loop helps to understand the evolutionary process of this reduction in the spatial extent of Arctic sea ice (Weiss 2008). According to the latest IPCC report (IPCC 2019), the proportion of multi-year ice that is at least five years old decreased by about 90% between 1979 and 2018.

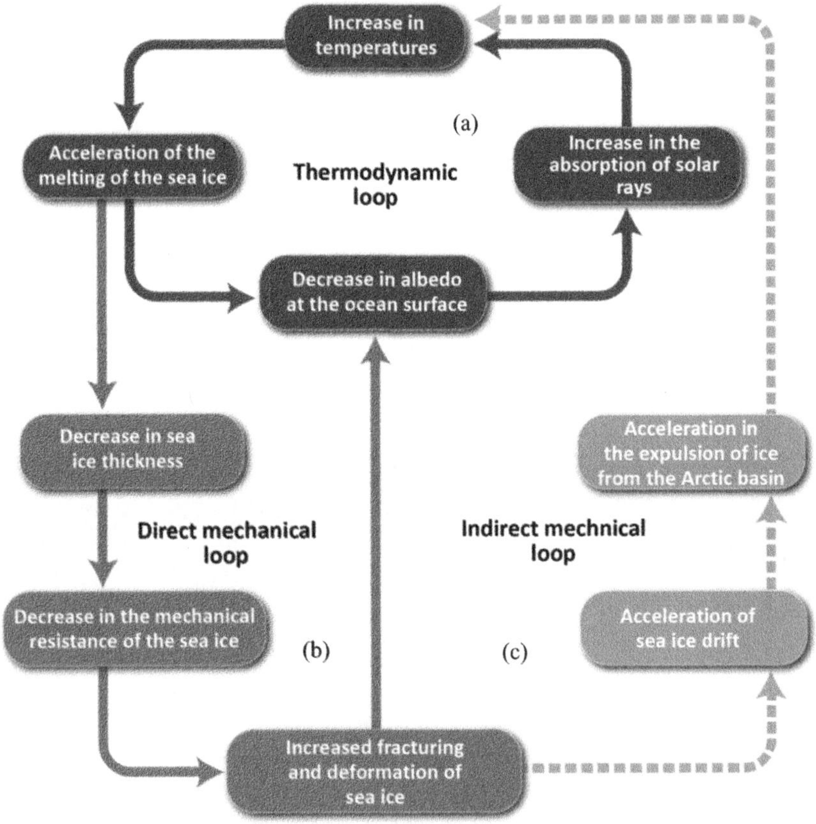

Figure 2.4. (a) *Thermodynamic feedback loops; (b) direct mechanics; and (c) indirect mechanics, all explaining the melting of the Arctic sea ice (source: modified from Weiss 2008). For a color version of this figure, see www.iste.co.uk/mercier/climate.zip*

2.3.2. *Antarctic sea ice*

On the other hand, the dynamics and evolution of the sea ice surrounding the Antarctic continent are of a different nature.

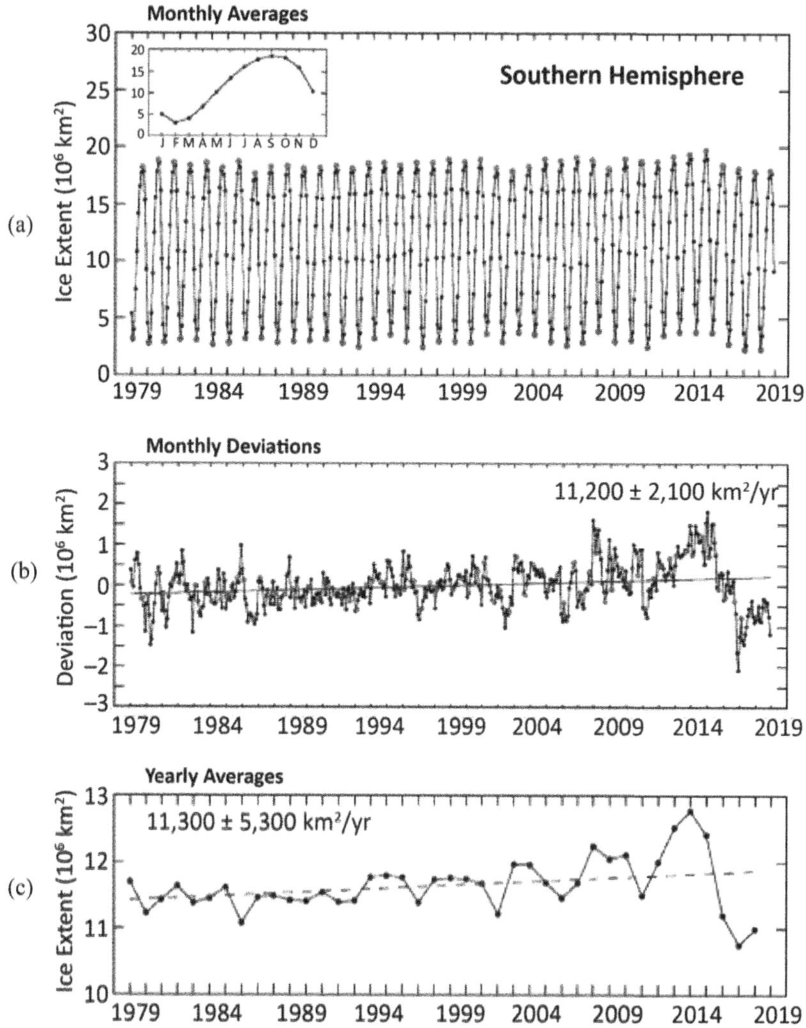

Figure 2.5. *Antarctic sea ice extents derived from satellite data DMSP Nimbus7 and NASA (source: Parkinson 2019). For a color version of this figure, see www.iste.co.uk/mercier/climate.zip*

COMMENT ON FIGURE 2.5.– *(a) Mean monthly sea ice extent for the southern hemisphere from January 1979 to December 2018. The February ranges are shown in red, the September ranges in green and all other ranges in black. (Box) The 40-year average annual cycle. (b) Monthly deviations determined from the monthly mean data of (a), with the same monthly color coding and with the line of least squares. (c) Mean annual sea ice extent and trend.*

The spatial extension of the Antarctic sea ice is symmetrical around the continent from the South Pole and reveals a double spatial astronomical and therefore thermal logic. In addition, sea currents and winds circulate continuously around the continent in an hourly direction. They then act as a thermal barrier surrounding the continent. Unlike the Arctic Basin, the absence of a land boundary around the continent allows Antarctic sea ice to float freely towards mid-latitudes where warmer waters cause it to melt. As a result, most of the sea ice that forms during the southern winter melts during the summer season. During the winter, the Antarctic sea ice reached an average of 18 million km^2 between 1981 and 2010 with a maximum extension in September. In September 2014, it even reached 20 million km^2. Its minimum extension was still registered in February with less than 5 million km^2 (see Figure 2.5; NSIDC 2019; Parkinson 2019).

Since record levels in 2014 and despite the significant declines in recent years in the spatial extent of the Antarctic sea ice, the trend for the period 1979–2018 still remains positive with an increase of 11,300 km^2 per year (Parkinson 2019). The causes of this increase in the spatial extent of the Antarctic sea ice have not yet been agreed (Parkinson 2019). On the other hand, the impact of this variability in the extension of the sea ice around Antarctica determines the extent of pre-precipitation for the continent. Between a reduced extension and a vast extension of the sea ice, the difference in precipitation is estimated at 102 Gt per year (Wang *et al.* 2020).

2.4. Melting of the Earth's cryosphere

2.4.1. *Melting ice sheets*

2.4.1.1. *The melting of Greenland*

Between 2006 and 2015, the Greenland ice sheet lost ice mass at an average rate of 278 ± 11 billion tons per year (IPCC 2019). This melting represents an equivalent of 0.77 ± 0.03 mm per year in global sea level rise. Melting mainly affects the surface of the ice sheet (see Figures 2.6 and 2.7).

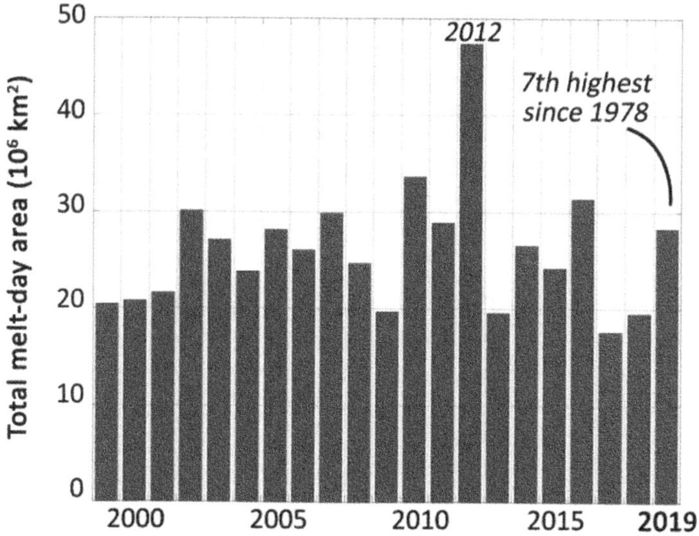

Figure 2.6. *Total extent of melt day, or sum of daily melt area during the 1999–2019 melt season in Greenland (source: NSIDC). For a color version of this figure, see www.iste.co.uk/mercier/climate.zip*

Since 2000, the Greenland ice sheet has experienced a general increase in melting, with a melt day area for 2019 totaling 28.3 million square kilometers. Melting has been observed over nearly 90% of the island on at least one day, even reaching the Summit station and much of the high-altitude areas. It was particularly intense along the northern edge of the ice cap, where, compared to the average from 1981 to 2010, melting occurred on an additional 35 days. The number of melt days was also slightly above average along the western flank of the ice cap, with about 15 to 20 more melt days than average. In the south and southeast, the melt was slightly below average within a few days.

This melting is caused by warm, moist air flows associated with active summer and winter cyclogenesis, from the south or southeast. These mechanisms result in increased cloud cover, low-level liquid and snowy precipitation at high elevations in the ice sheet, better absorption of long-wave radiation, and a decrease in albedo in the south and near the coast, accelerating the melting of the snowpack (Oltmanns *et al.* 2019).

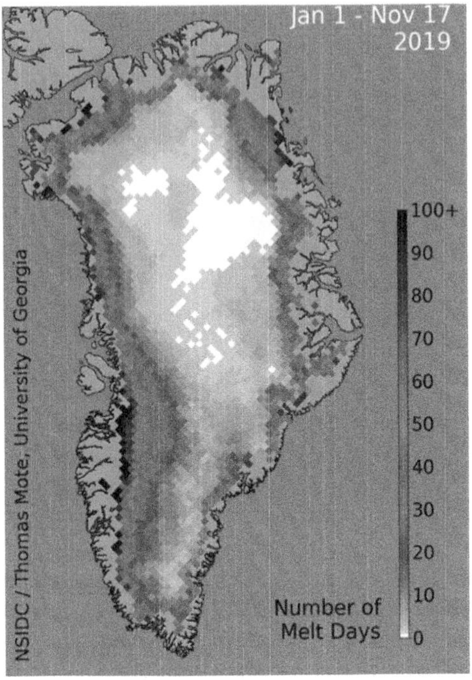

Figure 2.7. *Number of days of melt at the surface of Greenland's ice sheet between January 1 and November 17, 2019 (source: NSIDC, Thomas Mote, University of Georgia). For a color version of this figure, see www.iste.co.uk/mercier/climate.zip*

2.4.1.2. *The melting of the Antarctic*

Between 2006 and 2015, the Antarctic ice sheet lost ice mass at an average rate of 155 ± 19 Gt per year (0.43 ± 0.05 mm per year), mainly due to the rapid thinning and retreat of large glaciers downstream of the continent draining the West Antarctic ice sheet, especially in the Amundsen Sea and Wilkes Land in the eastern part (IPCC 2019).

The total mass loss of the Antarctic ice sheet is increasing every decade. It increased from 40 ± 9 Gt per year in 1979–1990 to 50 ± 14 Gt per year in 1989–2000, 166 ± 18 Gt per year in 1999–2009 and 252 ± 26 Gt per year in 2009–2017 (Rignot *et al.* 2019; see Figure 2.8).

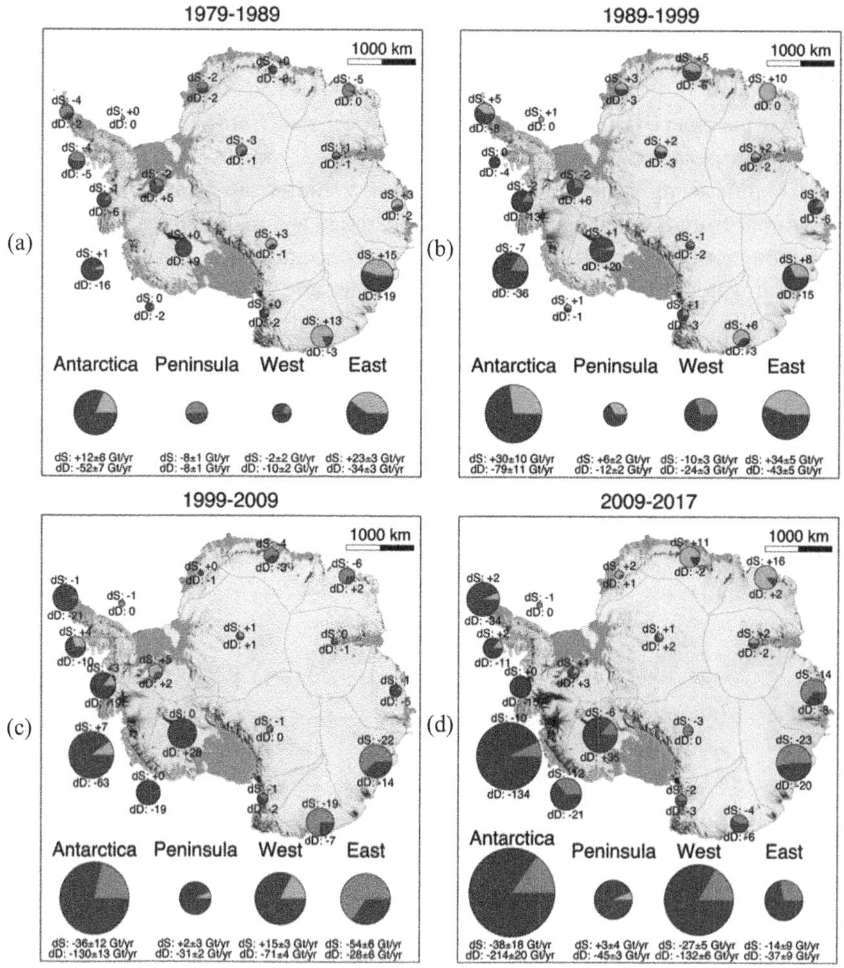

Figure 2.8. *Mass balance of the Antarctic ice sheet. For a color version of this figure, see www.iste.co.uk/mercier/climate.zip*

COMMENT ON FIGURE 2.8.– *The size of the circle is proportional to the absolute magnitude of the anomaly in D (dD = SMB1979-2008 - D) or SMB (dSMB = SMB - SMB1979-2008). The color of the circle indicates a loss in dD (dark red) or dSMB (light red) relative to a gain in dD (dark blue) or dSMB (light blue) in billions of tons (1,012 kg) per year. The dark color refers to dD; the light color refers to dSMB. The graphs show totals for Antarctica, Antarctic Peninsula, West Antarctica and East Antarctica. The bottom is the total mass balance distributed over the catchments with a color code ranging from red (loss) to blue (gain) (source: Rignot* et al. *2019).*

Between 2009 and 2017, mass loss was dominated by the Amundsen and Bellingshausen Sea sectors in West Antarctica (159 ± 8 Gt per year), Wilkes Land in East Antarctica (51 ± 13 Gt per year), and the Western and Northeastern Peninsula (42 ± 5 Gt per year). The contribution to sea level rise from Antarctica averaged 3.6 ± 0.5 mm per decade with a cumulative 14.0 ± 2.0 mm since 1979, of which 6.9 ± 0.6 mm from West Antarctica, 4.4 ± 0.9 mm from East Antarctica and 2.5 ± 0.4 mm from the Peninsula (i.e. East Antarctica is a major contributor to mass loss). Throughout the period, the mass loss was concentrated in the closest to deep circumpolar water (DCW) that is warm, salty, subterranean, which is consistent with the enhanced polar winds pushing DCW towards Antarctica to melt its floating ice shelves, destabilize glaciers and raise sea level (Rignot *et al.* 2019).

2.4.2. *The melting of mountain glaciers*

The World Glacier Monitoring Service (WGMS) provides standardized statistical data, such as ice front variations and mass balances[1].

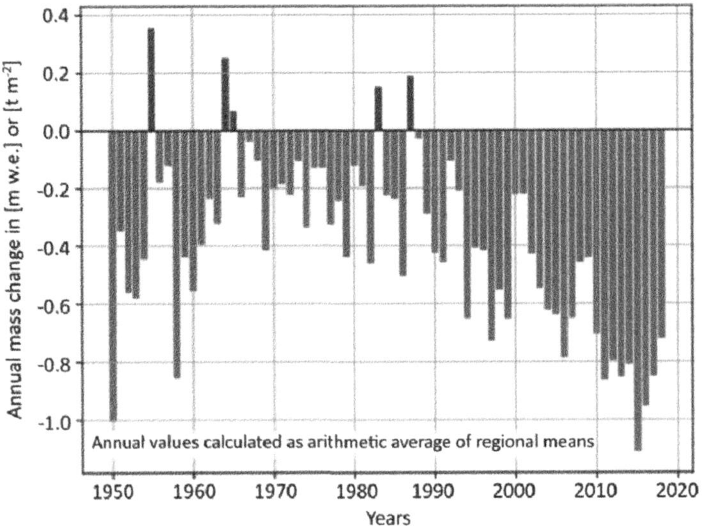

Figure 2.9. *Annual mass balance of reference glaciers with more than 30 years of glaciological measurements from 1950 to 2018. The values of annual mass change are given on the y-axis in water equivalent (w.e.) per meter of water, which corresponds to tons per square meter (t/m^2) (source: Zemp* et al. *2017). For a color version of this figure, see www.iste.co.uk/mercier/climate.zip*

1 https://wgms.ch/.

In particular, this graph (Figure 2.9) shows that seven of the ten most negative mass balance sheet years were recorded after 2010. A value of –1.0 m water equivalent per year represents a mass loss of 1,000 kg per square meter of ice cover or an annual loss of glacier-wide ice thickness of about 1.1 m per year, since the density of ice is only 0.9 times the density of water (Zemp *et al.* 2017).

Data at regional scales show that all of the world's glaciated mountain areas have been melting over the last few decades. Glaciers in North America and Central Europe are suffering the greatest losses. Between 2006 and 2015, the world's other glaciers melted at an average rate of 220 ± 30 billion tons per year, equivalent to 0.61 ± 0.08 mm per year in sea level rise (IPCC 2019).

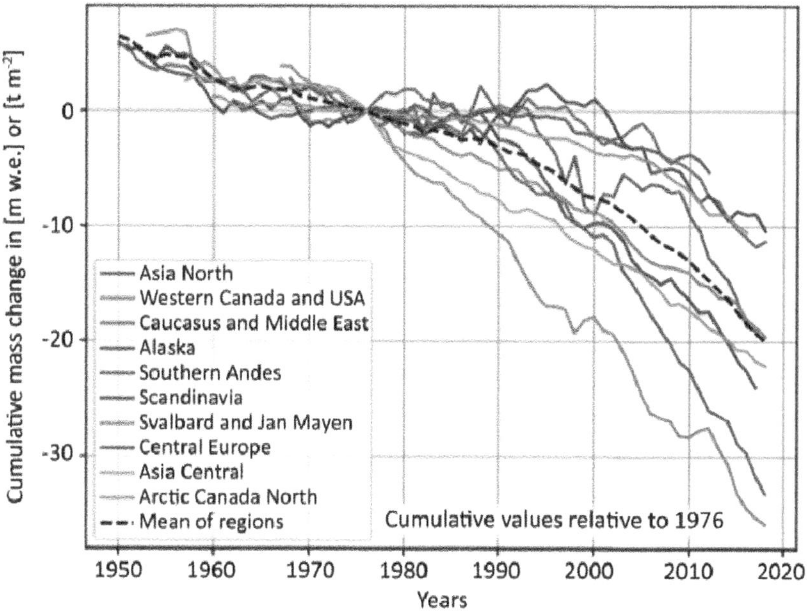

Figure 2.10. *Cumulative mass change from 1976 for regional and global averages based on reference glacier data (source: Zemp et al. 2017). For a color version of this figure, see www.iste.co.uk/mercier/climate.zip*

Recent studies are trying to determine the date of disappearance of some glaciers. For example, the evolution and characteristics such as volume, area, ice thickness, runoff and duration and mode of disappearance of glaciers have been projected for the Austre Lovénbreen glacier on the Brøgger peninsula in northwestern Spitsbergen in the Svalbard archipelago (Wang *et al.* 2019; see Figures 2.11 and 2.12). Based on the

21st Century warming trend of the Arctic in the IPCC Fifth Assessment Report, glacier evolution was simulated under three hypothetical climate scenarios: pessimistic, high probability and optimistic. The results predicted that the glacier would retreat until it disappeared under all three scenarios, and its time to disappear would likely be about 111 years, that is by 2120.

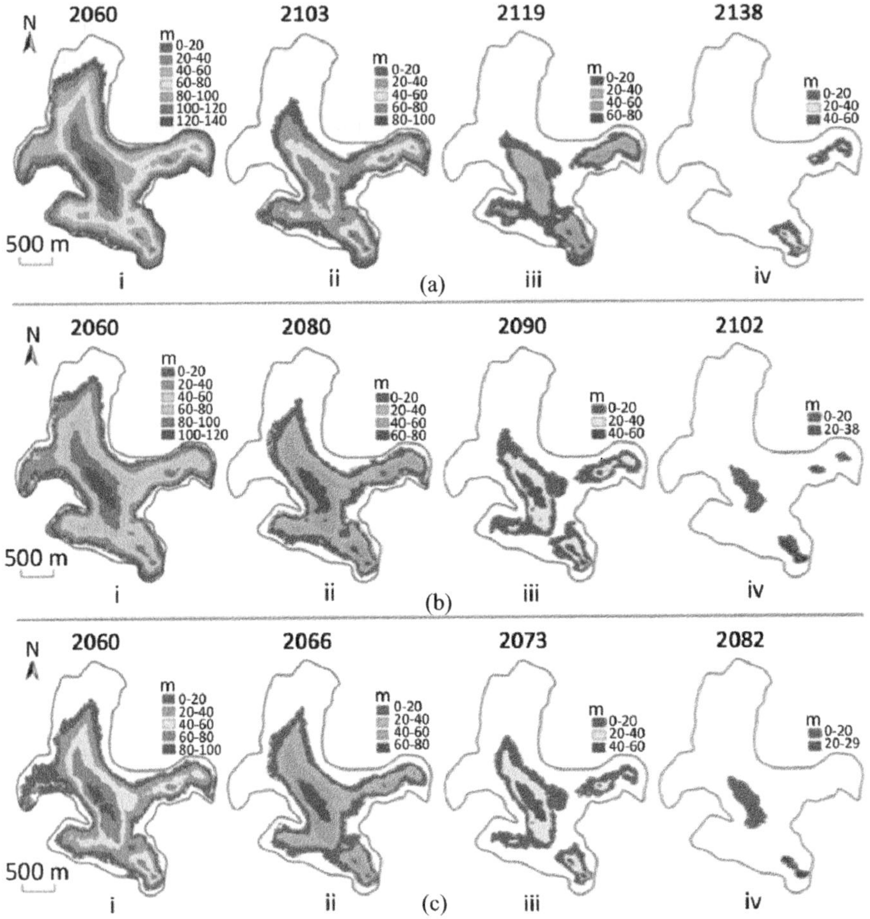

Figure 2.11. *Area and thickness of the Austre Lovénbreen glacier (Svalbard) simulated for different years: (i) 50th year (2060); (ii) disappearance of the western tributary; (iii) break between the main current and the eastern tributary; (iv) late decomposition according to the scenarios (a) optimistic; (b) high probability; (c) pessimistic (source: Wang* et al. *2019). For a color version of this figure, see www.iste.co.uk/mercier/climate.zip*

Figure 2.12. *The Austre Lovénbreen glacier in northwestern Spitsbergen (Svalbard) in the background. The flat space in front of the glacier corresponds to the space freed by its melting since the beginning of the 20th Century. In the foreground, the prograding deltas on the Kongsfjorden are fed with sediments by the meltwater runoff from the glacier (source: © photo by D. Mercier taken on August 24, 2017). For a color version of this figure, see www.iste.co.uk/mercier/climate.zip*

2.4.3. *Decreasing permafrost*

The IPCC Cryosphere Synthesis (IPCC 2019) provides information on the increase in permafrost temperatures since the 1980s. From 2007 to 2016, permafrost temperatures increased by an average of 0.29°C ± 0.12°C in the polar and high mountain regions of the world. The intensity of climate warming in the Arctic is exacerbated by this melting of Arctic and boreal permafrost, which may eventually release between 1,460 and 1,600 Gt of organic carbon, nearly twice the carbon in the atmosphere (IPCC 2019).

2.4.4. *Melting snow*

The decrease in land snow cover extent in June for the Arctic was 13.4 ± 5.4% per decade between 1967 and 2018, a total loss of approximately 2.5 million km^2, mainly due to the increase in surface air temperature (IPCC 2019).

In almost all high mountain regions, the depth, extent and duration of snow cover has decreased in recent decades, especially at low altitudes in relation to rising temperatures and the rising rain-snow limit (IPCC 2019).

2.5. Consequences of the melting cryosphere

2.5.1. *On a global scale: rising sea levels*

At this scale, the most important consequence of the melting of the cryosphere is sea level rise. In addition to the thermal expansion of the oceans, the main sources of this sea level rise are the melting of the Greenland ice sheet and the Antarctic ice sheet, the contribution of mountain glaciers and permafrost. It was 18 cm during the 20th Century, and the various IPCC scenarios envisage a rise of around 60 to 100 cm by the end of the 21st Century (IPCC 2019).

However, we should not think in terms of this deadline alone, but rather that the rise of the seas and oceans will continue over the coming centuries as part of the melting of continental ice that has begun since the beginning of the Holocene interglacial period in which we live.

Thus, an increase (rise) of 5 m will surely be recorded by 2300. The consequences for low-lying coastal areas such as estuaries, tidal marshes, deltas, etc. will affect the economic activities and human occupation of millions of citizens (see Chapter 4).

A recent assessment by Zemp *et al.* (2019) shows that glaciers alone lost more than 9 billion tons of ice between 1961 and 2016, raising water levels by 27 millimeters (see Figure 2.13).

With more than 3,000 Gt, the Alaska Glaciers (ALA) have contributed the most to sea level rise. The glaciers of Southwest Asia (ASW, green circle) were the only ones to record an increase in mass.

Glaciers in the European Alps, the Caucasus mountain range and New Zealand have also suffered significant ice loss. However, because of their relatively small glacial areas, they have played only a minor role in sea-level rise (Zemp *et al.* 2019).

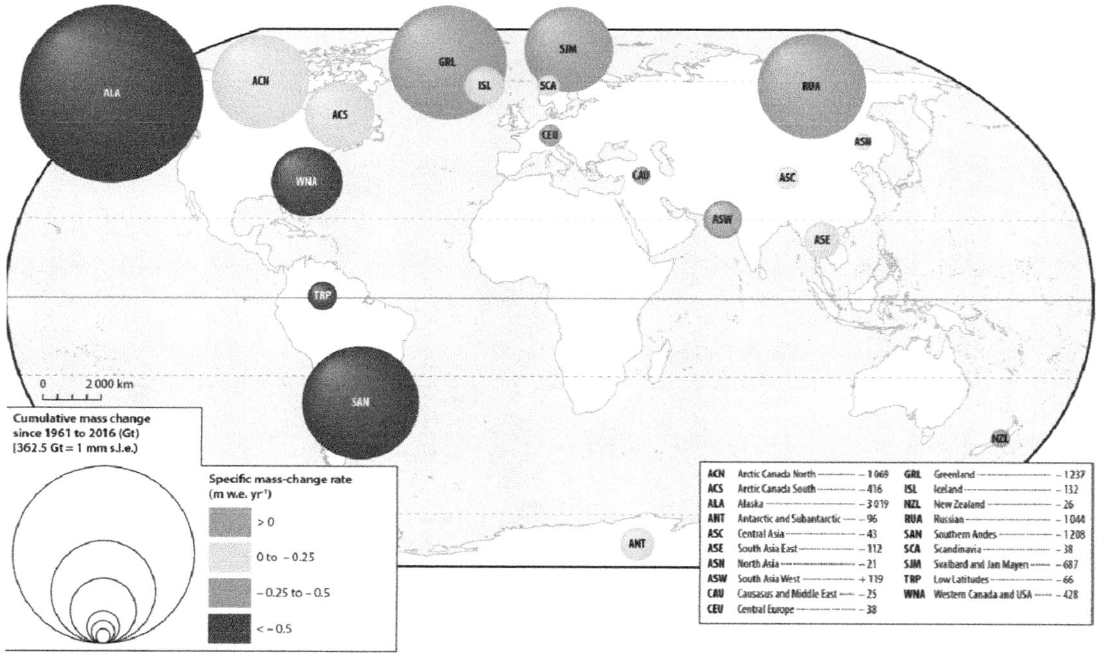

Figure 2.13. *Regional share of glaciers in sea-level rise from 1961 to 2016. The cumulative change in regional and global glacier mass (in gigatons, 1 Gt = 1,000,000,000 tons) corresponds to the size of the circles. The synthesis is based on 19,000 glaciers worldwide (source: modified from Zemp* et al. *2019). For a color version of this figure, see www.iste.co.uk/mercier/climate.zip*

2.5.2. *Regionally: paraglacial risks*

At the scale of mountain ranges, the melting of glaciers and areas with permafrost has consequences for the paraglacial sedimentary cascade (see Chapter 5), periglacial dynamics (see Chapter 6), hydrology of Arctic rivers (see Chapter 7), and water supply in Central Asia (see Chapter 9).

On the other hand, the shrinking of glaciers also brings with it paraglacial risks for the populations living on the margins of these glaciated areas.

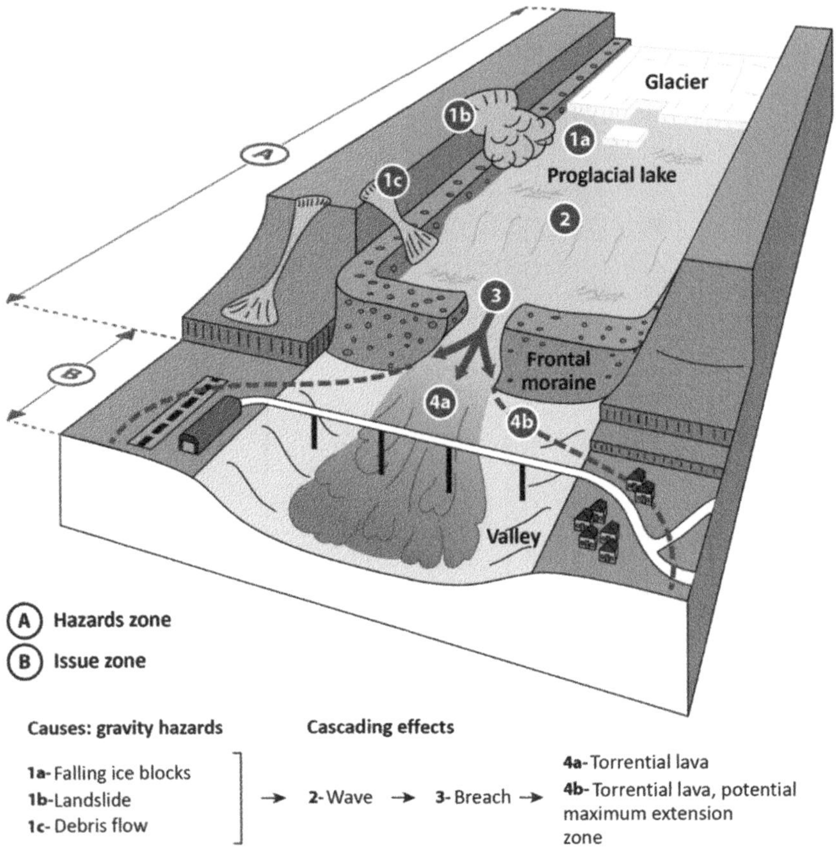

Figure 2.14. *Paraglacial hazards induced by melting glaciers (source: design D. Mercier; drawing F. Bonnaud, Sorbonne University, 2019). For a color version of this figure, see www.iste.co.uk/mercier/climate.zip*

Glacier melt induces the formation of lakes in the immediate periphery of glaciers, with meltwater often retained by natural dams formed by frontal moraines (see Figure 2.14). This glacial melting and these lakes feed rivers that flow downstream in valleys to the seas and oceans. When glaciers calve and large areas of ice fall into lakes (see Figure 2.14, step 1a), waves can form (see Figure 2.14, step 2) and breach the frontal moraines (see Figure 2.14, step 3). The flows then become torrential with a mixture of water and sediment, resulting in highly turbid, torrential lava flows that can affect localized issues at the periphery of the flow area (see Figure 2.14, steps 4a and 4b). As a result, infrastructure such as roads, bridges, airstrips, hangars, dykes, houses, etc., can be damaged by these lava flows (see Figure 2.14, Zone B).

The problem of draining water pockets remains a threat in the Alps since the dramatic accident in Saint-Gervais in 1892, which claimed 175 victims downstream from the Tête Rousse glacier. These Glacial Lake Outburst Floods (GLOFs) are present in many mountains, in the Andes, in the Himalayas[2] (Westoby *et al.* 2014). In Iceland, these floods, called jökulhlaups (literally "the running glacier") are associated with the melting of glaciers under the effect of global warming but can also be exacerbated by sub-glacial volcanic activity. In North America, the Alaska Climate Adaptation Science Center (AK CASC) is funding research on the Mendenhall Glacier to better understand its dynamics and the risks induced by its flash floods in order to model its dynamics and predict flooding by monitoring, and among other things, the current water level in the lake, its spatial extent and its bathymetry.

The melting of glaciers also induces landslides on the slopes because the pressure exerted during the glacial sequence is replaced by a decompression of the walls (Mercier 2016). Thus, rock volumes can move down the slopes and end their course in proglacial lakes (see Figure 2.14, step 1b), potentially generating waves, breaches and torrential lava, or directly affecting the infrastructures below the uplift zones. Melting permafrost in the walls, or increased rainfall, may also cause debris flows that may also end their course in the proglacial lake, leading to the same chain of events (see Figure 2.14, step 1c). These gravitational hazards are therefore all linked directly or indirectly by melting glaciers and are potential hazards to downstream urbanized areas (see Figure 2.14, zones A and B).

2 See Chapter 9 on Central Asia and particularly section 9.2.3.

2.6. Conclusion

The contemporary melting of the terrestrial and marine cryosphere is a reality observed at high latitudes as well as in the mountains. Although this melting has had many precedents in the history of our planet, the consequences of this loss of ice and snow cover have repercussions on different spatial scales, from the global to the local level.

Some of the consequences will be global, such as sea level rise related to the melting of the Earth's cryosphere. Although the melting of the marine cryosphere does not induce sea level rise, the melting of the Arctic sea ice does have an impact on the North Atlantic thermohaline circulation and thus has implications for the general atmospheric circulation. Locally, the disappearance of glaciers induces changes in the morphogenic dynamics that cause hazards, potentially dangerous for populations, and in plant colonization.

Moreover, climate predictions for the late 21st Century and for the centuries to come converge to affirm that the melting of the various components of the cryosphere will continue (IPCC 2019). This long-term evolution is doubly logical, on the scale of the current global warming and on the scale of the interglacial period in which humanity has been living for thousands of years.

2.7. References

AMAP (2017). *Snow, Water, Ice and Permafrost in the Arctic (SWIPA)*. Arctic monitoring and assessment programme, Oslo.

Francou, B. and Vincent, C. (2011). *Les Glaciers à l'épreuve du climat*. IRD, Paris.

IPCC (2019). Special report on the ocean and cryosphere in a changing climate [Online]. Available at: https://www.ipcc.ch/srocc.

Lageat, Y. (2019). *Les variations du niveau des mers*. Presses Universitaires de Bordeaux, Pessac.

Mercier, D. (2016). L'Arctique face aux crises géomorphologiques paraglaciaires. In *L'Arctique en mutation*, Joly, D. (ed.). EPHE, Paris.

NSIDC (2019). National Snow & Ice Data Center [Online]. Available at: https://nsidc.org/.

Oltmanns, M., Staneo, F., Tedesco, M. (2019). Increased Greenland melt triggered by large-scale, year-round cyclonic moisture intrusions. *The Cryosphere*, 13, 815–825 [Online]. Available at: https://doi.org/10.5194/tc-13-815-2019.

Parkinson, C.L. (2019). A 40-y record reveals gradual Antarctic sea ice increases followed by decreases at rates far exceeding the rates seen in the Arctic. *PNAS*, 116(29), 14414–14423.

Rignot, E., Scheuchl, B., van den Broeke, M., van Wessem, M.J., Morlighem, M. (2019). Four decades of Antarctic Ice Sheet mass balance from 1979–2017. *PNAS*, 116, 1095–1103.

Wang, Z., Lin, G., Ai, S. (2019). How long will an Arctic mountain glacier survive? A case study of Austre Lovénbreen, Svalbard. *Polar Research*, 38, 3519.

Wang, H., Fyke, J.G., Lenaerts, J.T.M., Nusbaumer, J.M., Singh, H., Noone, D., Rasch, P.J., Zhang, R. (2020). Influence of sea-ice anomalies on Antarctic precipitation using source attribution in the Community Earth System Model. *The Cryosphere*, 14, 429–444 [Online]. Available at: https://doi.org/10.5194/tc-14-429-2020.

Weiss, J. (2008). Petite tectonique des plaques de banquise. *Pôles Nord & Sud*, 1, 68–81.

Westoby, M.J., Glasser, N.F., Brasington, J., Hambrey, M.J., Quincey, D.J., Reynolds, J.M. (2014). Modelling outburst floods from moraine-dammed glacial lakes. *Earth-Science Reviews*, 134, 137–159.

Zemp, M., Nussbaumer, S.U., Gärtner-Roer, I., Huber, J., Machguth, H., Paul, F., Hoelzle, M. (2017). Global glacier change bulletin No. 2 (2014–2015) [Online]. Available at: https://wgms.ch/downloads/WGMS_GGCB_02.pdf.

Zemp, M., Huss, M., Thibert, E., Eckert, N., McNabb, R., Huber, J., Barandun, M., Machguth, H., Nussbaumer, S.U., Gärtner-Roer, I., Thomson, L., Paul, F., Maussion, F., Kutuzov, S., Cogley, J.G. (2019). Global glacier mass changes and their contributions to sea-level rise from 1961 to 2016. *Nature*, 568, 382–386.

3

Between Warming and Globalization: Rethinking the Arctic at the Heart of a Stakes System

Éric CANOBBIO
University of Paris 8 Vincennes – Saint-Denis, Paris, France

3.1. Spatial impacts of climate change in the Arctic

It now seems scientifically and politically apparent that the spatial impacts of climate change should be considered in the geo-physical context of the Arctic vastness. No other region in the world, except Antarctica, has such global influence on its future, which is strongly linked to the rise in temperatures in the cryosphere and its effects on the northern hemisphere and the rest of the world. Images and maps of the summer ice pack or of the retreat of the Arctic glaciers and the Greenland ice sheet have thus become a true veritable visual piece that has been used since the Conferences of the Parties (COPs) of the United Nations Framework Convention on Climate Change (UNFCCC) began, in a scientific and militant mediation, contributing to the rounds of international negotiations on the reduction of greenhouse gases. The current scientific consensus is that by 2100, global warming will be at least twice as great in the northern-Arctic zone as in the rest of the northern hemisphere. The halving of the area and thickness of the pack ice has been observable in its average trajectories since 1979, with a gradual disappearance of

Spatial Impacts of Climate Change,
coordinated by Denis MERCIER.

multi-year ice[1]. Satellite observation of the Greenland ice cap, based on data provided by Grace 1, estimates the weight of ice lost each year to be 290 million tons (with significant annual variations) and annual records of summer temperatures at the North Pole confirm the occurrence of above-freezing temperatures.

However, while the warming Arctic has gradually become one of the proofs of the tangible nature of global climate change over the last two decades, the analysis of contemporary polar issues in their social, political, economic and strategic evolution calls for other global contexts, in particular the effects of globalization and its influencers on new polar spatial representations acting on the notions of "openness", "access", "integration" or "mobility" in boreal spatial inventories. This globalized interpretation of new polar opportunities, indexed on climate models, confirms without nuance the meltdown of polar isolates.

The central objective of this contribution will be to propose an analysis of the interplay between spatial impacts, essentially understood in their political translation in terms of doctrines and actions, and climate change and globalization.

Its approach lends itself to the geography of the development of the polar regions, questioning the long-term production of political, socio-economic and cultural models, which animates "stakeholders" of major polar issues at the regional and national levels or in original forms of inter-regional or international cooperation.

Indeed, the study of current Arctic dynamics shows an increasingly sophisticated cooperation between the traditional stakeholders of northern administration, northern policy making and sovereignty, and the groups of stakeholders involved in the analysis of boreal transformations: non-polar states such as China, regional integration bodies such as the European Union, strategic cooperation mechanisms such as NATO, non-governmental environmental organizations that produce an activist polar strategy such as Greenpeace, or expertise and mediation organizations such as the Word Wildlife Foundation (WWF) or the International Union for Conservation of Nature (IUCN), and finally the integrative approaches of multinationals to certain polar resources.

3.1.1. *Clarifying the terms of the subject in their polar contexts*

It may be useful to briefly clarify the key terms constituting the proposed sub-topic – impact, space and climate change – in their polar contexts of use and their interpretation trajectories within the framework of our study.

1 See Chapter 2 on melting of the cryosphere.

3.1.1.1. *The notion of "impact": standards and modeling*

The notion of "impact" is an ancient notion linked in its emergence in the 1970s with the deployments of the first boreal pioneer fronts, particularly in Alaska and Canada, and to the consubstantial production of the first specific "impact study" standards in the high latitudes. Initially centered on environmental assessments, these protocols will quickly tend to broaden their diagnoses to include socio-economic and cultural issues of northern land use projects affecting Indigenous communities within the planning regions. The spatial impact thus becomes systematically multi-dimensional in its assessment process and is both anthropogenic and spatially limited. It has also been, since its origins and in its normative translations, a relevant indicator of the legal and democratic relationship in the North and in particular to the corpus of Indigenous rights in the polar regions, where impact study protocols can be implemented.

In terms of climate challenges, the interpretation of the notion of impact has broadened its traditional application scales and favors three dimensions that can act in parallel but not always in dialogue:

– *the local impact* of Arctic communities and social-environmental and economic constructs, such as pastoral systems or fishing activities;

– *the regional impact*, which requires the choice of the target regionality: natural region, administrative region, cultural region or economic region. This issue of delimitation is contained in a large number of recent studies, in particular the second report "Arctic Human Development Report (AHDR), Regional Process and Global Linkages" published in 2014 by Norden[2], which chooses a specific delimitation of its statistical study areas (see Figure 3.1), now considered as a new viable delimitation of the "Arctic" (Nymand Larsen and Fondhal 2014);

– the impact assessed on a *global scale* sizes the entire physical Arctic region and sets its central objectives on the analysis of the evolution of the natural components, cryosphere, permafrost, biomes, flora and fauna, etc., which does not extract the capacities for political, or even strategic and economic interpretation of these impacts on the polar basin. In 2004, the publication of the report "Arctic Climate Impact Assessment, impacts of a warming Arctic" (ACIA), opened a new scientific period in the now systemic approach to the term "impact". This first international scientific work on the state of the Arctic region, commissioned by the

2 Norden refers to the Nordic Council created in 1951 involving Denmark, Sweden, Norway, Iceland and Finland (she joined the Council in 1956). The Norden is the oldest cooperation institution in the North and maintains active interregional cooperation activity alongside other bodies such as the Council of Baltic Sea States, the Barents Euro-Arctic Council or the Arctic Council.

Arctic Council, which itself contributed to its production based on the work of its two programs, the Arctic Monitoring and Assessment Programme (AMAP) and the Conservation of Arctic Flora and Fauna (CAFF), also inaugurates a new format for mediating the comprehensive polar analysis of the impacts of climate change to decision-makers, the media and public opinion.

This shift from local (or regional) to macro-regional assessment does not, however, quench the normative dimension initially contained in terms of impact. Thus, the impact of the summer reduction in sea ice and the extension of the period of navigability in the seas bordering the Arctic Ocean led to the initiation and adoption by the International Maritime Organization (IMO) on January 1, 2017, of a polar code[3] attempting to clarify the standards for the construction and equipment of ships and the skills of crews that could allow access to "Arctic waters", which made it necessary to define a new *polar* maritime spatial perimeter where these rules would henceforth be applied.

3.1.1.2. *Polar space, a multidimensional and dynamic geographical object*

The definitions of the constituent geographies of the polar basin are restricted to their "biophysical" or "geophysical" and political boundaries, and section 3.1.1.3 on the generation of polar issues will address issues related to the production of the new representations of the northern zone. This need for geographical precision nevertheless remains a dynamic subject, particularly because many climatic, economic or geopolitical issues require the reconfiguration of their areas of expertise beyond the traditional Arctic boundaries. The study of permafrost provides a useful example, but the analysis of polar landscapes also reflects the spatial variability of the major Arctic biomes. As geographer Marie-Françoise André stated,

> In mid-continent environments where summers are brief and sunny, the boundary between tundra and boreal forest extends beyond the 70th parallel, well north of the Arctic Circle. This is particularly the case in Alaska and Siberia. Conversely, in the maritime environments on the eastern seaboard of continents, such as southern Labrador, the polar tundra landscapes extend down to the 51st parallel, the latitude of Brussels! (André 2011)

In confrontation or in complementarity of representations, two Arctics thus coexist in high latitudes, the biophysical Arctic and the political Arctic.

3 Named by the IMO "International Code for Ships Operating in Polar Waters".

The biophysical Arctic is historically bounded by the Arctic Circle, 66° 33' north latitude, a circumterrestrial space of more than 21 million km^2, two-thirds of which is maritime, a complex grouping of an ocean and its coastal seas, islands, vast archipelagos and continental margins offering a wide variety of shapes and landscapes. Its early mapping in boreal exploratory processes invented the first polarocentric projection with the publication in 1595 of Gerard Mercator's "Septentrionalium Terrarum Descriptio", which describes an imaginary arctic, continentalized in its center, but integrating the coastlines and straits already discovered at the beginning of the 17th Century. This Arctic places the North Pole at its center and gives priority to a perennial Mediterranean-style representation of the high latitudes, opening up confusion about the regional realities of compartmentalization and enclaves.

The political Arctic is the product of the long historical processes of integration of the polar regions in the eight countries occupying high latitudes: Russia and Canada, the two great "northern" world nations (Dorion 2015), the United States with Alaska, Norway, Sweden, Finland, Denmark, which despite the increased autonomy of Greenland in 2008 retains sovereignty over the polar island and owns the Faroe Islands, and Iceland, a sub-Arctic country with some territorial fragments beyond the Arctic Circle and an "Arctic" exclusive economic zone. Since 1996, these eight states, known as the "A8", have been organizing multilateral polar governance, essentially based on the Arctic Council, a body of diplomatic representation that is supposed to provide an efficient framework for the resolution of certain regional issues, particularly those of an environmental or socio-economic nature.

The five countries bordering the Arctic Ocean – Russia and Canada, the United States, Norway and Denmark – named "A5" in 2008, have been filing particularly complex maritime claims since the 2000s, mainly centered on the extension of national sovereignty areas to extended continental shelves and the status of the Northeast and Northwest Arctic Passages. This process explicitly enhances the strategic dimension of A5 as an "Arctic power". In its southern margins, which correspond to the strongest gradient of structural facilities, population density and urbanization, the political Arctic draws heterogeneous and plural borders, corresponding for the most part to the limits of the administrative regions formed during the historical integration processes and originally designated by the idea of the "Great North" or "High North" in the wealth of national representations.

Figure 3.1. *The Arctic, a macro-region with variable spatialization, map proposed by the NORDREGIO Institute in 2015 (source: AHDR, AMAP and CAFF, analysis and design: J. Roto). For a color version of this figure, see www.iste.co.uk/mercier/climate.zip*

3.1.1.3. *Boreal spatiality and "nordicity", a fundamental re-interpretation of northern and Arctic regional identities*

Incomplete in accurately defining northern regions in terms of their human, environmental and economic identities, these spatial designations have been refined in Russia and Canada through the concept of "nordicity", which remains a particularly efficient tool for measuring the North when assessing the spatial impacts of climate change and globalization at high latitudes. As Quebec-based geographer Dorion notes:

> Nordicity is a relative and complex concept that Russian northern geographers G.A. Agranat and S.V. Slavine, as well as Quebec

geographer Louis-Édmond Hamelin, have seriously analyzed. These researchers proposed various indicators, both natural and socio-economic, to study nordicity. It follows that the degree of nordicity of different territories depends not only on the latitude at which they are located, but also on factors such as vegetation, the presence of transportation infrastructures and settlement dynamics. (Dorion 2015)

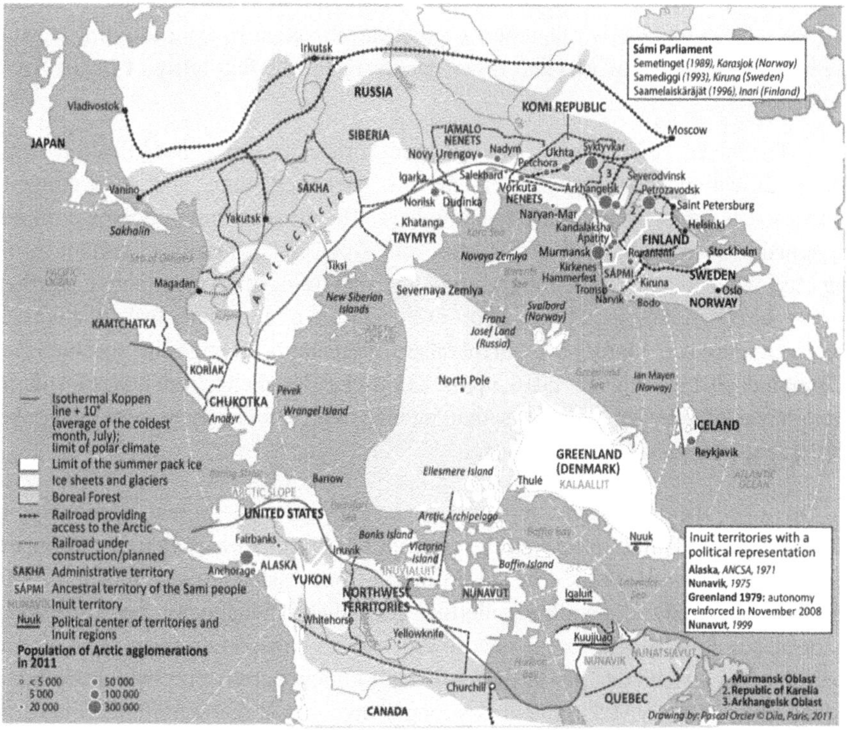

Figure 3.2. *The political Arctic, nations and territories (source: Canobbio 2011). For a color version of this figure, see www.iste.co.uk/mercier/climate.zip*

As early as the 1970s, in a context of pioneering fronts on northern resources, Hamelin thus detailed a nordicity indicator based on the addition of 10 criteria, each with a polar value (vapo), the maximum value being reached at the North Pole with 1,000 vapo. According to a simple quantitative method, the geographer takes as human elements: accessibility, the presence of air services, the resident or wintering population, the regional population density or the number of inhabitants in the agglomeration, the degree of economic activity (local or regional). Natural elements include latitude, summer heat, annual cold, local ice cover, precipitation and vegetation cover (open forest, tundra, stony deserts). The use of this indicator allows Russia and Canada to use two major space productions:

– To redefine the concept of the "High North" which becomes a regionality precisely delimited by an index of between 500 and 800 vapo, and which spatially succeeds the "Middle North", or in Russia the "near North" (Radvany 1990), whose nordicity is between 200 and 500 vapo. The Middle North, which can rise very high in latitude in Northern Europe or Russia according to the Hamelin nordicity criteria, defines a new strategic world zone for polar development by concentrating most of the infrastructures, networks and flows necessary for the feasibility of major Arctic projects.

– To place the North in dynamic processes of "nordification" and "de-nordification", based on elements of socio-economic and environmental analysis. The urban object is illuminating here. The urbanization of polar centralities equipped with a high level of facilities and services "denordifies" certain Arctic regions, whereas, on the contrary, processes of relegating unattractive northern spaces through a drop in public support or the closure of an activity that structures the local economy can "nordify" a boreal space, particularly through the enclave effect (Canobbio 2011; Vaguet 2016, pp. 125–134). Nordicity reminds us of the importance of relevant indicators that interact with development dynamics when attempting to assess polar mutations.

On this theme, which remains a major issue in boreal analyses, the NORDREGIO Institute proposed the concept of "Northern and Sparsely Populated Areas" (NSPA) at the request of the European Commission at the end of the 2000s, based on the study of population density (less than 8 inhabitants per km^2) which made it possible to delineate a new low-density Northern European regional area (less than 4.9 inhabitants per km^2) involving 14 administrative regions. Recognized by the OECD, the NAPS concept is strongly involved in the reconfiguration of the analysis of current or prospective boreal transitions. It implies a reorientation of public policies in support of specifically northern socio-economic issues, with priority given to urban networks, the enhancement of boreal amenities and the improvement of inter-regional material or immaterial flows (OECD 2016).

3.1.1.4. *Climate change and polar weather, the past and the revolution*

Climate change is calling for a new relationship with weather in the Arctic. Over the past two decades, it seems to have irreversibly abrogated the relationship to the historical geographical determinism of the Arctic regions, particularly in the analysis of a periphery whose isolated nature was more due to its "extreme" climatic characteristics than to its spatial remoteness from the centers of globalization. When, in the early 2010s, the international media announced the "opening of Arctic sea routes" on the assumption of increased navigability due to the retreat of the summer ice pack, the interest of these routes was essentially measured by the time saved between the nodes of globalization, implying an Arctic representation due to its geographical proximity. The publication of a new map of polar inclusion in a new regional space presented as "functional" thus confirms without nuance the transition from geographical marginality to integration in globalization.

While in-depth academic or sectoral expertise on international maritime transport attempts to demonstrate the realities of Arctic transits and their characteristics, the geographical implication of these hypotheses is producing a real revolution in the way the Arctic space is integrated into the world system (Lasserre 2010).

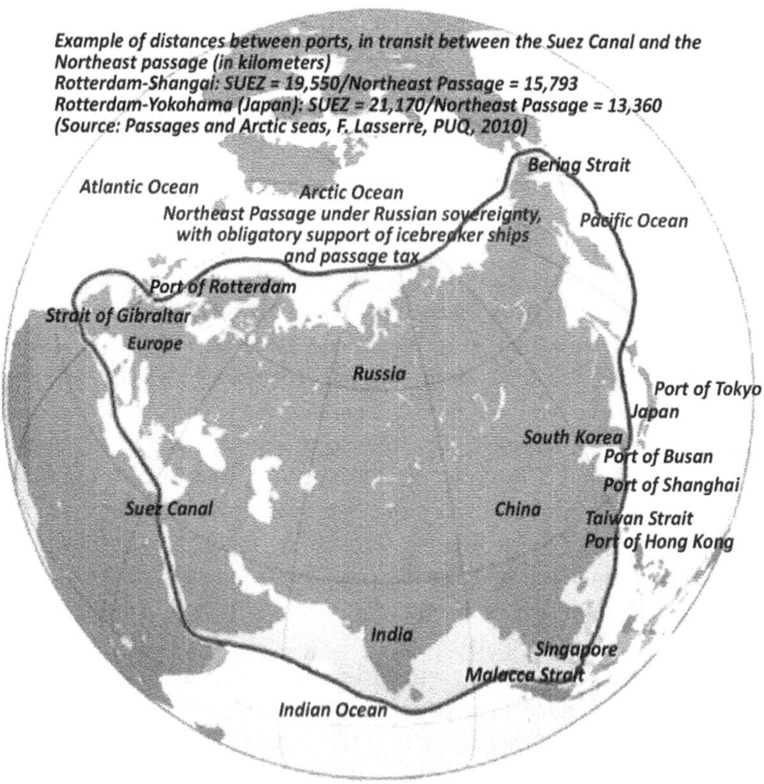

Figure 3.3. *A tenacious representation: Arctic integration through the routes of globalization (source: Canobbio 2019, based on a map by the Baltic shipowner Tschudi, in the 10th Annual Russian-Norwegian Oil and Gas Conference, Oslo, January 25, 2012). For a color version of this figure, see www.iste.co.uk/mercier/climate.zip*

This new global polar imagination of a collapse of Arctic space characteristics, in the confusion between reality and prospective, is sustained by the evocative power of climate modeling in the cryosphere at horizons within the century. In an attempt to clarify the real effects on their strategic dimensions, the U.S. Navy adopted three assessment timeframes in a 2014 framework document: the short term 2014–2020, the medium term 2020–2030, and the long term post-2030 (U.S. Navy 2014).

This hybridization of boreal weather implied by climate change between *polar present time* – social time, the time of indigenous activities (Lavrillier and Gabishev 2017), the time of human development, adaptation to change and crises, the polar seasons, the astronomical realities of solar radiation and sunshine duration – and *the prospective and scripted time* through the effect of modeling, has brought about a fundamental breakthrough in the interpretation of polar issues.

3.2. The manufacture of polar issues, between global warming and globalization

The 2000 decade was an arena for the emergence of the major contemporary polar issues, prefiguring the relationship between "opportunities", "risks" and "responsibilities" that now shapes most of the political doctrines centered on Nordic-Arctic issues. The two contexts for assessing polar issues, climate change and the effects of globalization on the cold zone, represent a radical break with the post-Soviet context of Arctic environmental emergency which had produced inter-regional scientific and political cooperation based on issues of security, denuclearization and assessment of polar contaminants and their impacts on Indigenous populations. Thus, the establishment of the Arctic Environmental Protection Strategy (AEPS) in June 1991 on the initiative of Finland and Canada, through the signature of the Rovaniemi Process by the eight polar states, defined ethno-cultural, environmental and health issues as its subjects of competence and intervention. The creation of the Barents Euro-Arctic Council in 1993, a forum for intergovernmental cooperation involving the 13 Barents regions, pursues convergent objectives by focusing its initial actions on the environmental security of this essentially Norwegian and Russian region, which constituted both a strategic fishing zone in the Arctic and the epicenter of the Russian military system in the North from the Poliarny-Severomorsk complex. The first decade of the 21st Century affirmed a paradigm shift in the Arctic based on a dual assessment process that would produce a powerful media impact:

– A process for evaluating energy and mining potential, which is more often associated with the effects of global warming than with multinationals' new economic models that integrated the Arctic's potential into a global materials market under very strong upward pressure until the effects of the 2007 financial crisis. Its spread to developed economies will durably modify polar investment strategies, based on increased project selectivity, particularly due to strong fluctuations in raw material prices, the arrival of new oil and gas offers from American shales, which have had an impact on hydrocarbon prices, and the better documented production of risk/opportunity analyses. Total announced in 2012, as did Shell in 2013, that it would withdraw from oil extraction projects in Arctic sea areas subject to permanent or seasonal freeze-ups.

– A political and strategic process to clarify the Arctic maritime claims of the polar states, which had remained unresolved during the Cold War and which generated a collective momentum for resolution.

3.2.1. *Warming and space production, a decade of confusion off the Arctic coasts*

Without a doubt, the relationship between climate change and spatial impact has been primarily embodied in the Arctic on maritime issues, and this issue remains an active focus of global media interest. The expression "territorialization of the Arctic Ocean" has sometimes been used in international relations to interpret the relationship between climate change and spatial impact. This is a historic step in the geopolitical reconfiguration of the Arctic, which, after 20 years of confusion, is only just beginning to be resolved. The placement of a Russian Federation flag in the North Pole in August 2007 at a depth of 4,200 meters by the privately funded Russian Arktika expedition stirred a number of phantasms associated with a return from the "Cold War". The 12 years that separate us from this spectacular event have shown that the polar maritime issue is politically more sophisticated than a programmed confrontation between Arctic powers. In 2008, the "A5" agreed on the Ilulissat Declaration, which recognized the framework of the International Law of the Sea – and its bodies – as the framework for resolving all maritime disputes, then in 2010 Russia and Norway signed the resolution of an old maritime dispute in the Barents Sea. This first "modern" border agreement between two major Arctic maritime nations was taking effect at the heart of a highly strategic region that is valued for its gas potential, rich fishery resources and its importance in the transit of Russian ships using the Northern Sea Route to European markets. The purpose of this section is not to produce an exhaustive analysis of a dossier that requires a great deal of legal technicality on the international law of the sea, which did not include chapters dedicated to the specificities of the polar regions, other than a modest clause on the relationship between sea ice and environmental protection, known as clause No. 234 (Lasserre 2010).

The Arctic maritime issue is based on four main themes:

– the extension of the *sovereign* domains of coastal states beyond the exclusive economic zones (EEZs), which implies the resolution of "disputes" related to overlapping claims;

– clarification of the status of the two polar sea routes: the Northeast Passage on the Russian side, historically known as the Northern Sea Route (NSR), and the essentially Canadian Northwest Passage. These routes are not affected by maritime claims, being spatially in the sovereign domains close to the Russian and Canadian coasts, but they remain the centerpieces of regional development planning, particularly energy and mining. This planning became a reality in Russia in 2017 with the opening of the Yamal gas complex, which finalizes the progressive south-north exploitation of the Ob oil and gas basin up to its mouth and adjacent maritime areas (Vaguet 2016, pp. 125–134). The status of these two maritime routes pits Russia and Canada – who share a common approach on the intangible national character of these routes – against the rest of the A5, which assesses the "international" characteristics of these passages on the model of the Strait of Gibraltar. The current implicit status of routes "under sovereignty" remains under the control of Russian and Canadian logistical arrangements for surveillance and intervention, which implies a "polar maritime administration", and the payment of a passage fee for the NSR. Russia, which has included this seaway in its Arctic development project since the late 1920s, considers this route to be a major axis of its Eurasian development, particularly gas. The Yamal project requires a new fleet of 14 ice-breaking LNG carriers (under construction between 2017 and 2023) and consolidates the economic development of the only functional Arctic route with a real fleet of ice-breakers, including the *Sibir,* launched in 2017. This first of three ships of a new polar class will be capable of sailing up the estuaries of the great Siberian rivers. China, which has declared itself a "country close to the Arctic", is a partner in the Yamal project through the China National Petroleum Corporation (CNPC) and the Silk Road Fund, the sovereign fund of the Silk Road. China is thus integrating the NSR as a contributor to a boreal Europe-Asia axis that is strategic for its national development.

However, this assessment should not mask the low level of flows recorded by a few hundred transits per year despite an expected increase in the number of transits in the NSR in the coming years. On the Canadian side, the Northwest Passage remains a low-traffic route with poor reception infrastructure, awaiting the fully operational commissioning of a new deep-water port at Nanisivik in the northern part of the Nunavut Territory. Mostly used for the summer refurbishment of coastal Indigenous communities, the Northwest Passage is nevertheless seeing new entrants, mainly cruise ships, which are increasing in size. The passage during the two summers of 2016 and 2017 of the *Cristal Serenity*, a vessel of more than 800 feet and 1,600 tourists and crew members, was the subject of a wide-ranging debate on the risks taken by shipowners, the position of the major insurance companies and the capacity of the riparian countries to intervene and cooperate in the event of an accident or damage in areas lacking any logistical infrastructure. This cruise was not renewed after 2017.

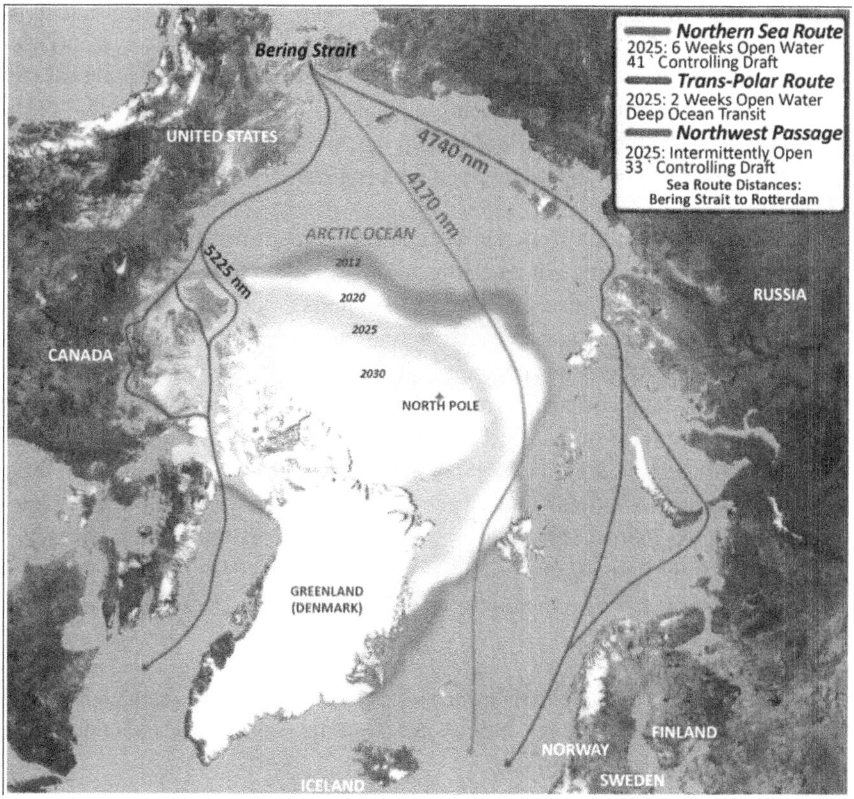

Figure 3.4. *Polar shipping routes (in nautical miles, nm) and the reduction of summer sea ice: mapping a new "transpolar route" incorporates the evolution of sea ice by 2030 (source: U.S. Navy 2014). For a color version of this figure, see www.iste.co.uk/mercier/climate.zip*

In spite of this decision, the increase in polar tourism and the projection of the effects of the drastic reduction in summer sea ice on a more navigable season continue to raise questions about the following:

– Canada's ability to regulate these new forms of traffic without a massive investment in a renovated polar fleet, which was announced more than 10 years ago but has had no effect;

– regulatory conditions for the transit of ships in "Arctic waters", and the setting of national and international standards: since 2017, the Polar Maritime Code has been the central element of the international legal system in the face of the lack of normative co-production by the A5;

– the conditions for the exploitation of offshore resources, in particular gas, today mainly in the form of liquefied natural gas (LNG)[4] and oil (however, off-shore remains limited to Russia). Both the prevention and management of ecological crises potentially linked to this exploitation and to polar transits, and the assessment of their environmental and anthropogenic impacts on the Arctic basin, require a transregional approach and a framework for circumpolar cooperation.

It was Russia that in 2001 opened the current round of maritime claims by filing a claim with the UN[5] setting out its claims to the extension of its sovereignty beyond its 200 nautical mile exclusive economic zone, on its continental shelf as far as the North Pole. In its text, Russia relied on the surveying of the Lomonosov and Alpha-Mendeleyev Ridges as the morphological and geological extensions of its continental shelf, in accordance with the obligations set out in Article 76 of the United Nations Convention on the Law of the Sea (UNCLOS). In 2002, Denmark, Canada, the United States and Norway questioned the Russian case and began programs of expertise in their underwater domains, often within the framework of bilateral scientific cooperation, as was the case during the 4th International Polar Year of 2007–2009 in the Canada-USA cooperation in the Beaufort Sea and on the Chukchi Plateau or in the cooperation between Canada and Denmark off Greenland and Ellesmere Island.

This beginning of the claims process remains within the framework of the international Law of the Sea, but it faces three pitfalls that give the resolution of this issue a multi-decennial time frame:

– the filing of claims is only legally viable within 10 years of ratification of the Law of the Sea Convention, which does not imply an order of recognition of claims (the United States has not ratified the treaty). In 2015, based on recent expert reports, Russia clarified its claims to a 1.2 million km² maritime domain including the North Pole, in accordance with the historical and symbolic delimitation of the "Russian Arctic" dating from 1926;

4 Norway and now Russia are increasing their share of LNG from the Barents production area and, since its start-up in 2017, from the Yamal complex, with the port of Sabetta allowing ice-breaking LNG carriers to transport LNG to European and Asian markets.

5 Russia submitted its claims to the Commission on the Limits of the Continental Shelf, which is competent to deal with all national claims relating to extended continental shelves.

– claims on extended continental shelves are particularly complex in the Arctic because of the geological evidence required;

– the Commission for Continental Shelves, producing reasoned opinions on the basis of the geological expertise provided, shall be competent to request additional studies for the purpose of claims.

The current assessment of "proven or probable" energy resources remains more than 90% within the territorial seas or exclusive economic zones of coastal states, so claims to extended continental shelves have no short- or medium-term impact on national economies and their extractive development patterns. However, they do demonstrate the early integration of the effects of climate change and the summer de-icing of the Arctic Ocean into the common will of coastal states to control the entire Arctic Ocean by extending their "sovereign" dominions, a term that must nevertheless be weighed by the complex legal characteristics of the law and use of these maritime zones. The dynamic of ongoing claims did not mobilize international public opinion on the unprecedented consequences of the almost total appropriation of a zone of the high seas, the "common good" of humanity, by five states. Only the European Union, in October 2008, attempted through a first resolution of its Parliament in favor of a balanced development policy for the Arctic region, to propose the sanctuarization of vast maritime zones and their uninhabited environments.

3.2.2. *Three interacting contexts*

The proposed sketch delineates the main themes associated with the three contexts that interact today in the assessment of polar issues and in the evaluation of spatial impacts: the climatic context, the globalization context and the geopolitical context to be understood in their circumpolar realities and in multipolar productions. The symbolic entry into the Arctic Council in 2013, as "permanent observer members" without voting rights, of India, China, Singapore, South Korea and Japan alongside the "historical" observers of the Council since its creation in 1996 – France, Spain, Germany, Italy, Poland, the United Kingdom and the Netherlands – could be interpreted, not without some naivety, as a signal that polar diplomatic bodies are open to multilateralism. However, it should be pointed out that the Arctic Council develops its activities in direct relation to "sustainable development and environmental protection in the Arctic. It is not competent to deal with disputes relating to resources or the delimitation of territories or any other issue related to security" (UEA 2016).

This opening did not, however, concern the admission of the European Union, which was once again rejected, revealing the ambiguity of its relations with the polar nations and regions. As Russia's leading energy partner and increasingly well connected to Russian Arctic oil and gas terminals from Varandey-Murmansk (and now Yamal-Sabetta, linked by the Nord Stream 1 pipeline network and in 2020 by Nord Stream 2), the EU is also an important player in supporting the polar regions through several programs with specific funds such as Kolarctic.

Despite Greenland's withdrawal from the European Community by referendum in 1985 (although it retains the status of an Overseas Country and Territory, or OCT), several European compensation mechanisms benefit the Arctic Island's fisheries and education sectors. The EU's "Northern Dimension", adopted in 2009, also marks a policy in favor of northern development.

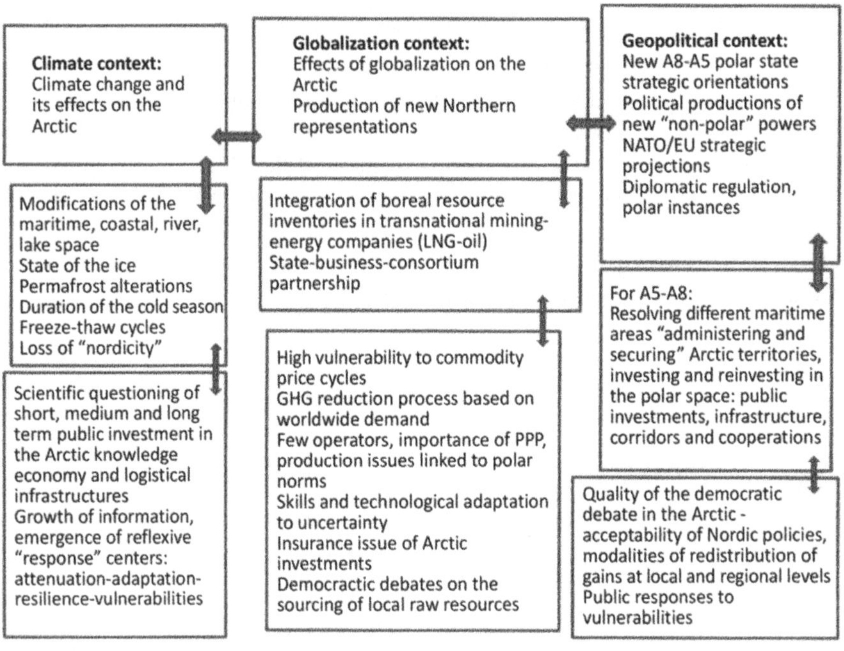

Figure 3.5. *Schematic representation of interaction games between climate change, globalization and geopolitical recomposition issues (source: Canobbio 2019). For a color version of this figure, see www.iste.co.uk/mercier/climate.zip*

Since 2016, the EU has had a real project in favor of the Arctic regions, consolidated by the adoption of an "integrated policy" that breaks with sectoral approaches in the Arctic and which must perpetuate its intervention programs in the EU's community and extra-community border polar regions. However, since 2009 and on the basis of a renewed policy of banning the import of seal products denounced by the Inuit leaders of Canada and Greenland, the EU has still not managed to become a full-fledged "player" in Arctic processing (Canobbio 2014).

3.3. The production of polar doctrines: rhetoric and frameworks for action

Between 2006 and 2018, all the Arctic states, some provinces and territories such as Quebec, which has jurisdiction over the economic development of its northern-Arctic space, non-Arctic powers such as China or France, and the European Union published a set of framework documents specifying the foundations of their "Arctic policies" (Kopra 2013; UEA 2016).

For the polar states, these Arctic policies are both directed towards their polar regions – in the administrative and political sense of their boreal regionalities, but also involve the entire Arctic space as a functional dimension of their assessments based on the issues of governance, multilateral "polar" and international cooperation, particularly in its scientific dimensions (Polar Research and Policy Initiative 2009; RFPA 2009; DGF 2011; SSA 2011; FSAR 2013; The White House 2013; NOP 2014; Norwegian Ministry of Foreign Affairs 2014).

This paragraph proposes a simplified analysis of the converging and differentiating factors that can be extracted from this heterogeneous corpus of national policies; some analysts have recently proposed comparative reviews of these policies by intervention sector or for ethical and environmental assessment purposes (Schulze 2017). It seems necessary to underline the fact that to date there is no "Arctic policy" at the scale of the polar basin, in particular no binding "Arctic climate policy" that would make it possible to unify all the mechanisms for mitigating the industrial risks linked to extractive projects. All these national Arctic policies share a twofold functional reality: they do not fit into a framework of legal or normative constraints and must therefore be understood in their declarative dimensions. However, they do define areas of intervention, objectives or priorities, depending on the terms chosen, which must produce sectoral public policies, particularly in their regional dimensions, which will essentially act on public territories. Private ownership is residual in the Arctic in favor of forms of concessions, "categories of land and resources" in the territories of Aboriginal agreements such as in Alaska (ANCSA 1971), Quebec (CBJNQ 1975) or Nunavut (Entente 1999), or "land usufruct for livestock farming" in Lapland, a model often associated with transfers of land rights to the regional managing authority as in the Norwegian Finmark Act of 2005.

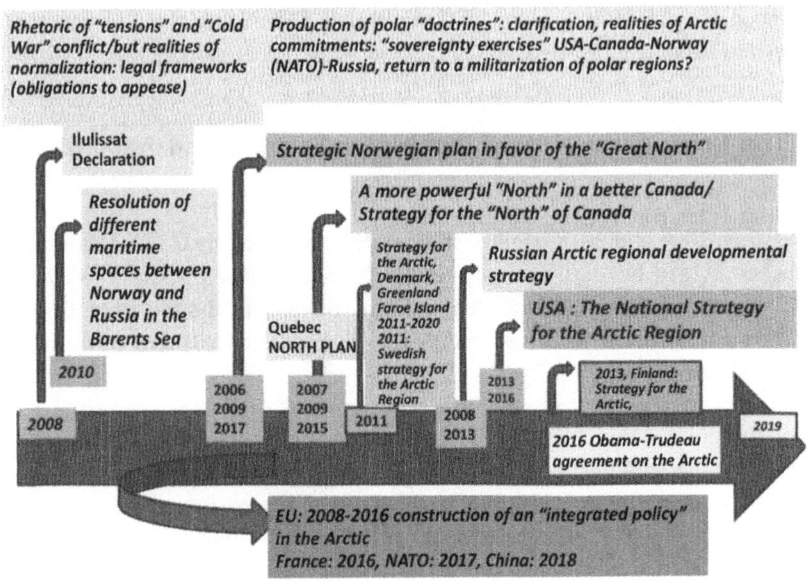

Figure 3.6. *Schematic representation of the production of Arctic policies during the period 2008–2018 (source: Canobbio 2019). For a color version of this figure, see www.iste.co.uk/mercier/climate.zip*

3.3.1. *Factors of convergence and consensus*

These factors are in the majority and form a common basis for Arctic policies. They can be grouped into six sectors of public priority:

– The affirmation of climate change in the Arctic (the use of the "High North" remains highly prevalent), and of its environmental impacts, which require a fundamental rethinking of national strategies in favor of Arctic development models. The notions of vulnerabilities/opportunities/adaptabilities/uncertainties frame all reflexive processes.

– The affirmation of the emergence of a polar regional economy that must be a growth driver for the 21st Century, a new "value-creating" space. This new Nordic-Arctic economy is detailed in its traditional components: extractive industries, minerals with high added value, oil and gas, land resources and tourism. According to a less conventional approach to the polar economy, industrial innovation linked to the increase in Arctic development projects, technological progress and industrial processes are still mainly concentrated in shipbuilding (methane tankers or polar-class ore carriers), or in the design of new industrial infrastructures, which prefigure the key sectors of the 21st Century. The regulatory economy (environmental standards, risk management, industrial standards and social standards), which is part of an emerging specification of an "Arctic economy of sustainable development", and the knowledge economy, which requires the development of resource centers on the Norwegian Svalbard model must accompany this transition. But this boreal "new economy" requires massive structural investments, such as the setting up of monitoring networks and satellite programs.

– The need to produce an attractive economic development environment, which must be ensured by three key words in Arctic policies: stability, cooperation (regional and international), and peace.

– The involvement of Indigenous institutions, sensitive in Arctic policies but low prioritiy except in the northern policy of Quebec and Canada (Polar Research and Policy Initiative 2009; Gouvernement du Québec 2015). Greenland holds a unique place here, particularly since the 2008 referendum on reinforced self-government and the associated transfer of new powers to the Greenland Government over the development of the island's resources.

– The nodal issue of cross-border cooperation is clarified both in its classic economic dimensions of regional structural investment (development corridors, e.g. the northern European "Arctic corridor") or political and cultural interactions, but also with regard to cooperation, security and rescue.

– A prospective dimension is called for in the need to adapt Arctic policies to "real" environmental, geopolitical and economic developments in the polar space. Implicit in this aspect is the great vulnerability of national polar ambitions to global economic cycles in terms of energy and minerals, but also the economic questions that weigh in the medium term (2030–2040) on the effects in the Arctic of a transition of industrial economies towards their CO_2 reduction objectives, taken in 2015. No doubt this paradoxical context, which is shrouded in uncertainty and national ambitions for the cold zone, explains a new process for analyzing boreal issues, in which the geostrategic, geopolitical, economic and now social dimensions of the Arctic are being brought together in a search for coherence.

3.3.2. *Differentiation factors*

The first factor that differentiates Arctic policies is of course the maritime dimension that underpins the A5 and increases the strategic dimension of the Arctic policies of the bordering states. Sweden and Finland nevertheless remain two Baltic tropic nations that are particularly sensitive to developments in cooperation or conflict that could alter economically strategic inter-regional relations in their Arctic territories from Lapland. For obvious geopolitical reasons dictated by its Russian neighbor, Finland, the only polar country to defend an "open" multilateral approach to the Arctic (called A8+), remains at the forefront in promoting a model of integrated cooperation, based on a region pacified by the development of new Arctic regional economies, which is contained in the concept of a "new Arctic era" set out from 2009.

The second differentiating factor is expressed in the relationship between polar states and their boreal Indigenous populations, which reveals the asymmetry of regional ethno-cultural integrations in the Arctic. Canada and, more moderately, the United States involve dialogue with Indigenous populations as part of the foundation of new Arctic policies. This common will led in March 2016, under the Obama administration, to the creation of Canada-U.S. Arctic marine protected areas and in Alaska to the invention of the *Northern Bering Sea Climate Resilience Area* for the protection of indigenous coastal fisheries. In Canada, the vast territory of Nunavut, created in 1999 and 85% Inuit, allows for greater institutional and political integration of the Canadian Arctic in support of sovereignty claims, in particular a "Canadian" status for the Northwest Passage that runs through the territory of Nunavut and is therefore considered part of Canada's internal waters.

3.3.3. *The strategic dimensions of Arctic policies, the complex issue of polar militarization*

On August 8, 2019, the probable explosion of a nuclear propellant of a new type of missile at Nyonoska in the Western Russian Arctic during a ballistic test near the city of Severodvinsk served as a reminder of the existence of a vast military-industrial complex in this highly urbanized Arctic region between Murmansk and Arkhangelsk. The question of the militarization of the Arctic was once again revived in the media by this dramatic event, which should not hide the complexity of understanding the strategic polar theater and its inclusion in multipolar issues that are broader than the Arctic dimension itself.

The nationalist rhetoric of the polar states on defending their Arctic interests and the powerful media coverage of major joint maneuvers in the northern space, particularly the "Nanook" operations in Canada since 2007, the unprecedented deployment of Russian troops alongside Chinese troops invited to take part in Operation Vostok 2018 in Eastern Siberia in 2018 has contributed to re-establishing the geostrategic dimension of the polar basin as one of the major impacts of the effects of climate change and its consequences on national representations (Connolly 2017). This dimension appeared as early as the Second World War from the Arctic routes used to transport the war effort from North America to Russia. As early as 1946, the sociologist and geographer André Siegfried described the forces acting on the "Arctic front" as "ice-water imperialism" (Siegfried 1946). For nearly half a century, the Cold War will thus fix the polar space as a strategic glacis organized around chains of advanced operational bases of the Russian and American warning systems (Besnault 1992).

Polar military geography then recognizes the three dimensions of the Arctic Basin: the air dimension, which also includes civil transpolar and transcontinental routes, the land and sea "surface" dimension, and the sub-glacial dimension. The region was also an important area for Russian atmospheric nuclear testing in New Zealand and underground testing in the Aleutian Alaskan Islands.

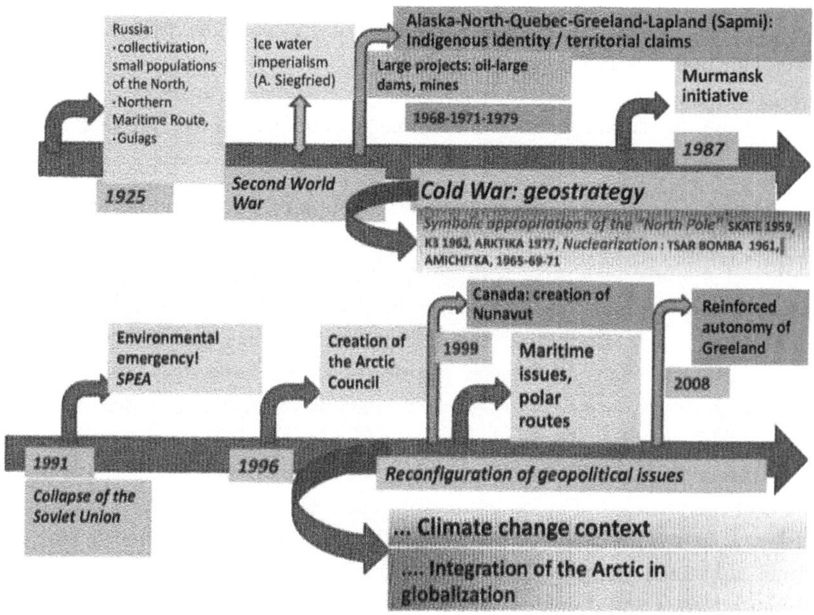

Figure 3.7. *Simplified schematic approach to historical trajectories of national recompositions and global contexts in the Arctic during the period 1925–2008 (source: Canobbio 2019). For a color version of this figure, see www.iste.co.uk/mercier/climate.zip*

The current process of militarization seems to be inscribing the Arctic into a new dramaturgy of confrontation. However, it remains essentially analyzed on the basis of the Russian Arctic policies of 2009 and 2013, which plan to redeploy by 2020 the of task forces in contact with the Northern Sea Route. The opening of several bases in the Russian Arctic and the increase in human and military logistical resources were announced and publicized by Moscow as one of the key elements of its Arctic development policy. Geostrategic analysts thus confirm the rearmament of the Northern fleet, the resumption of Russian strategic submarine activity in the North Atlantic and the Arctic, the reopening of six old or new bases in New Assembly, in the Franz Joseph Archipelago, at Cape Schmidt and on Wrangel Island, the reactivation of 13 military airfields and the creation of "Arctic brigades" (RFPA 2009; Gatollin 2013; Laruelle 2015; Connolly 2017). But the Arctic policies proposed by the eight polar states since 2006, including those formalized after Russia's annexation of Crimea in 2014, collectively propose a strategic reinvestment and a strengthened relationship with polar sovereignty. They also affirm in an apparent paradox that the Arctic, as a "zone of low conflict intensity", must be preserved from crises by regional cooperation mechanisms (MacDonald 2015; Norwegian Ministries 2017; U.S. Navy 2019).

This singular situation, which seems to contradict the Russian strategic assessment, must be interpreted in its two main dimensions:

– the first is involved in the operational function of troops and military assets in the Arctic, which remains the sole regional security force, including in the face of scenarios of major civil disasters or environmental crises. The Russian, Canadian, American and Norwegian doctrines place the question of securing the Arctic regions and maritime routes (including the fishing and tourism sectors) as a mission involving military means as a matter of priority, in the same way as the defense of national sovereignty. Norway, in the latest versions of its Arctic policy dating from 2014 and 2017, specifies the role of armed land, naval and air troops in the Arctic in terms of rescue and search forces, but also in the control of Arctic fishing zones. The Arctic is at the center of a complex interplay between the economy, geopolitics and geostrategy, particularly in Norway and Russia, which combine in the same doctrinal approach security issues, the deployment of capabilities to defend sovereign interests and domains, and the implementation of economic development projects. In the coastal and maritime arctic areas of these two countries, the gradual increase in gas and oil production migrating towards the arctic zones, mainly in the Barents and Kara Sea regions, and in the lower Ob basin as far as the Yamal Peninsula, objectively contributes to a long-term logistical and operational re-commitment. The security of extractive infrastructures and induced maritime traffic, including against a terrorist risk, thus directly contributes to the credibility of northern industrial projects developed through the constitution of international consortiums. A major regional issue, the Russian Arctic now accounts for 20% of national GDP and 25% of its exports (Gattolin 2013);

– the second dimension articulates the polar basin in a much larger strategic area that involves both the evolution of the Euro-Atlantic area and international relations, in particular between Russia, the United States and NATO. The question of the deployment of the American "defensive" anti-missile shield in Europe has prompted a Russian response that is just as much a defensive one, which places the Arctic space in its historical vocation as a "contact zone" and not as a central regional issue in terms of its capacity to produce its own forms of belligerence.

This multidimensional context of the Arctic's strategic identity has nevertheless been deeply affected by Russia's annexation of Crimea in 2014, opening specifically in Norway, and outside NATO, in Sweden and Finland, a new cycle of prudential relations with Russia, without giving up the preservation of the Arctic as a peaceful region. Former Norwegian Foreign Minister Jonas Ghar Støre[6] points out that "on 27 March 2014, shortly after Russia's annexation of the Crimea, Moscow reached an agreement with Denmark on the modalities for delimiting their continental shelves in the Arctic" (Gahr-Støre 2016). NATO, in a recent paper on Arctic security published in October 2017, implies this complex conceptual confrontation of an Arctic region that is simultaneously an area of cooperation, an area of economic development and an area militarized by objective rearmament (Connolly 2017).

3.4. Geography of a new system of stakeholder relations in the Arctic

The regional integration processes outlined above have redefined the stakeholder landscape in the high latitudes. Figures 3.8 and 3.9 propose to specify the contours and simplified fields of interaction based on two simple objectives: to establish the landscape of polar stakeholder groups and to propose the relational dynamics that act between these stakeholders.

These two diagrams present in particular new stakeholders who are involved in polar issues alongside the traditional players (Arctic nations and territories). The major public companies, which are very active in boreal development, remain key players alongside the polar states. The role of the Norwegian company StatoilHydro (now Equinor) and the Russian company Gazprom was mentioned in the resolution in 2010 of cross-border conflicts in the Barents Sea, at the heart of a vast gas exploration zone that is gradually bringing its economically viable fields into production.

---

6 Jonas Ghar Støre was Norway's foreign minister. He participated in the signing in 2010 of a Norway-Russia agreement on the delimitation of the maritime border in the Barents Sea, putting an end to a 40-year-old maritime dispute.

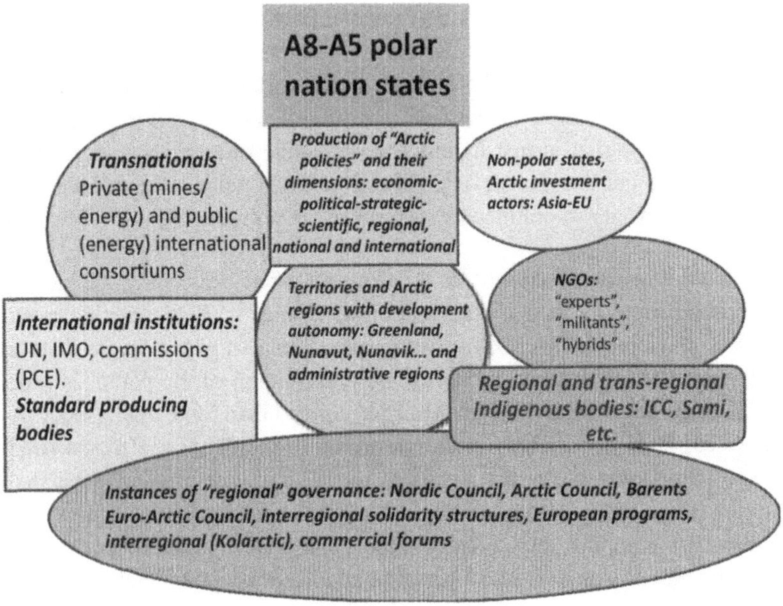

Figure 3.8. *Schematic representation of "stakeholder" actors in the Arctic (source: Canobbio 2019). For a color version of this figure, see www.iste.co.uk/mercier/climate.zip*

This "industrial diplomacy" must thus be evaluated in its capacity for parallel action. The rise of non-polar states in development projects, mostly represented by industrial consortiums combining large state-owned companies with private or parastatal multinationals, is a recent development that places China and its investment capacities in the Arctic as a major player in current energy and mining developments. The role of NGOs and public opinion in the current awareness of the effects of climate change on the Arctic must not be underestimated in the acceptability of offshore energy projects, particularly oil projects. A key player in the stakeholder landscape is not often called upon in this respect, namely the major reinsurance companies such as Lloyd's or Munich RE, which are now investing their *expertise in the risk* of polar developments and influencing the production of new standards and prohibitions linked to the risk/benefit balance sheet. In this respect, the report "Arctic Opening, Opportunity and Risk in the High North" published by Lloyd's in 2012 demonstrates the selective and particularly prudential approach of the major players in the insurance industry to the challenges of polar development (Emmerson 2012; Munich 2018). It is indeed the economic and environmental assessment of the spatial impacts of climate change that is now carried out by these major stakeholders in the world economy. The announcement in 2019 by the three container ship majors MSC, CMA-CGM and Hapag-Lloyd not to integrate the Arctic routes "as is" because of the environmental impacts of container ship traffic on the environment, seems to consolidate this prudential approach, which nevertheless remains sectoral and not systemic.

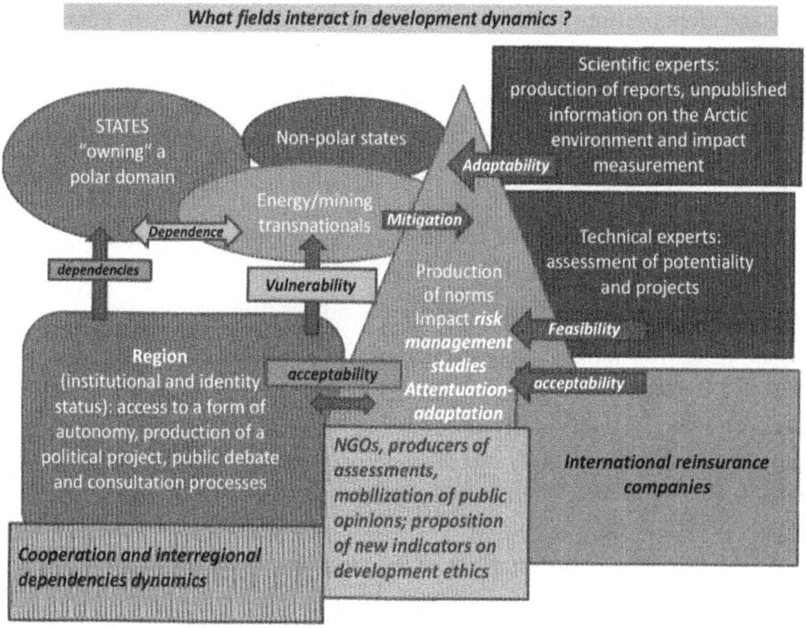

Figure 3.9. *The relational dynamics between "stakeholders" in northern-Arctic developments (source: Canobbio 2019). For a color version of this figure, see www.iste.co.uk/mercier/climate.zip*

3.5. Conclusion: polar metamorphisms

This chapter attempted to explain how "polar space", in the diversity of its regional components, and "polar time", in the plurality of its study dimensions (*present time, planned time and modeled time*), interact in the assessment of contemporary polar issues. The Arctic region, considered by simplification as a homogenous and indivisible space, appears as a global model for prefiguring the impacts of climate change. However, the regional "metamorphosis" processes demonstrate the extreme diversity of polar situations: polar regions already under development by globalized investors, such as certain areas of the Russian Arctic, northern Europe or the Canadian midwestern basins, regions awaiting development such as Greenland, and vast territories that seek to believe in an economic future that will bring social progress for communities when announcing new polar opportunities. This fragmented landscape in the fragility of its cumulative changes – climate and environmental, social and cultural, economic and political – is neither an *El Dorado* nor an indifferent fringe to the great upheavals of the world; it has become one of its centers of study. The analysis of the impacts of the interactions between climate change and globalization is all the more complex to address and to disentangle at the heart of a system of issues that calls upon the Arctic as one of the epicenters of environmental vulnerabilities and contemporary geopolitical uncertainties.

3.6. References

André, M.-F. (2011). Où commencent les régions polaires ? In *Cap sur les Pôles*, Lemarchand, A., André, M.-F., Rémy, R. (eds). Omnisciences, Montreuil.

Besnault, A. (1992). *Géostratégie de l'Arctique*. Economica, Paris.

Canobbio, É. (2011). Mondes arctiques miroirs de la mondialisation. *La Documentation photographique*, 8080.

Canobbio, É. (2014). La zone froide européenne au défi des intégrations régionales, éléments objectifs de diagnostic. In *La régionalisation du monde, construction territoriale et articulation global local*, Gana, A., Richard, Y. (eds). IRMC Karthala Tunis, Paris.

Connolly, G.E. (2017). L'OTAN et la sécurité dans l'Arctique. Report, Commission politique/Assemblée parlementaire de l'OTAN [Online]. Available at: https://www.nato-pa.int/download-file?filename=sites/default/files/2017-11/2017%20-%20172%20PCTR%2017%20F%20r%c3%a9v.%201%20fin%20-%20ARCTIQUE.pdf.

DGF, Denmark, Greenland and the Faroe Islands (2011). Kingdom of Denmark Strategy for the Arctic 2011–2020 [Online]. Available at: http://library.arcticportal.org/1890/1/DENMARK.pdf.

Dorion, H. (2015). La nordicité. In *Québec, Canada, Russie : 100 miroirs*, Dorion, H. (ed.). Presses de l'Université Laval, Quebec.

Emmerson, C. (2012). Arctic Opening: Opportunity and risk in the high north. Chatham House. Available at: https://www.chathamhouse.org/sites/default/files/publications/0412arctic.pdf.

FSAR, Finland's Strategy for the Arctic Region (2013). Artiken strategia 2013 [Online]. Available at: http://vnk.fi/documents/10616/334509/Artiken+strategia+2013+en.pdf.

Gahr Store, J. (2019). Worried about increased tension in the Arctic [Online]. Available at: https//www.highnorthnews.com.

Gatollin, A. (2013). Arctique : préoccupations européennes pour un enjeu global. Information report, Commission des affaires européennes [Online]. Available at: https:// www.senat.fr/notice-rapport/2013/r13-684-notice.html.

Gouvernement du Québec (2015). Le Plan Nord à horizon 2035 [Online]. Available at: https://plannord.gouv.qc.ca/wp-content/uploads/2017/05/Synthese_PN_FR_IMP.pdf.

Government Offices of Sweden (2011). Sweden's strategy for the Arctic region [Online]. Available at: https://openaid.se/wp-content/uploads/2014/04/Swedens-Strategy-for-the-Arctic-Region.pdf.

Kopra, S. (ed.) (2013). China's Arctic interest. In *Arctic Yearbook Heinimen*, 1, 107–124 [Online]. Available at: http://www.arctic.arcticyearbook.com/images/Articles_2013/KOPRA_AY13-FINAL.pdf.

Laruelle, M. (2015). *Russia's Arctic Strategies and the Future of the Far North*. Routledge Taylor/Francis Gray, London/New York.

Lasserre, F. (2010). Passages et mers arctiques : géopolitique d'une region en mutation. Presses de l'université du Québec, Quebec.

Lavrillier, A. and Gabyshev, S. (2017). *An Arctic Indigenous Knowledge of Landscape, Climate, and Human Interactions: Evenki Reindeer Herders and Hunters*. Kulturstiftung Sibirien, Fürstenberg/Havel.

MacDonald, A. (2015). La militarisation de l'Arctique : nouvelles réalités, exagération et distraction. *Revue Militaire Canadienne*, 15(3), 18–28.

Munich RE (2018). Corporate responsibility report [Online]. Available at: https://www.munichre.com/en/company/corporate-responsibility.html.

Norwegian Ministries (2017). Norway's Arctic strategy between geopolitics and social development [Online]. Available at: https://www.regjeringen.no/contentassets/fad46f0404e14b2a9b551ca7359c1000/arctic-strategy.pdf.

Norwegian Ministry of Foreign Affairs (2014). Norway's Arctic policy: Creating value, managing resources confronting climate change and fostering knowledge. Development in the Arctic concerns us all [Online]. Available at: https://www.academia.edu/16302611/norways_arctic_policy.

Nymand Larsen, J. and Fondhal, G. (eds). (2014). Arctic human development report: Regional processes and global linkages. Nordic Council of Ministers [Online]. Available at: http://norden.diva-portal.org/smash/get/diva2:788965/FULLTEXT03.pdf.

OECD (2016). Territorial reviews. Northern sparsely populated areas, policy highlights. OECD Publishing, Paris [Online]. Available at: http://www.oecd-ilibrary.org/regions.

Paet, U. and Pietikäinen, S. (2016). UE Rapport sur une politique intégrée de l'Union européenne pour l'Arctique (2016/228(INI). Report, Commission des affaires étrangères/commission de l'environnement, de la santé publique et de la sécurité alimentaire [Online]. Available at: https://www.europarl.europa.eu/doceo/document/A-8-2017-0032_FR.html.

Polar Research and Policy Initiative (2009). Canada's northern strategy: Our North, our heritage, our future [Online]. Available at: http://polarconnection.org/canadas-northern-strategy-north-heritage-future/.

Radvany, J. (1990). *L'URSS, régions et nations*. Masson, Paris.

RFPA (2009). Basics of the state policy of the Russian Federation in the Arctic for the period till 2020 and for a further perspective [Online]. Available at: http:/www.arctis-search.com/Russian+Federation+Policy+for+the+Arctic+to+2020.

Schulze, V.G. (2017). Arctic strategies round-up 2017 [Online]. Available at: https://www.arctic-office.de/fileadmin/user_upload/www.arctic-office.de/PDF_uploads/Arctic_Strategies_EN_10.11.17.pdf.

Siegfried, A. (1946). Le front Arctique du Canada. *Revue de Paris*, 7.

The White House (2013). USA. National strategy for the Arctic region [Online]. Available at: https://obamawhitehouse.archives.gov/sites/default/files/docs/nat_arctic_strategy.pdf.

U.S. Navy (2014). Arctic roadmap 2014–2030. Report, Navy Task Force Climate Change [Online]. Available at: https://www.navy.mil/docs/USN_arctic_roadmap.pdf.

U.S. Navy (2019). Strategic outlook for the Arctic. January [Online]. Available at: https://www.navy.mil/strategic/Navy_Strategic_Outlook_Arctic_Jan2019.pdf.

Vaguet, Y. (2016). Les formes et les enjeux de l'urbanisation en Arctique. In *L'Arctique en mutation*, Joly, D. (ed.). EPHE, Paris.

Xinhua (2018). China's Arctic policy: The state council information. Office of the People's Republic of China [Online]. Available at: http://www.xinhuanet.com/english/2018-01/26/c_136926498.htm.

4

Coastlines with Increased Vulnerability to Sea-level Rise

Axel CREACH
Sorbonne University, Paris, France

4.1. Introduction

In 2010, the President of the Republic of Maldives announced the creation of a sovereign fund, a function of which could be to acquire land in foreign countries in order to build the "New Maldives". This island state is located entirely on a coral reef formed by the growth of corals, which is in fact, by definition, a low-lying territory (80% of the Maldives are located less than 1 m above the current sea level). It is therefore directly exposed to contemporary sea-level rise, which is accelerating under the intensification of climate change effects. The double constraint of a low-lying territory built on coral reefs and the rise in sea level makes it necessary to consider, among other solutions, the medium- or long-term displacement of the population and the Republic of Maldives, which had about 400,000 inhabitants in 2017. This is a paradoxical situation for a country where a large part of the current economy is based on tourism, which takes advantage of its insularity and the coral reef: beaches, turquoise waters, hotels built close to the water, multiple possibilities of activities related to the presence of coral reefs, all characteristics that meet tourist expectations associated with coastal areas.

The example of the Maldives bears witness to the twofold antagonistic dynamic facing coastal areas: these spaces at the interface between land and sea are directly

Spatial Impacts of Climate Change,
coordinated by Denis MERCIER. © ISTE Ltd 2021.

affected by climate change and, first and foremost, the rise in sea level, which will redefine the boundaries of this interface in the future; at the same time, coastal areas are an attractive territory for populations and are the site of multiple activities that have been greatly strengthened in the past century and should continue with respect to the coastalization and the maritimization of societies (Miossec 2012). Societies are nowadays dependent on coastlines, since 90% of the world's goods traffic transits through ports (Rodrigue 2020) and they are one of the world's main tourist destinations. In this respect, societies, their facilities and their economies can be considered vulnerable to the constraints caused by sea level rise and climate change.

It is necessary to reflect on the future of coastal societies and, by extension, global society.

This chapter looks at the dual dynamics of the constraints of sea-level rise and the recent increase in populations and activities. The redefinition of these territories will therefore require the management and adaptation of societies in order to survive and maintain key activities. In the light of these dynamics, the different risk management strategies will be exposed, with the relocation of a state such as the Maldives to other territories being one possibility among others.

4.2. Coastlines under the influence of sea-level rise

4.2.1. *The pressures of climate change on coastlines*

4.2.1.1. *Sea-level rise*

The current rise in global temperatures is causing a general rise in sea level by two main mechanisms:

– on the one hand, increasing air temperatures are causing continental ice to melt, returning water in liquid form to the global ocean. This is glacio-eustatism (see Chapter 2);

– on the other hand, the increase in the temperature of the atmosphere leads to an increase in the temperature of the surface water. The increase in water temperature causes the water to expand. Thus, an increase of 1°C in a layer of water measuring 200 m causes it to rise by 20 cm (Paskoff 2001). This is the steric effect or thermal expansion.

According to the IPCC report (IPCC 2019), during the decade 2006–2015, when the global ocean level rose by 3 cm, 47% of this rise was due to thermoeustatism,

20% to the melting of mountain glaciers, 27% to the melting of the ice sheets of the Greenland ice sheet and 14% to the melting of the Antarctic ice sheet.

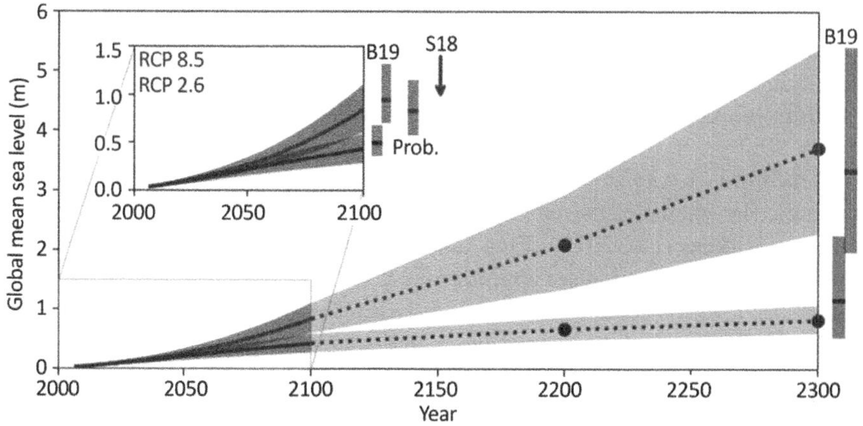

Figure 4.1. *Projections of sea-level rise between 2000 and 2300 for two greenhouse gas emission scenarios: RCP 2.6, "optimistic" scenario; RCP 8.5, "pessimistic" scenario (source: IPCC 2019). For a color version of this figure, see www.iste.co.uk/mercier/climate.zip*

The average sea-level rise is currently around 3 to 4 mm/year (IPCC 2019). The IPCC predicts that this rate will increase to between 4 and 10 mm/year in the coming decades (see Figure 4.1). Thus, by 2050, the sea level is expected to have risen by 0.24 m to 0.32 m depending on the scenario. By 2100, this rate could accelerate, and the sea level could reach between 0.43 m and 0.84 m by that time. Beyond that, projections are more uncertain. By 2300, the sea level could be higher than today's levels by between 2.3 and 5.4 m.

These figures represent the average sea level rise due to eustatism (or global mean sea level, GMSL). Locally, these variations may be counteracted or exacerbated by isostatic phenomena. It is therefore the relative sea level rise (RSL), a combination of global sea-level rise and local adjustments, that will directly affect coastlines.

Isostasy refers to the equilibrium of the lithosphere. Continents may be stable, uplift or subsidence, for natural reasons (rebound and glacio-isostasis adjustments) or anthropogenic reasons (subsidence linked to the weight of urbanization, pumping of groundwater). Thus, in the case of a rise in sea level (positive eustatism), three scenarios determine a relative sea level (see Figure 4.2). If the continents are stable and eustatism is between 0.43 and 0.84 m by 2100, the relative sea level rise will be of the same nature. If the continents rise by 1 m and eustatism is between 0.43 and 0.84 m by 2100, then the relative sea level will drop between 0.16 and 0.57 cm. This scenario occurs on northern coasts that were glaciated during the cold sequences of the Pleistocene and which are still experiencing a glacio-isostatic rebound (e.g. in Scandinavia and northern Canada).

In the third case, relative sea level rise may be exacerbated by coastal subsidence (sinking). For example, the city of Bangkok is now subsiding by an average of 3 cm/year (Erkens *et al.* 2015). If this rate of subsidence continues, it could be about 3 m deeper by 2100 even as sea level is projected to have risen by 0.43 to 0.84 m. The relative sea level rise could therefore be between 3.43 and 3.84 m. This is the case for all subsiding coastlines around the world, including the deltas of major rivers where large cities are located, which are thus doubly vulnerable to sea-level rise.

4.2.1.2 *Impact of climate change on tropical- and extra-tropical cyclones*

Climate change is likely to change the frequency of tropical- and extra-tropical cyclones, causing extreme sea-level events and morphogenic pressures on coastlines. As far as cyclones are concerned, the IPCC concludes that the intensity of events is likely to increase but remains very cautious about an increase in the number of cyclones (IPCC 2019). The increase in surface temperatures is also likely to widen the formation areas of tropical cyclones. As far as mid-latitude storms are concerned, no significant trend has yet been identified.

In any case, the extreme sea levels generated by the surges associated with tropical or extra-tropical cyclones are expected to mechanically increase and intensify as a result of sea-level rise.

For example, in February 2010, storm Xynthia generated an exceptional sea level of 4.51 m in the port of La Rochelle (Pineau-Guillou *et al.* 2010). The storm surge was measured at 1.53 m in this port. This has the likelihood of a returning period of less than 100 years, or less than a one in 100 chance of occurring each year (Pineau-Guillou 2012). In the future, if the sea level actually rises by 0.84 m by 2100, then the level measured at Xynthia could be reached by a surge of 0.69 m, which occurs on average once a year. On the other hand, if the surge generated by Xynthia were to occur in 2100, then the sea level reached could be 5.35 m.

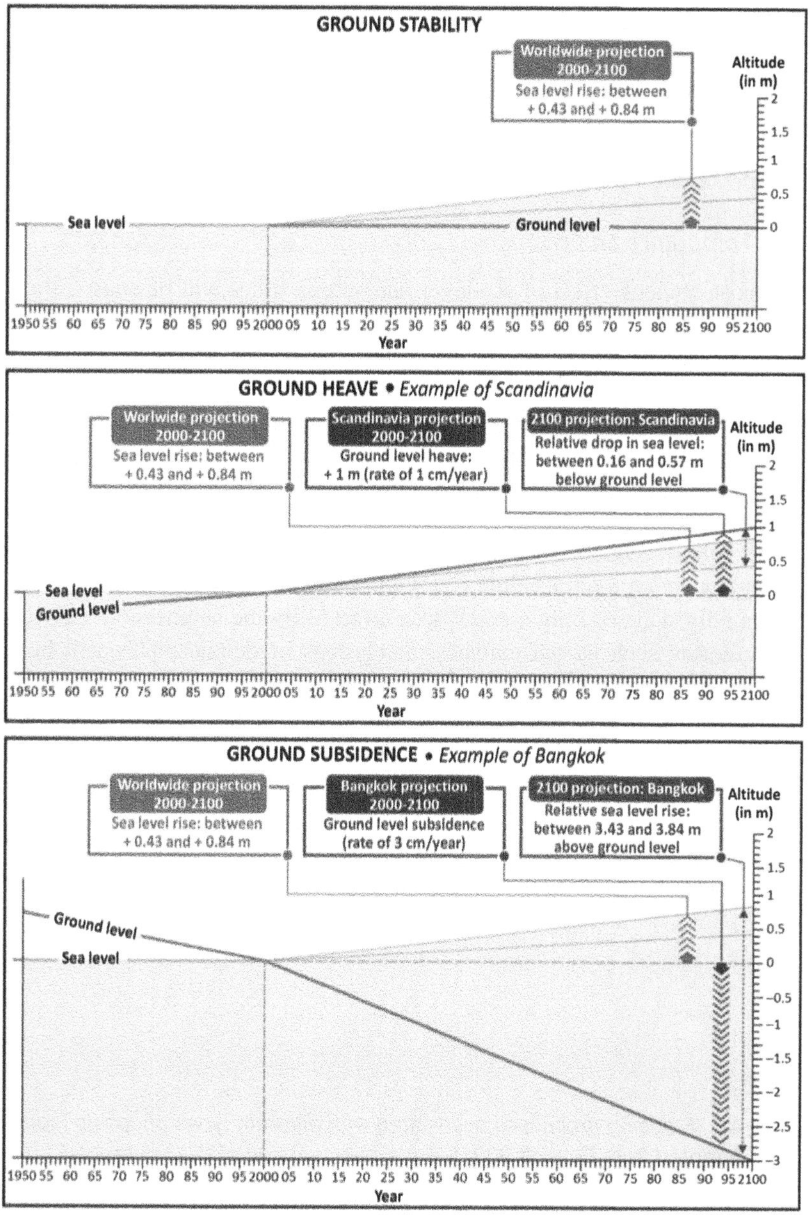

Figure 4.2. *The concept of relative sea level on coastlines as a function of different isostatic behaviors (source: adapted from Paskoff 2001). For a color version of this figure, see www.iste.co.uk/mercier/climate.zip*

If the Xynthia event were to recur in 2100, an additional 13% of the territory of two regions mainly affected in 2010 could be flooded (CETMEF 2009). In 2010, 41 victims of the marine submersion were recorded in those two regions.

4.2.2. *Consequences of sea-level rise on coastlines*

4.2.2.1. *Temporary and permanent submergence of low-lying coasts*

The most obvious effect of sea-level rise on coastlines will be coastal flooding. This submersion will take place in two ways: on the one hand, temporary submersion could be more frequent due to the impact of meteorological and marine phenomena, which should be less intense for a similar impact than today (see section 4.2.1.2); on the other hand, the inexorable rise in sea level will lead to the permanent submersion of low coastlines located below this level if these coasts are no longer protected.

These submergences will affect the low-lying coastlines at different rates depending on the type of coastline (see Figure 4.3). Low-lying coastlines that gain altitude rapidly will see only a small area affected by the submersion. Conversely, low-lying coasts such as sea marshes, mangroves or deltaic plains will be more exposed to permanent submersion. For example, an average sea level rise of 1.5 m could submerge 22,000 km² of the deltaic plains of Bangladesh and affect 17 million people (Fitzgerald *et al.* 2008; Webb *et al.* 2013).

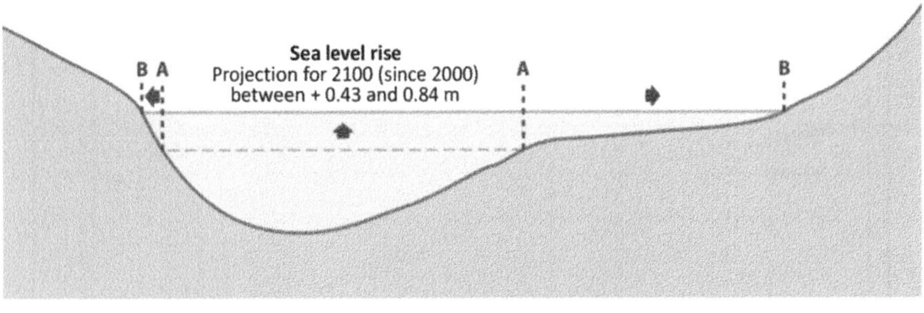

Figure 4.3. *Flooding processes associated with different types of coasts (source: adapted from Paskoff 2001). For a color version of this figure, see www.iste.co.uk/mercier/climate.zip*

Today, 76 million people live in an area that can be submerged by an extreme rise in sea level (100-year event), or 1.3% of the world population (Muis *et al.* 2016), and 110 million people currently live below sea level (Climate central 2019).

By 2100, this figure could rise to 200 million depending on the projected sea level (RCP scenario 4.5). In addition, by 2050, 300 million people could be living in an area that could be exposed to annual flooding.

Logically, the vast low areas are the most exposed. For example, Asia and its large delta areas are home to 151 million people in areas that could be permanently flooded by 2100 (Climate Central 2019), while the relative rise in sea level could be exacerbated by subsidence.

In Europe, the sea level could be 0.29 to 0.59 m higher according to the optimistic IPCC scenario, and 0.60 to 1.10 m higher according to the pessimistic scenario (AGE 2019). In 2300, the sea level could even reach 5.4 m in the worst-case scenario. Thus, extreme sea levels would increase, mainly as a result of average sea level rise rather than a change in the frequency or intensity of storms (Vousdoukas *et al.* 2017). The 100-year sea level extremes would thus be 0.57 m (optimistic scenario) to 0.81 m (pessimistic scenario) higher in 2100.

The most exposed territories are thus the great plains from northern France to north-western Germany and Denmark, the English coasts and especially the east of the country, the Atlantic and Mediterranean coasts of France and the northern part of the Adriatic Sea and especially the Venice region. A total of 9.8 million Europeans now live between 0 and 1 m above the current sea level, and 17.5 million between 0 and 6 m (AGE 2019). In addition, 5 million Europeans would be exposed to a 100-year flood (Vousdoukas *et al.* 2017).

4.2.2.2. *Coastline erosion and retreat*

The coastline, which represents the boundary between land and sea, can be seen as a barrier or protection against rising sea levels. However, the sea-level rise could lead to shoreline retreat.

Erosion can affect two types of coastlines: rocky coasts, made of bedrock and destined to retreat, and sandy coasts formed by the accumulation of loose materials. The latter may experience alternating phases of erosion and accretion due to the occurrence of morphogenic phenomena or a shortage/abundance of sediment. Accumulation coastlines, in the form of dune belts, are important protections for large low areas located behind. Their dynamics are therefore essential in a context of rising sea levels: erosive dynamics will accelerate submersion, while accretion dynamics will – temporarily – compensate for flooding.

These accumulation coasts represent 31% of the world coastline, excluding ice-covered areas (Luijendijk *et al.* 2018). Over the period 1984–2016, it has been shown that 24% of the world's sandy coastlines have been retreating, 28% have gained and 48% have remained stable. However, the volumes committed are not commensurate. Over the period 1984–2015, 28,000 km² were eroded, the equivalent of a country such as Haiti, while 14,000 km² were gained by accretion (Mentaschi *et al.* 2018). These losses are mainly related to anthropogenic extractions, changes in sediment flows due to irrigation and large dams, and the destruction of mangroves.

Sea-level rise is expected to increase erosion as a result of various processes (Ranasinghe 2016) although erosion is primarily related to the effect of extreme sea levels (Mentaschi *et al.* 2018). Indeed, in the face of a steady rise in sea level, the coastline can remain relatively stable (Duvat 2019), which is particularly true for coral coasts (which are not, strictly speaking, accumulation coasts).

The coastline is therefore exposed to rising sea levels. It is, moreover, a space densely occupied by societies.

4.3. Increasingly attractive coastlines for societies

4.3.1. *The coastalization process*

From an empty coastline mentioned by A. Corbin to designate the occupation of the coast until the end of the 18th Century (Corbin 1990), R. Paskoff spoke of a "grasping" coastline (Paskoff 1993) to designate the contemporary coastline where conflicts of use have become recurrent in the face of the multiplication of activities in these areas. This formula illustrates the consequences of the change in the way Western societies look at coastlines and the process of coastal development that is taking place today.

Coastalization refers to the attractiveness of contemporary societies for settlement near the sea. Whereas the coastline was previously little occupied and was used for purely functional purposes (food, energy, defense), two main processes have reversed this trend and given rise to coastalization: the development of the leisure society and globalization (Miossec 2012).

The birth of the leisure society, starting in the 19th Century (Merckelbagh 2009) gave rise to a change in the way we look at the coastline. This change of outlook was first expressed by the phenomenon of haliotropism, which can be defined as the attraction for the sea and the coastline. As a counterpoint to the urbanization of Western societies, coastlines offer an "airy" and healthy space. The 19th Century was the period of the birth of sea bathing, initially reserved for a certain elite. In a second phase, heliotropism was added to this first trend. It expressed the attraction of societies towards sunny regions. This manifested itself in particular through the development of seaside tourism, representing a seasonal activity, but has also resulted in the settlement of a permanent population, often retired (Merckelbagh 2009).

At the same time, the development of globalization has led to a very sharp increase in international trade, mainly through maritime transport. The regularity and practicality of maritime transport (linked to the containerization of goods transport) and the economy of scale made possible by the gigantic size of ships make it the most widely used means of transport on the surface of the globe (Rodrigue 2020). The volume of goods transported by sea thus doubled between 1990 and 2012 (UNCTAD 2013). Today, 90% of the world's transported volumes are carried by sea (Rodrigue 2020). These volumes transit via ports, at the interface between land and sea and at the point of load breakage, which involve multiple sectors linked to logistics, storage, transport by land or air, processing, as well as decision-making centers.

This phenomenon of coastline development is likely to continue, given the continuing strong attraction for the coastline. In 2006, Columbia University projected that the share of the world's population living within 100 km of the coast would increase by 35% between 1995 and 2025 (The Earth Institute at Columbia University 2006). Similarly, in 2014, 50% of people in France preferred to live by the coast; this figure was around 30% in 2006 (SOES *et al.* 2017). This power of attraction of the coastline is also found in the first results of the GlocalMap survey (Grasland *et al.* 2019): for French cities, those appearing most attractive to respondents are mainly those in the Atlantic arc.

Spatially, the phenomenon of coastline development is reflected in two ways: by a densification of activities on the coasts and by moving closer to the sea.

4.3.2. *A densification of activities on the coastlines*

Compared to the pre-19th-Century "territory of emptiness" (Corbin 1990), the first consequence of the phenomenon of coastalization is a multiplication and densification of traditional activities and the appearance of seaside tourism in coastal areas.

In practice, this process is reflected in the size of the world's coastal population. Analysis of population densities on a global scale illustrates the difference between coastal and non-coastal regions: the average density is 112 inhabitants/km^2 on the "coast" (understood as an envelope 100 km wide and less than 100 m in altitude) compared to 44 inhabitants/km² in the rest of the world (Small and Nicholls 2003). Thus, 23% of the world's population is located in this envelope.

Figure 4.4 shows the high concentration of the population in low-lying and near-coastal areas. In particular, the figure shows the importance of the millionaire cities near the sea (top left) as well as the density of urban spots larger than 10 km² in the same area (bottom right). A total of 15 of the 20 cities with more than 10 million inhabitants are located on the coasts. According to MacGranahan *et al.* (2007), in 2000, 10% of the population lived below 10 m high, or 2% of the world's surface.

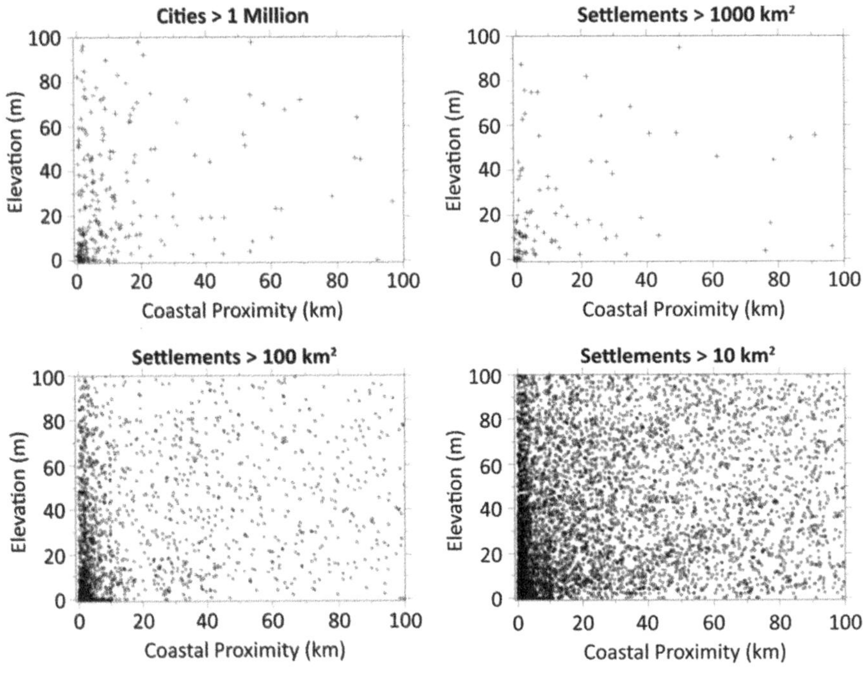

Figure 4.4. *Distribution of world population (1999) as a function of distance (x-axis) and elevation (y-axis) from the coast for cities with more than 1 million inhabitants (top left), urban patches of more than 1,000 km² (top right), 100 km² (bottom left) and 10 km² (bottom right), (source: Small and Nicholls 2003). For a color version of this figure, see www.iste.co.uk/mercier/climate.zip*

This mass of cities close to the coasts can partly be explained by the fact that the sea represents one of the major vectors of the globalization process. The hub of this trade by sea is of course the ports, which are places where cargo is transferred between ships and land or air transport (having much lower carrying capacities), resulting in the creation of employment and concentration of capital. In 2005, 13 of the 20 most populated cities in the world were ports (Hanson *et al.* 2011).

At the same time, the development of seaside activities represents a significant consumption of space. Compared to the national average, French coastal communities have a population density that is 2.4 times higher and a land artificialization 2.6 times higher (Colas 2017). Housing density is 2.7 times higher than the national average. Over the period 1990–2009, the population of these municipalities increased by 9.4%, even though they represent only 4% of the national territory (SOES *et al.* 2017). Over the equivalent period 1990–2012, these municipalities accounted for 12% of housing construction, with part of the new construction being for second homes or tourist accommodation (SOES *et al.* 2017). Thus, these municipalities offer an estimated 7 million tourist beds, which is 16.5 times higher than in the rest of the country (Colas 2017).

Finally, many other activities have gained ground on the coast. The energy sector is currently present and should continue to grow. Currently, 30% of the oil and 27% of the gas consumed comes from offshore sources, which account for 20% of the world's oil reserves and 25% of its gas reserves, respectively (Merenne-Schoumaker *et al.* 2015). Nuclear energy is also present on the coasts in order to benefit from a cooling source for reactors. This explains the presence of 30% of French nuclear reactors on the coast. Finally, the development of marine renewable energies could cover up to twice the estimated global needs for 2050 (Merenne-Schoumaker *et al.* 2015). Most of these energy sources require installations close to the coast: tidal dams in estuaries, tidal turbines in straits, wind turbines in shallow waters and the installation of onshore stations for connection to the electricity grid. Finally, food-producing activity also makes its mark on the coast, marked by the presence of fishing ports. Salt marshes, but especially the development of aquaculture are also the hallmarks. About 40% of the world's fish production currently comes from farming, installed close to the shore or even on land. Only coastal agriculture is experiencing a decline, mainly in favor of urbanization. In France, the share of agricultural land has decreased 2.5 times faster in coastal communities than in the rest of the country over the period 1970–2010 (Colas 2017).

4.3.3. *A closer approach to the sea*

In addition to this reinforcement of the occupation of the coasts, a logic of implantation as close to the sea as possible, or even on the sea, can be observed. Two activities particularly illustrate this phenomenon: seaside tourism and the development of ports.

The development of mass tourism from the second half of the 20th Century onwards led to the spreading of seaside resorts in order to benefit from the “sea view” and privileged access to beaches. This logic did not always prevail since, at the beginning, the installations necessary for this activity were established either in continuity with the existing ports, or in the extension of villages which were initially set back from the coast as was common at that time. This led to a “doubling” of the coastal municipalities (Meur-Férec and Morel 2004), with the old center being inland, the seaside resort on the coast (see Figure 4.5).

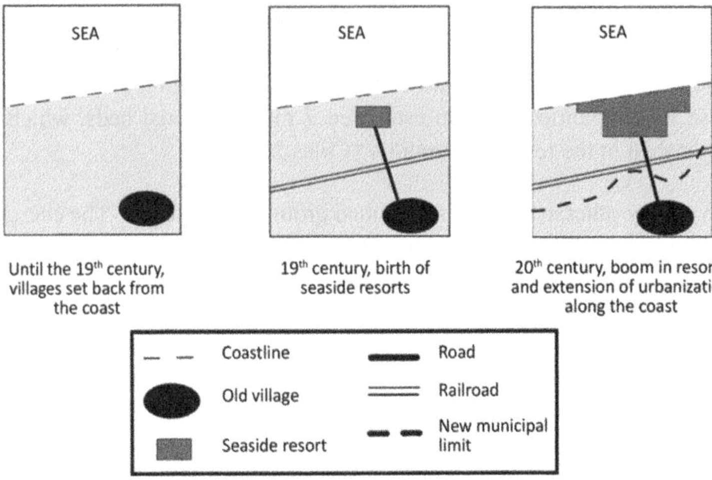

Figure 4.5. *Schematic illustration of moving of seaside resorts as close to the sea as possible (source: Meur-Férec and Morel 2004). For a color version of this figure, see www.iste.co.uk/mercier/climate.zip*

The artificialization of the coastal strip illustrates this trend. In metropolitan France, on a strip from 0 to 500 m from the sea, 29% of the territory is artificialized while the national average is 5% (SOES *et al.* 2017).

This artificialization of the coastline for the benefit, mainly, of seaside tourism, in some cases results in creations *ex nihilo* on the sea. The emblematic examples of the artificial peninsulas of Palm Jumeirah and Palm Jebel Ali in the emirate of Dubai in the United Arab Emirates are an illustration of this and testify that this phenomenon can be found on all tourist coasts.

This dynamic is not without consequences, since among the populations living on the world's coasts, 40% are concentrated in urban areas (population density per km² between 1,000 and 10,000) and 60% are spread out (Small and Nicholls 2003). This fragmentation symbolizes the spread of populations along the coasts.

The logic of the location of the ports is also interesting. The port has two main functions: to offer shelter to ships and to allow the breaking of load between maritime transport and other means of transport (Rodrigue 2020). For a long time, the search for a "natural" shelter presided over the installation of ports that were mainly "estuary" ports: good shelter coupled with the possibility of using the waterway to transport goods inland (ports of Rouen, Nantes or Southampton). The intensification of maritime transport since the 1950s and advances in civil engineering now lead to a different location logic: the need to accommodate increasingly large ships (up to 400 m in length for the largest container ships) with shorter call times (to reduce the number of calls) as well as the creation of vast storage and handling areas for containers are two factors explaining the migration of ports downstream, in a position known as "frontage" (the port of Le Havre is the counterpart of Rouen, Saint-Nazaire for Nantes or Portsmouth for Southampton). This implantation, which partly ignores the sheltered position, is compensated by the creation of massive structures: earthworks, embankments, dikes, etc. Increasingly, ports are even extending out to sea. The example of the port of Rotterdam is typical of this movement: while the old port was located in the estuary of the Rhine and Meuse, it gradually moved downstream until the inauguration in 2013 of the port of Maasvalkte 2, consisting of 2,000 hectares built entirely on the North Sea and allowing ships with a large draught to call without having to sail up the estuary.

4.4. Towards the necessary adaptation of coastal areas

4.4.1. *The coastline, an area at risk*

The coastlines are thus faced with a double constraint: on the one hand, a rise in sea level linked to climate change, which should lead to the temporary and definitive submersion of low-lying coastlines; on the other hand, attractive spaces for societies, resulting in a densification of people and activities on the coastline and an increasingly close proximity of these installations to the sea.

This makes coastlines risk areas where the likelihood of these risks is expected to increase in the future. Disasters resulting from meteorological and marine hazards are already among the most frequent on earth. For the year 2017, storms (in the sense of events of marine meteorological origin) were the leading cause of disasters in the world (127 events recorded), ahead of floods (126 events) (UCL *et al.* 2018). In the period 1998–2017, storms were the second most common cause of natural disasters (28.2% of events) after floods (43.4% of events, CRED 2018). The 2,049 storm-related disasters over this period concerned 16% of the world's disaster-affected populations (after floods for 45%), 17% of the victims (after earthquakes, 56%) but 46% of the economic damage, which testifies to the concentration of the issues at stake on the coasts. Thus, among the three most costly disaster years over the period 1998–2017, two were storm-related: the 2017 hurricane season (hurricanes Harvey, Irma and Maria) with $245 million in damage representing the most costly year, while the 2005 hurricane season (hurricanes Katrina, Rita and Wilma) ranks third (CRED 2018). The second year with the most losses was 2011, marked by the tsunami in Japan: while the origin of the hazard was geological, it was the coastline that was directly affected.

This situation is expected to worsen. Projections by the OECD show that, as a result of sea-level rise and persistent population growth in coastal cities, the population exposed to coastal risks could increase threefold by 2070 (Hanson *et al.* 2011) to reach 150 million people, mainly in Asia, or even 300 million by 2050 according to more recent work (Climate central 2019; Kulp and Strauss 2019). The global GDP affected by coastal risks could increase from 5% to 9%. Action is therefore becoming a necessity. The World Bank shows the need for adaptation: the potential losses linked to the risk of coastal floodings are currently estimated at $6 billion (in 2005) for the 136 largest coastal cities (Hallegatte *et al.* 2013; Kulp and Strauss 2019). With economic and demographic development alone, this figure would reach $52 billion in 2050. If sea-level rise (between 20 and 40 cm) is included, these potential losses would amount to $1 trillion in 2050. This would not be sustainable.

4.4.2. *Possible coping strategies*

There are multiple possible actions to cope with coastal hazards (IPCC 2019). Depending on whether one seeks to act on a particular factor responsible for a risk situation, these actions can be grouped into three categories:

– Acting on the hazard: if removing a hazard is not possible, different actions can reduce its impact on a given territory. These generally involve counteracting erosion or limiting the extension of areas likely to be flooded by the sea using different techniques (IPCC 2019; see Figure 4.6). The range of possible actions is wide: hard protection, sediment-based protection, ecosystem-based adaptation, which involves taking advantage of the protective role of coral reefs, sea marshes and mangroves, or coastal advance, which involves reclamation or embankment to gain land from the sea. These different actions have very different costs, effectiveness and environmental impact. The choice of one action or another will depend on the issues to be protected.

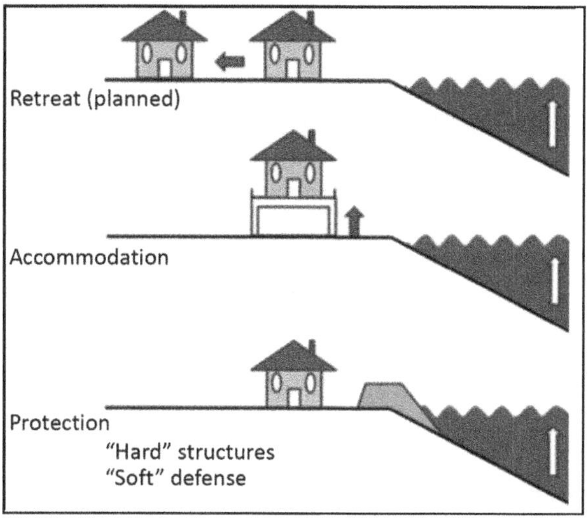

Figure 4.6. *Three possible strategies for dealing with the consequences of sea level rise (source: Creach 2015, adapted from Nicholls 2011). For a color version of this figure, see www.iste.co.uk/mercier/climate.zip*

– Acting on the issues: removing or limiting the issues is one way of reducing risk situations. The IPCC (2019) distinguishes between two types of intervention: planned retreat and forced displacement, depending on the degree of anticipation. Given the long time required to implement this type of policy and its irreversible nature, relocation is particularly suitable in a context of erosion where the retreat of the coastline can be anticipated.

– Acting on vulnerability: this is an in-between where the stakes are maintained and where the main actions concern their modification/adaptation in order to limit damage in the event of a hazard, whether through structural adaptation of buildings or networks or by strengthening warning systems in order to promote preventive evacuation of exposed populations (IPCC 2019). This type of measure may prove useful for dealing with one-off events where the main concern is to be able to "cope" in the event of a crisis while ensuring the quickest possible return to normal.

These three main strategies obviously benefit from being implemented jointly (Creach 2019).

4.4.3. *The example of the Netherlands*

The Netherlands is home to three of Europe's major estuaries: the Rhine, the Meuse and the Scheldt. About 65% of the territory is likely to be flooded, by sea or by rivers (Gueben-Venière 2014) where many issues are at stake: 56% of the Dutch live in an area exposed to the risk of flooding; Rotterdam, Europe's leading port for goods traffic, is located there; 65% of the country's GDP would be likely to be directly impacted in the event of flooding (2008 values).

Due to its configuration, the Netherlands has experienced a great deal of flooding by the sea in the past. This has led the country to develop a centuries-old tradition of protection and reclamation through the construction of "offensive" dikes (Wagret 1959).

As a result, almost 17% of the country's surface area has been artificially removed from the sea (Gueben-Venière 2014). In 1953, 1,836 people died in the marine submersion associated with the storm of January 31 when more than 150 breaches were recorded in the protective dikes facing the sea (Gerritsen 2005). Following this event, the Dutch state implemented an environmental protection plan, the Delta Plan, supplemented by further complementary actions.

4.4.3.1. *Actions for hazard reduction*

The Delta plan consisted mainly of reinforcing the hard protection by closing three estuaries *with* storm surge barriers sized to contain a flood of decamillennial return period, as well as maintaining and consolidating the existing dikes (Gerritsen 2005). However, these actions have not been systematic: 64% of the Dutch coastline is represented by dune massifs, so periodic sand nourishment is common.

One of the emblematic examples is the "sand engine" started in the region of the Hague (Gueben-Venière 2014) where it is no longer simply a question of recharging the foot of the sand dune but of depositing a large volume of sand on the foreshore and taking advantage of the effect of coastal drift to protect, by extension, a 10 km stretch of coastline where the aesthetic challenge linked to the presence of seaside activity is essential.

4.4.3.2. *Action on issues/follow-up*

The Netherlands has also started a process of de-polderization where some structures in which breaches had occurred have not been repaired (Goeldner-

Gianella 2013). This process can be seen as a form of retreat since it accepts the submersion and thus the damage to certain structures that may eventually have to be moved.

4.4.3.3. *Actions on vulnerability reduction*

Protecting is not enough to eliminate the risk. One of the stated objectives of the Delta Plan is to guarantee long-term safety on different scales (Gerritsen 2005) by strengthening warning and prevention. A wide range of actions are implemented in this sense, from the permanent monitoring unit at the national level to the implementation of crisis management plans at the local level. In addition, the Netherlands is also acting on the issues at stake to make them less vulnerable to marine submersion. Prototype neighborhoods composed of amphibious houses have been built in Rotterdam and Amsterdam, providing a solution to land pressure and better security for the inhabitants.

4.5. Which coastline for tomorrow?

Coastal areas are subject to antagonistic pressures that make them risky areas:

– climate change-related sea-level rise is expected to aggravate erosion and submersion, temporary and permanent. Natural forcing is thus exerted from sea to land (see Figure 4.7);

– the coastalization and maritimization of societies result in a densification and multiplication of activities and populations as close to the coastline as possible. A form of anthropogenic forcing is thus exerted from the land to the sea (see Figure 4.7).

This double pressure at the land-sea interface inevitably results in an increase in the risks linked to marine submersion if the frequency and intensity of the hazards increase (natural forcing) in an area with increasingly high stakes (anthropogenic forcing).

Adaptation is therefore a necessity: never before has such a concentration of issues been exposed to coastal risks, and never before has such a diversity of possible strategies been offered to societies. The question of adaptation therefore arises in terms of technical feasibility, economic viability and social acceptability over a long-time scale.

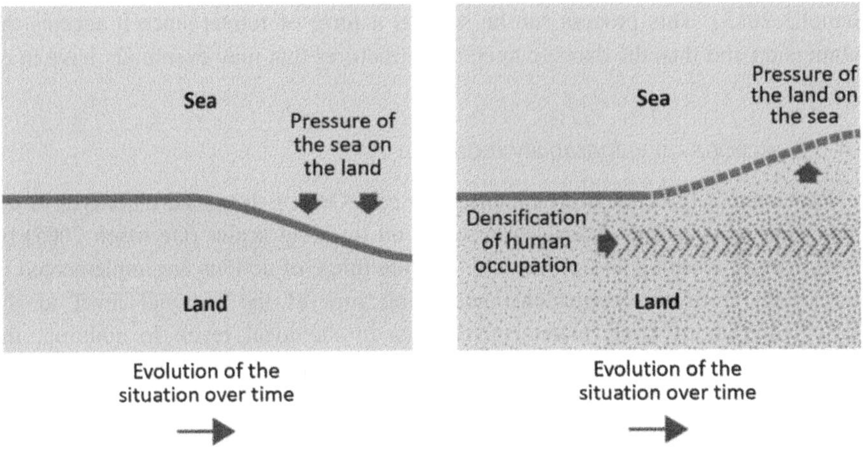

Figure 4.7. *Schematic illustration of the increase in coastal risks due to natural forcings exerting pressure from the sea to land (left) and anthropogenic forcings marked by a densification of populations and activities over time (green dots) and an overhang on the sea (right) (source: Creach 2020, adapted from Meur-Férec and Morel 2004). For a color version of this figure, see www.iste.co.uk/mercier/climate.zip*

The diversity of solutions presented is proof that the technique can enable adaptation to the consequences of climate change on coastlines. The range of possibilities for protection could make it possible to protect all coasts. Technical feasibility does not seem to be an obstacle, but at what cost?

The economic dimension of adaptation alone illustrates the challenge facing our societies and suggests that it may not be possible to protect everything: the protection of all land exposed to coastal risks would amount to between \$25 million and \$270 billion per year for a hypothesis of sea-level rise of between 0.5 m and 2 m between 2005 and 2100 (Nicholls *et al.* 2010). The construction of new dykes alone would amount to \$80 billion to \$120 billion for a sea level rise of 0.5 meters (Anthoff *et al.* 2010). In addition, raising a sea wall to keep pace with sea level rise is exponentially expensive (Jonkman *et al.* 2013). Economic viability is therefore a crucial aspect of the adaptation process.

On the other hand, the social acceptability of accommodations is essential. If adaptation is unavoidable, it is a matter of not undergoing the process but of anticipating it. This anticipation requires acceptance of the process by all stakeholders. A *status quo* that would lead to a postponement of adaptation measures could lead to a risk of "maladaptation" (Magnan and Duvat 2018), the consequences of which could be additional costs in projects or the possible shift of vulnerability to other risks (Meddtl 2011). The risk of maladaptation is exacerbated when the adaptation process is part of the urgency of managing an event and when adaptation is a response to an observed shortcoming. This "reactive" and undergone adaptation often ignores consultation due to a lack of time.

Thus, all scenarios are possible as long as they are anticipated, prepared and accepted. The words of the President of the Republic of Maldives say little else: "Yes, my people can move, but they will do so with dignity."

4.6. References

AGE, Agence européenne pour l'environnement (2019). Global and European sea-level rise [Online]. Available at: https://www.eea.europa.eu/data-and-maps/indicators/sea-level-rise-6/assessment.

Anthoff, D., Nicholls, R.J., Tol, R.S.J. (2010). The economic impact of substantial sea-level rise. *Mitigation and Adaptation Strategies for Global Change*, 15, 321–335.

Cetmef, Cete Méditerranée, Cete Ouest. (2009). *Vulnérabilité du territoire national aux risques littoraux*. CETMEF/DLCE, Compiègne.

Climate Central (2019). Flooded future: Global vulnerability to sea level rise worse than previously understood [Online]. Available at: https://www.climatecentral.org/news/report-flooded-future-global-vulnerability-to-sea-level-rise-worse-than-previously-understood.

Colas, S. (2017). *Des pressions plus fortes en bord de mer, surtout dans les territoires ruraux et périurbains*. MTES, Paris.

Corbin, A. (1990). *Le Territoire du vide : l'Occident et le désir du rivage, 1750–1840*. Flammarion, Paris.

Creach, A. (2015). Cartographie et analyse économique de la vulnérabilité du littoral atlantique français face au risque de submersion marine. PhD thesis, Université de Nantes, Nantes.

Creach, A., Bastidas-Arteaga, E., Pardo, S., Mercier, D. (2019). Adaptation of residential buildings to coastal floods: Strategies, costs and efficiency. In *Climate Adaptation Engineering*, Bastidas-Arteaga, E., Stewart, M. (eds). BH-Elsevier, Oxford.

CRED, UNISDR (2018). Economic losses, poverty and disasters (1998–2017) [Online]. Available at: https://www.preventionweb.net/files/61119_credeconomiclosses.pdf.

Duvat, V.K.E. (2019). A global assessment of atoll island planform changes over the past decades. *Wiley Interdisciplinary Reviews: Climate Change*, 10.

Erkens, G., Bucx, T., Dam, R., de Lange, G., Lambert, J. (2015). Sinking coastal cities. *Proceedings of the International Association of Hydrological Sciences*, 372, 189–198.

Fitzgerald, D.M., Fenster, M.S., Argow, B.A., Buynevich, I.V. (2008). Coastal impacts due to sea-level rise. *Annual Review of Earth and Planetary Sciences*, 36, 601–647.

Gerritsen, H. (2005). What happened in 1953? The big flood in the Netherlands in retrospect. *Philosophical Transactions of the Royal Society A*, 363, 1271–1291.

Goeldner-Gianella, L. (2013). *Dépoldériser en Europe occidentale. Pour une géographie et une gestion intégrées du littoral*. CNRS, Paris.

Grasland, C., Didelon-Loiseau, C., Brennetot, A., Pecout, H., Pistre, P., de Ruffray, S., Berroir, S. (2019). Premiers résultats de l'enquête GlocalMap. Dossiers no. 8, CIST, Paris.

Gueben-Venière, S. (2014). Vers une gestion renouvelée du littoral nord-ouest européen : des ingénieurs néerlandais, anglais et français de plus en plus "verts" ? PhD thesis, Université Paris Panthéon-Sorbonne, Paris.

Hallegatte, S., Green, C., Nicholls, R.J., Corfee-Morlot, J. (2013). Future flood losses in major coastal cities. *Nature Climate Change*, 3, 802–806.

Hanson, S., Nicholls, R., Ranger, N., Hallegatte, S., Corfee-Morlot, J., Herweijer, C., Chateau, J. (2011). A global ranking of port cities with high exposure to climate extremes? *Climatic Change*, 104, 89–111.

IPCC (2019). Special report on the ocean and cryosphere in a changing climate [Online]. Available at: https://www.ipcc.ch/srocc/.

Jonkman, S.N., Hillen, M.M., Nicholls, R.J., Kanning, W., van Ledden, M. (2013). Costs of adapting coastal defenses to sea-level rise – New estimates and their implications. *Journal of Coastal Research*, 29, 1212–1226.

Kulp, S.A. and Strauss, B.H. (2019). New elevation data triple estimates of global vulnerability to sea-level rise and coastal flooding. *Nature Communications*, 10.

Luijendijk, A., Hagenaars, G., Ranasinghe, R., Baart, F., Donchyts, G., Aarninkhof, S. (2018). The state of the world's beaches. *Scientific Reports*, 8, 6641.

Magnan, A.K. and Duvat, V.K.E. (2018). (Mal)adaptation au changement climatique : commencer par bien faire ce que l'on fait mal. In *L'adaptation au changement climatique, une question de sociétés*, Euzen, A., Laville, B., Thiébault, S. (eds). CNRS, Paris.

McGranahan, G., Balk, D., Anderson, B. (2007). The rising tide: Assessing the risks of climate change and human settlements in low elevation coastal zones. *Environment and Urbanization*, 19, 17–37.

Meddtl (2011). Plan national d'adaptation de la France aux effets du changement climatique 2011–2015 [Online]. Available at: https://www.ecologique-solidaire.gouv.fr/sites/default/files/ONERC_PNACC_1_complet.pdf.

Mentaschi, L., Vousdoukas, M.I., Pekel, J.-F., Voukouvalas, E., Feyen, L. (2018). Global long-term observations of coastal erosion and accretion. *Scientific Reports*, 8, 12876.

Merckelbagh, A. (2009). *Et si le littoral allait jusqu'à la mer ! La politique du littoral sous la Vème République*. Quae, Paris.

Mérenne-Schoumaker, B. (2015). Les ressources énergétiques et minérales de la mer. In *Géographie des mers et des oceans*, Deboudt, P., Meur-Férec, C., Morel, V. (eds). Armand Colin, Paris.

Meur-Férec, C. and Morel, V. (2004). L'érosion sur la frange côtière : un exemple de gestion des risques. *Natures Sciences Sociétés*, 3, 263–273.

Miossec, A. (2012). *Dictionnaire de la mer et des côtes*. Presses Universitaires de Rennes, Rennes.

Muis, S., Verlaan, M., Winsemius, H.C., Aerts, J.C.J.H., Ward, P.J. (2016). A global reanalysis of storm surges and extreme sea levels? *Nature Communications*, 7, 11969.

Nicholls, R.J. (2011). Planning for the impacts of sea level rise. *Oceanography*, 24, 144–157.

Nicholls, R.J., Brown, S., Hanson, S., Hinkel, J. (2010). Economics of coastal zone adaptation to climate change [Online]. Available at: http://documents.worldbank.org/curated/en/229791468159607825/Economics-of-coastal-zone-adaptation-to-climate-change.

Paskoff, R. (1993). *Côtes en danger*. Masson, Paris.

Paskoff, R. (2001). *L'élévation du niveau de la mer et les espaces côtiers. Le mythe et la réalité*. Institut océanographique, Paris.

Pineau-Guillou, L. (2012). Statistiques des niveaux marins extrêmes des côtes de France (Manche et Atlantique). Report, SHOM/CETMEF.

Pineau-Guillou, L., Lathuilière, C., Magne, R., Louazel, S., Corman, D., Perherin, C. (2010). Caractérisation des niveaux marins et modélisation des surcotes pendant la tempête Xynthia. Report, *XIèmes Journées nationales génie côtier – génie civil*. Les Sables d'Olonne.

Ranasinghe, R. (2016). Assessing climate change impacts on open sandy coasts: A review. *Earth-Science Reviews*, 160, 320–332.

Rodrigue, J.-P. (2020). *The Geography of Transport Systems*. Routledge, Abingdon-on-Thames.

Small, C. and Nicholls, R.J. (2003). A global analysis of human settlement in coastal zones. *Journal of Coastal Research*, 19, 584–599.

SOES, Agence française pour la Biodiversité, Ifremer, Cerema (2017). Les données clés de la mer et du littoral. Synthèse des fiches thématiques, Observatoire, ministère de l'Environnement/Agence française pour la biodiversité/Ifremer, Paris.

The Earth Institute at Columbia University (2006). It's 2025. Where do most people live?. *ScienceDaily* [Online]. Available at: https://www.sciencedaily.com/releases/2006/07/060718090608.htm.

UCL, CRED, USAID (2018). Natural disasters 2017: Lower mortality, higher cost [Online]. Available at: https://www.preventionweb.net/publications/view/60351.

UNCTAD (2013). Review of Maritime Transport 2012. Report, United Nations, New York.

Vousdoukas, M.I., Mentaschi, L., Voukouvalas, E., Verlaan, M., Feyen, L. (2017). Extreme sea levels on the rise along Europe's coasts. *Earth's Future*, 5, 304–323.

Webb, E.L., Friess, D.A., Krauss, K.W., Cahoon, D.R., Guntenspergen, G.R., Phelps, J. (2013). A global standard for monitoring coastal wetland vulnerability to accelerated sea-level rise. *Nature Climate Change*, 3, 458–465.

5

The Consequences of Climate Change on the Paraglacial Sedimentary Cascade

Denis MERCIER[1] and Étienne COSSART[2]
[1]*Sorbonne University, Paris, France*
[2]*Jean Moulin University Lyon 3, France*

5.1. The paraglacial sedimentary cascade: elements of definition

5.1.1. *General principles of the concept of a paraglacial sedimentary cascade*

The term "paraglacial" means "beside the ice". It is composed from the Greek prefix "*para*" (beside) and the Latin "*glacies*" (ice) and has two meanings:

– It refers to non-glacial processes directly conditioned by deglaciation (Church and Ryder 1972). It refers to all proglacial processes and active processes operating around and within the formerly ice-bounded margins that are the result of the former presence of ice. Proglacial processes in action rework sedimentary sources inherited from past glaciations (moraines, rock walls stiffened by glacial erosion, etc.) and may take the form of mudflows or debris flows. They are also expressed through runoff, and are responsible for the establishment of steep torrential cones (>20°, alluvial cones), alluvial fans, lacustrine and offshore sedimentation, as well as fluvioglacial spreading plains (alluvial outwash plains, sandars) in which terraces can then be cut by the river incision according to variations in the base level.

– The term paraglacial is also used to describe the time sequence corresponding to the hinge sequence following glaciation during which paraglacial processes are particularly active. The pattern proposed by Church and Ryder (1972) in their seminal paper illustrates this sequence where erosion rates are higher than in the previous glacial sequence and in the post-glacial sequence following it (see Figure 5.1).

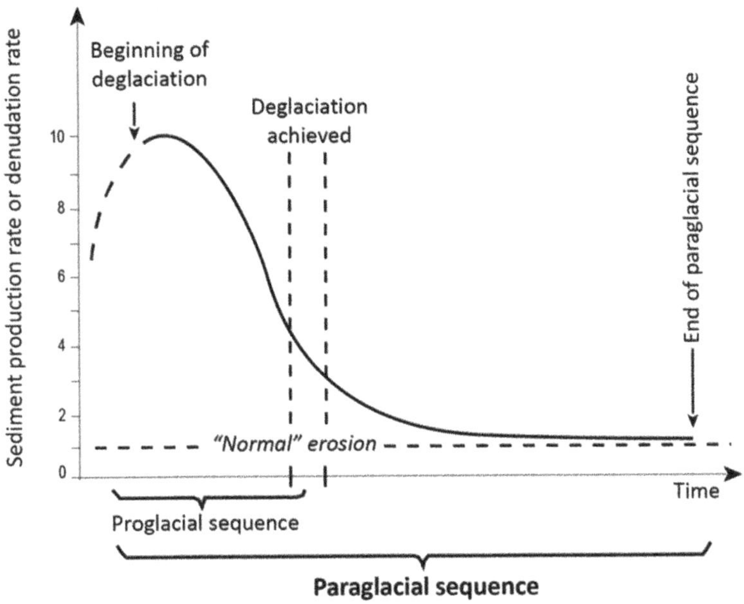

Figure 5.1. *The life span of the paraglacial sequence (source: modified from Church and Ryder 1972)*

The presence of abundant and unconsolidated sedimentary stocks of morainic origin and active processes, such as runoff, is one reason why denudation rates are so high during these paraglacial periods. This paraglacial period, which initially referred to late Pleistocene deglaciation, was later extended to all periods of glacial retreat that may have a temporal extension beyond the restricted deglaciation phase (Church and Ryder 1989). The paraglacial sequence thus has a life span that depends on several factors:

– the amount of sedimentary stocks to be reclaimed;

– the speed of the processes and therefore the climatic parameters, such as the length of the snow and ice melt season;

– the geographical location of the catchment area;

– of the post-glacial vegetation cover;

– the size, orographic configuration and geological nature of the watersheds in which paraglacial processes are expressed.

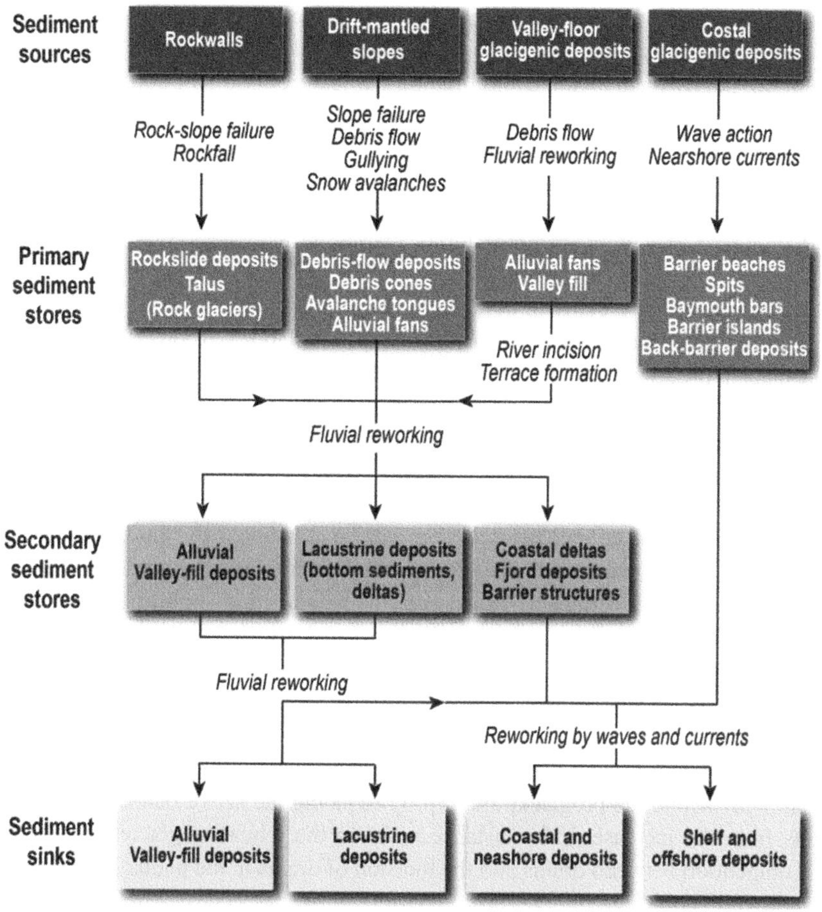

Figure 5.2. *Paraglacial sedimentary cascade by compartment from sedimentary sources to sinks, through deposits with highly variable lifetimes (source: modified from Ballantyne 2002)*

Logically, paraglacial denudation rates decrease as watersheds become larger. All other things being equal, rates of paraglacial remobilization are greater the more contrasting the relief. Lithological control is also emphasized by Church and Ryder (1972). The granites and gneisses of Baffin Island are more resistant to glacial and river erosion, and therefore rates of paraglacial denudation remain modest compared to those calculated in the metamorphic and sedimentary terrains of southern British Columbia. In contrast, weak rocks such as shales favor the dismantling of sedimentary stocks and landforms inherited from past glaciations. The post-glacial geomorphological evolution of the Queyras Massif (Southern French Alps) shows a rapid transition during the Holocene towards torrential processes not conditioned by glaciation, but by local tectonic and lithological features (Cossart 2005).

The concept of paraglacial periods as defined by Church and Ryder (1972) makes it possible to insist on the fact that paraglacial periods represent sequences of readjustments of environments, of the passage from one state of glaciation to another state, with rapid rates of evolution representing morphogenic crises.

Thirty years after the article by Church and Ryder (1972), a new definition of the concept of paraglacial periods is proposed by Ballantyne (2002), who refers to "non-glacial processes on the earth's surface, sedimentary accumulations, forms, systems and landscapes that are directly conditioned by glaciation and deglaciation." The paraglacial sedimentary cascade was formalized by Ballantyne (2002) to show how glaciated environments respond to climate change (see Figure 5.2).

This paraglacial sedimentary cascade can be illustrated by the example of the Midtre Lovénbreen catchment area in North-Western Spitsbergen (Mercier 2011) (see Figure 5.3). Sediments from lateral, frontal and bottom moraines are remobilized by concentrated proglacial runoff fed by the melting of the Midtre Lovénbreen glacier, which has retreated 1,200 meters since the beginning of the 20th Century when contemporary climate warming exacerbated in the Arctic. The concentration of the hydrographic network certainly allows the remobilization of sediments by a better competence of the flow drains, but on the other hand, it neglects spatial areas within the proglacial margin by reducing the active band. This reduction results from the increase in the distance between the retreating glacier front and the sea, competition between drains and the incision of drains in the plains, which are thus progressively colonized by vegetation (Moreau *et al.* 2008). These sedimentary additions continue beyond the intramorainic plain delimited by the frontal moraine (showing the maximum advance of the Midtre Lovénbreen glacier dated from the beginning of the 20th Century by photographs taken by the Isachsen expedition in 1907). The construction of the vast fluvio-glacial accumulation plains (sandurs) ends in prograding deltaic cones at the expense of the space occupied by the fjord in the foreground (see Figure 5.3). When these deltas are well fed by the paraglacial sedimentary cascade, they prograde, whereas they erode when the intramoraine competition of the flow drains does not allow the maintenance of terrigenous inputs (Mercier and Laffly 2005; Strzelecki *et al.* 2018). The work of Bourriquen *et al.* (2018) showed an overall trend towards coastal progradation on the south bank of the Kongsfjorden in North-West Spitsbergen from 1966 to 2017 (see Figure 5.4), with a mean value of +0.14 m/year, and a maximum side-draft advance value of +87 m from 1966 to 2017 at delta 5. However, this overall trend can be subdivided into three phases: a first phase of net progradation from 1966 to 1990, followed by a period of erosion from 1990 to 2016, and then again a tendency towards progradation, but only observed over one year of measurement.

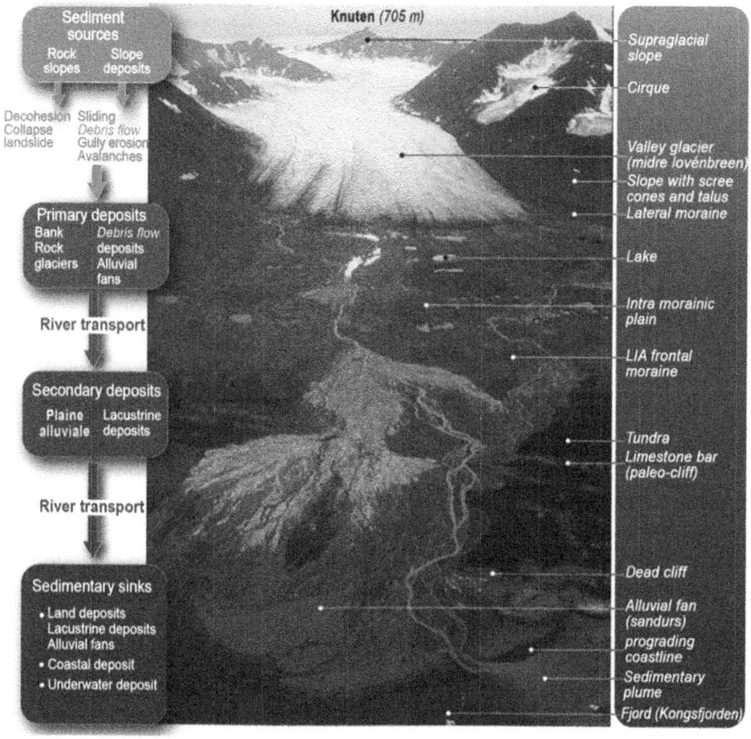

Figure 5.3. *Paraglacial sedimentary cascade for a small partially ice-covered polar watershed in northwestern Spitsbergen, Svalbard archipelago. The Midtre Lovénbreen glacier has been retreating since the beginning of the 20th Century as a result of contemporary climate warming in the Arctic (source: Mercier 2011). For a color version of this figure, see www.iste.co.uk/mercier/climate.zip*

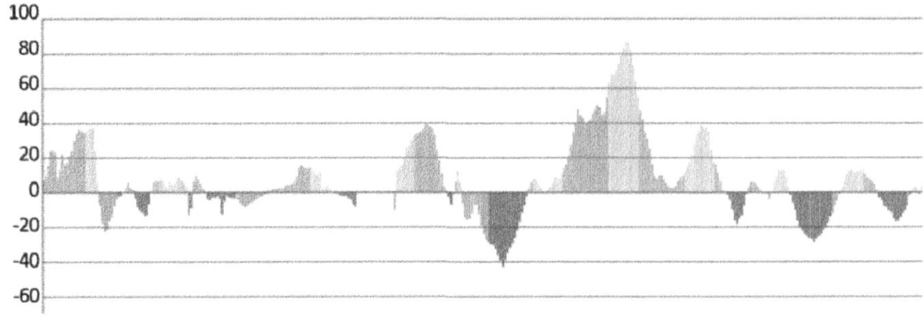

(a) Evolution of the coastline in meters from 1966 to 2017

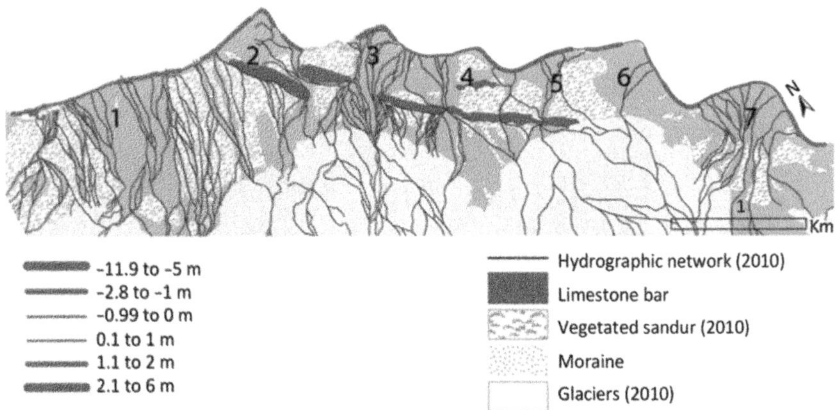

(b) Change in coastline in meters per year from 2016 to 2017

Figure 5.4. *Prograding and/or coastal erosion of the south bank of the Kongsfjorden in connection with the river system on the margins of the Loven glaciers in North-Western Spitsbergen (source: based on Bourriquen* et al. *2018). For a color version of this figure, see www.iste.co.uk/mercier/climate.zip*

5.1.2. *Paraglacial spatial boundaries*

The concept of paraglacial ice covers a very broad spatial reality and can be applied to all spaces on the Earth's surface that have been glaciated and have undergone a deglaciation sequence. This includes all surfaces that were glaciated during the Pleistocene cold sequences and their peripheries, and which are no longer glaciated today, that is about 40% of the Earth's surface, plus offshore sedimentation spaces. In addition, about 10% of the land surface remains ice-covered, a fraction of which is likely to undergo a deglaciation sequence in the current context of global warming (see Figure 5.5).

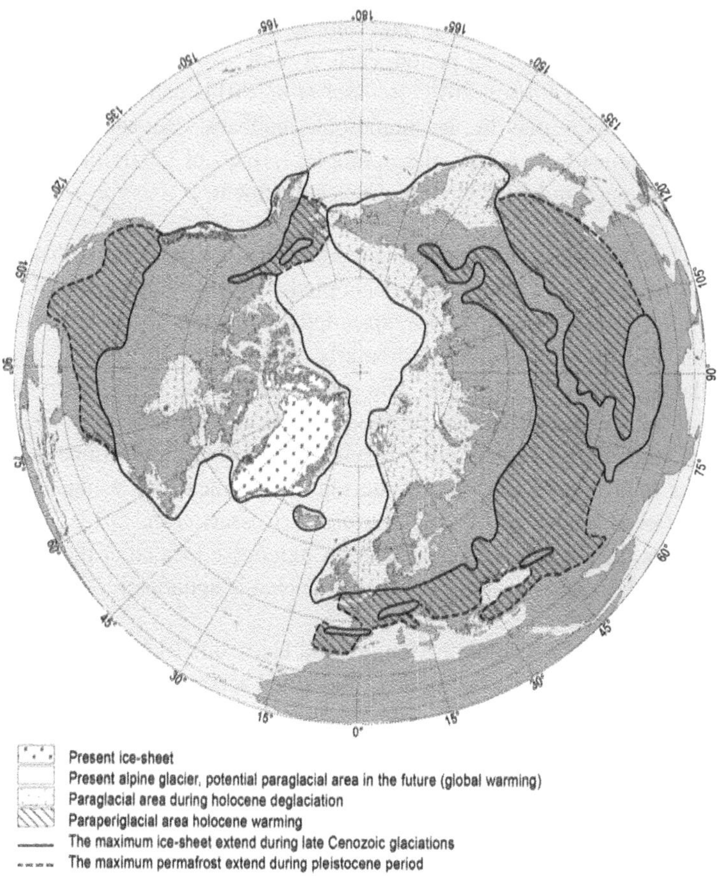

Figure 5.5. *Map of active paraglacial, paleoparaglacial and potential paraglacial (source: based on Mercier 2008)*

5.1.3. *The temporal limits of the paraglacial sedimentary cascade*

Since the work of Ballantyne (2002), the current debate also focuses on the temporal limits of the paraglacial period. This debate is particularly fueled by the study of mountain areas undergoing deglaciation, which are open-air laboratories in which a post-LGM (Last Glacial Maximum) paraglacial transition and a post-Little Ice Age (LIA) transition takes place, during which the direct imprint of glacial processes decreases over time. A recent synthesis by Knight and Harrison (2018) highlights the variability of meltwater and sediment flows during this paraglacial period; in particular, there is not necessarily parallelism or even linearity between water availability and sediment availability. In other words, hydrological and sedimentary signals are decoupled, and should be studied independently. Indeed, throughout the entire paraglacial period, each of these two signals is characterized by transient (Knight and Harrison 2018; Roussel *et al.* 2018) and even chaotic (Cossart *et al.* 2018) behavior due to the numerous feedback sequences linked to the paraglacial readjustment, and the complex role of extreme events (low frequency/high magnitude; LF/HM) in this readjustment.

Indeed, low frequency and high magnitude events animate the hydrological and sedimentary cascade in a sometimes-decoupled way (see Figure 5.6). On the one hand, from a hydrological standpoint, spasmodic behavior can be linked to a series of ice jams and breakup events, particularly when lakes are formed upstream of moraine dams (Roussel *et al.* 2016). The geomorphological impact then depends on the quantity of sediment that can be reworked along the flow axis. On the other hand, from a geomorphological point of view, sedimentary pulsations are linked to slope inputs, and in particular to classic mass movements of readjustment of the trough walls over-steepened by glacial erosion (Cossart *et al.* 2013). This mass of sediment clogs the sediment cascade if the watercourse located at the bottom of the valley is not itself in a paroxysm allowing the removal of sediments.

In terms of time, these findings (Knight and Harrison 2018) show that meltwater availability and sediment availability function according to their own self-organizations (see Figure 5.6). They can, however, coincide (when a hydrological event and a geomorphological event are synchronous) to activate sediment flows. This model does not correspond to the classic paraglacial model, which would imply a gradual decrease in sediment export after glacial retreat. Nevertheless, the paraglacial period prepares for sediment purges through sediment accumulation, while high magnitude hydrological events sporadically provide the necessary transport capacity for sediment export. The role of these events (including all LF/HM floods) has so far been underestimated (Cossart *et al.* 2018; Roussel *et al.* 2018), even though they are the ones that allow hydrological and sedimentary signals to act together.

In conclusion, the average of paraglacial sediment export rates calculated over one or more millennia (Church and Ryder 1972; Jackson *et al.* 1982) should not obscure the spasmodic nature of the functioning of paraglacial cascades. The large sediment supply caused by the disappearance of glaciers does not necessarily result in a peak in transfer and then export of sediment at the beginning of deglaciation. The authors now agree that there is a succession of spasmodic pulsations linked to the (de)phasing of hydrological and sedimentary signals. However, as the low-frequency and high-magnitude events typical of the paraglacial period are triggered (e.g. moraine dam breaks, gravitational readjustments of the slopes), threshold effects are crossed during this transition period, leading to a progressive resetting of the hydrological and sedimentary signals. In concrete terms, this means that the hydrosedimentary cascade, segmented by multiple breaks in connectivity, is progressively organized: hydrosedimentary connectivity increases each time a LF/HM event causes the dismantling of a dam or recreates a connection (Cossart *et al.* 2018). The end of the paraglacial phase could thus occur as soon as objects (moraines, mass-movement deposits) or processes (ice jams) no longer create a decoupling between hydrological and sedimentary signals (Knight and Harrison 2018).

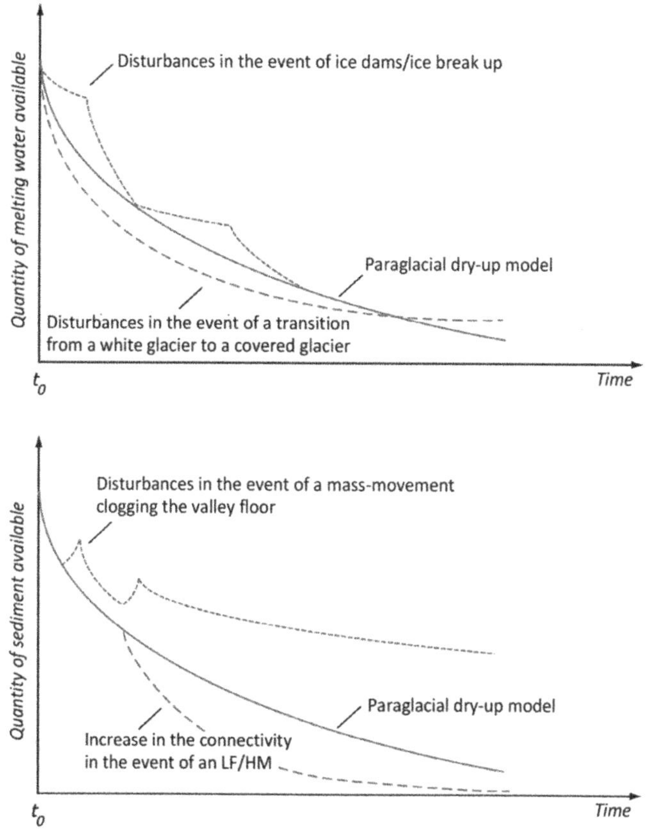

Figure 5.6. *Decoupled evolution of hydrological and sedimentary signals in a paraglacial context (source: modified from Knight and Harrison 2018). For a color version of this figure, see www.iste.co.uk/mercier/climate.zip*

5.2. Sediment inputs to the paraglacial sedimentary cascade

In the context of climate change in glaciated areas, three stages mark the paraglacial sedimentary cascade, which is illustrated on different time scales. The first corresponds classically to ablation and feeds the sedimentary cascade with sediments from the walls and slopes within the catchments. The second is illustrated by the remobilization of slope and surficial deposits. The third and final stage involves the deposition of sedimentary deposits as load breaks occur.

5.2.1. *Landslides*

Paraglacial landslides typically occur at the beginning of the deglaciation sequence during adjustments related to postglacial decompression. They represent the largest volumes of sediments mobilized since the retreat of the glaciers (Cossart *et al.* 2017). Landslides identifiable today therefore belong to two distinct phases of landscape history. The first corresponds to the landslide initiation at the end of the great cold sequence of the Weichselian during the global warming event that began around 15 ka and continued up to the very beginning of the Holocene. The second phase is the one observed in the still ice-covered regions of our planet with the post-LIA landslides. While the second sequence is well documented by archives and testimonies of mountain dwellers, the first requires the use of indirect dating methods to know the timing of the formation and the control factors.

The example of work carried out in Iceland in recent years helps to illustrate this phenomenon (Mercier *et al.* 2013; Cossart *et al.* 2014; Coquin *et al.* 2016; Decaulne *et al.* 2016; Peras *et al.* 2016; Mercier *et al.* 2017). The first-order control factor is above all global warming, which will exhaust glacier stocks and lead to the rapid deglaciation of Iceland between 15 ka and 10 ka (see Figure 5.7). Thus, deglaciation will lead to debuttressing, oversteepening, glacio-isostatic rebound and consequently paraglacial seismicity, all four of which will contribute to the triggering of landslides and deep-seated gravitational slope deformations (DSGSD). Landslide dating in Iceland is based on the use of geomorphological stacking of deposits such as glacioisostatic rebound beaches that are dated and fossilized by landslide masses (see Figure 5.8). They thus provide a very valuable pre-Holocene, pre-slide chronology. In addition, within the landslide deposits, depressions were colonized by plants at the beginning of the Holocene climate warming. Their fossilization allowed their conservation and macro remains were found during soundings and carbon-14 dated at 8 ka for the remains of birch trees. In addition, intra-slide depressions have also trapped volcanic ash during major eruptions in Iceland's history, such as those of the Hekla volcano. The use of tephrochronology thus makes it possible to go back as far as 10.2 ka with the tephra named Saksunarvatn from the

Grímsvötn volcano, dated at 10,200 ± 60 cal. yr BP (Decaulne *et al.* 2016). These sedimentary traps have also made it possible to refine the timing of these landslides through the use of age-depth models. Thus, 17 Icelandic landslides have been dated since the beginning of deglaciation in northern Iceland and before 10 ka, 76% of which occurred before 12 ka (Mercier *et al.* 2017).

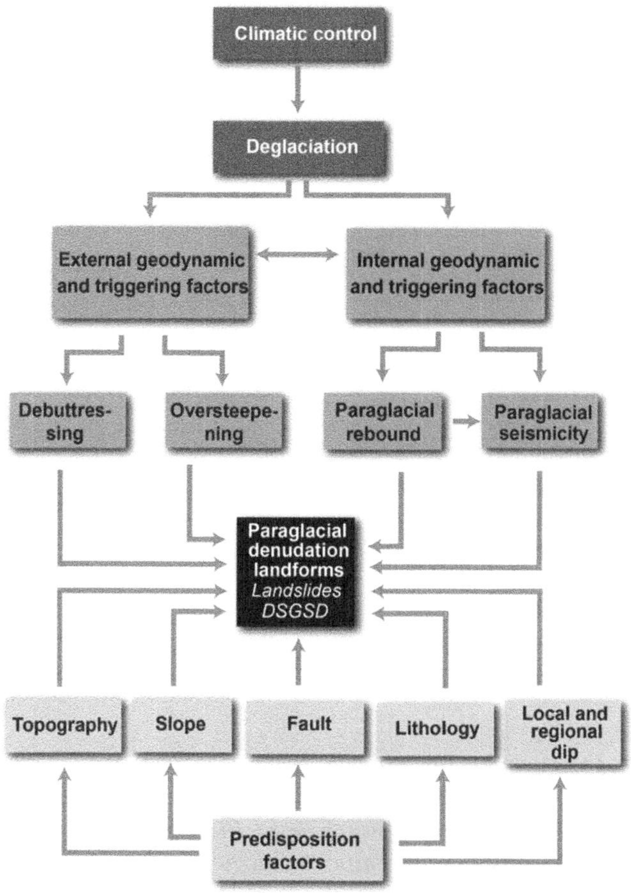

Figure 5.7. *Predisposing and triggering factors of gravity collapses and paraglacial landslides (source: based on Mercier* et al. *2017)*

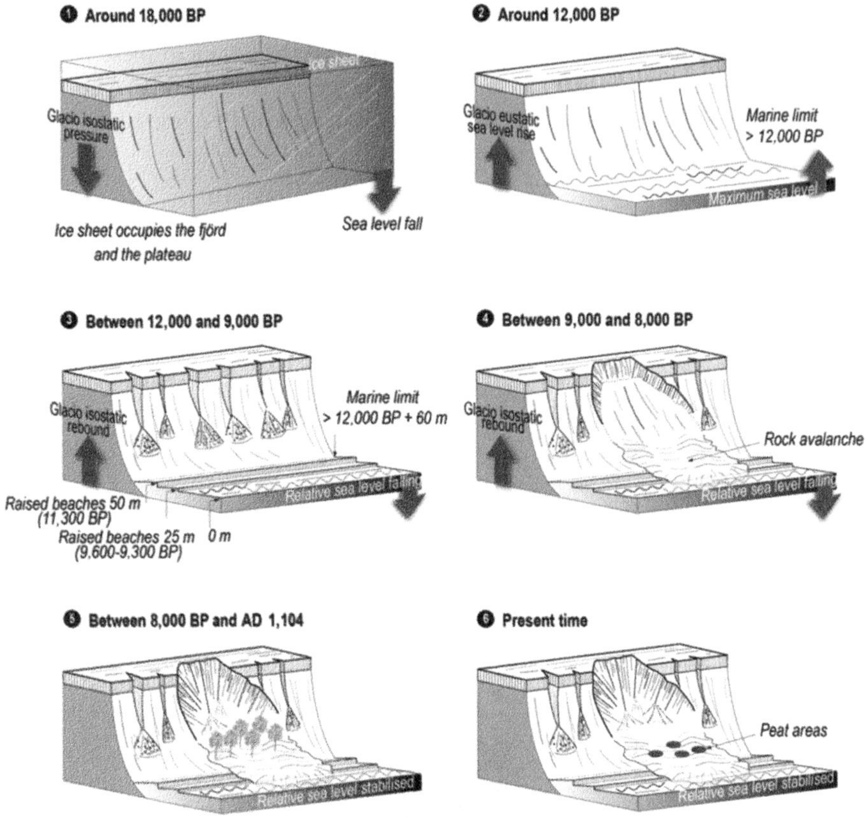

Figure 5.8. *Estimated dating of the Icelandic Höfdahólar landslide in Skagafjörður (65° 59' 43" N-19° 23' 10" W).*

COMMENT ON FIGURE 5.8.– *This estimate proposes a time range of 10,200 to 8,195 ± 45 cal. yr BP. These dates are obtained from geomorphological stacking such as raised beaches, the most recent of which dated 10,200 BP is partially fossilized by the landslide deposits, the use of tephrochronology (the oldest tephra layer found at this site is 4,211 ± 31 cal. yr BP from the Hekla volcano), the carbon-14 dating of plant remains such as birch trees trapped in the landslide depressions and dated at 7,975 ± 45 cal. yr BP and the use of an age-depth model that allows the entire sedimentary section to be encompassed beyond the dated elements and allows the beginning of sedimentation to be set between 8,113 ± 380 and 8,195 ± 45 cal. yr BP (source: from Mercier* et al. *2013).*

5.2.2. *Remobilization of slope deposits*

Slope deposits such as lateral moraines are masses of sediments inherited from the glacial sequence that will be remobilized by paraglacial processes (such as runoff).

5.2.2.1. *Runoff, gullying and debris flows*

The example of the evolution of the slope of Colletthøgda Mountain in the inner part of Kongsfjorden (78° 55' N, 12° 35' E) (see Figure 5.9) illustrates the role of runoff in the evolution of lateral moraines during the contemporary paraglacial sequence. The slope has a command of 610 m. The lower part of the slope was covered by the Kronebreen ice during the Little Ice Age to a height of 140 m. The Kronebreen watershed covers 690 km^2 and its front ends in an ice cliff in the waters of the Kongsfjorden (see the foreground in Figure 5.9).

Figure 5.9. *The northern slope of the Colletthøgda Mountain in the inner part of the Kongsfjorden (78° 55' N, 12° 35' E). For a color version of this figure, see www.iste.co.uk/mercier/climate.zip*

COMMENT ON FIGURE 5.9.– *The foreground is the seracs of the Kronebreen Glacier, which ends its course as an ice cliff in the waters of the Kongsfjorden with sedimentary plumes. The gray mass above the glacier corresponds to the lateral moraine of the LIA with sub-surface dead ice. This area is surmounted by overglacial scree cones from Holocene age. The right side of the slope has been glacially retreating since the 1970s and the lateral moraine is notably gullied by runoff fed by melting dead ice (source: oblique aerial photograph taken from a plane on August 24, 2017, by D. Mercier).*

This slope has undergone a rapid evolution linked to the retreat of the sea-front glacier by about 3 km since the early 1970s. The paraglacial sequence materializes by the abandonment of the dead ice at the base of the slope, the stripping of the dead ice under the action of translational slides of the supraglacial scree, the appearance of elementary gullies on the surface of the scree or dead ice (rill), followed by an installed runoff (gully), the dissection of the base of the slope, and the definitive melting of the dead ice preserved under the scree. It illustrates the major role of the correlative runoff of dead ice melt water in the paraglacial evolution of the slope. The evolutionary cycle is completed in only a few decades (Mercier *et al.* 2009).

5.2.2.2. *Paraglacial alluvial cones*

Alluvial cones are widely represented in paraglacial environments where they are set up following the dismantling of moraines (see Figure 5.10).

Figure 5.10. *Paraglacial Jutulhogget alluvial fans fed by the Kaldbekken stream at the outlet of the Kaldbekbotn valley in the Rondane massif in Norway (61° 53' N, 9° 47' E, 1,167 m above sea level). For a color version of this figure, see www.iste.co.uk/mercier/climate.zip*

COMMENT ON FIGURE 5.10.– *Sediment accumulates in a lake depression (Rondvatnet) in the foreground. In the background, the snow-covered slopes of the Vesleranden, which peaks at 1,473 m (source: photograph taken by D. Mercier on July 5, 1989).*

For example, Ballantyne and Benn (1996) showed their rapid establishment, in less than 200 years, in the Fabergstolsdalen valley in Norway (see Figure 5.11). Glacial retreat after the LIA allowed this sediment remobilization with considerable cone accretion rates of the order of 8 to 30 mm.yr^{-1}. These cones are now vegetated. The sedimentological signature of the three sequences is recognizable. Glacial sedimentation (morainic till) is disturbed in a second paraglacial phase characterized by debris flows (coarse heterometric facies), then fossilized by soil in the process of being built up.

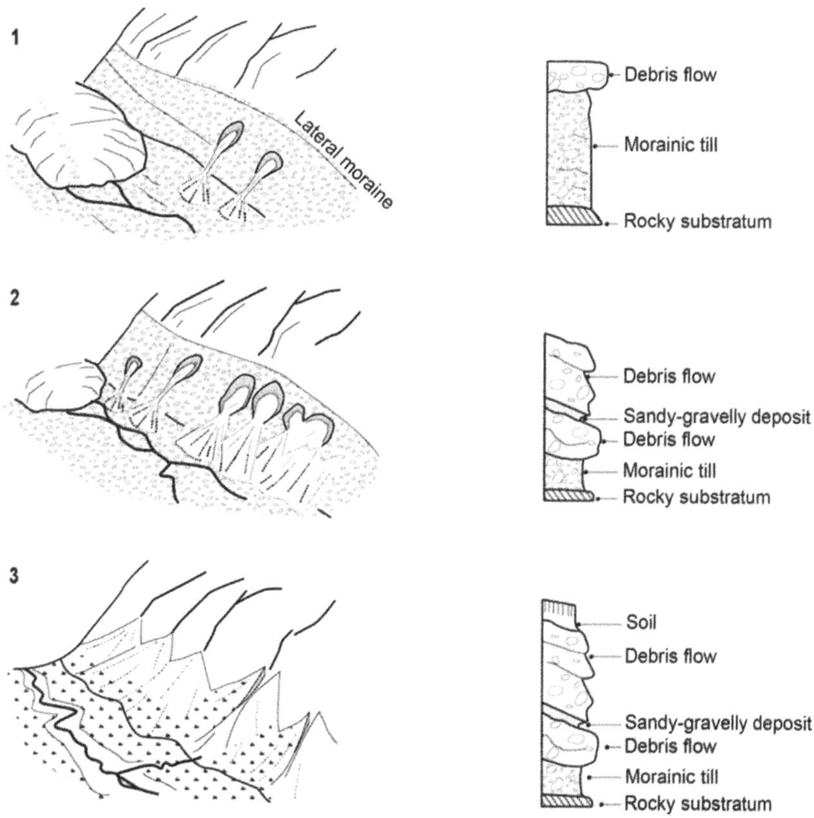

Figure 5.11. *Placement and fossilization of alluvial fans in the Fabergstolsdalen valley in Norway during a post-Little Ice Age paraglacial sequence (source: modified from Ballantyne and Benn 1996)*

5.3. Sediment fluxes within the paraglacial sedimentary cascade

The flows remobilizing sediments during the paraglacial sequences are essentially fed by the rivers themselves supplied by the melting of the glaciers.

5.3.1. *The evolution of ice margins on a decadal scale*

At fine time scales, well documented by post-LIA geomorphological reconstructions, the spasmodic character of paraglacial cascades and the pulsating evolution of ice margins are highlighted. This phenomenon is notably linked to the obturating role played by lateral and frontal moraines. The moraines constitute a local base level and hinder sediment transport from upstream to downstream during the first decades of deglaciation. As such, they induce a decoupling between the release of meltwater and geomorphological activity; this decoupling is at its peak when a lake forms upstream of the moraine (Roussel *et al.* 2016; Roussel *et al.* 2018; Knight and Harrison 2018).

Only special events such as ice jam floods (Roussel *et al.* 2016), low-frequency and high-magnitude hydrological events (Cossart *et al.* 2018) can effectively redesign or even dismantle this dam by widening and deepening the channels crossing the moraine. More specifically, a three-stage evolution of the proglacial stream can be reconstructed. In the first phase, sediment and water flows increase, which promotes the export of sediment downstream. Frontal moraines are then the main source of sediment and are gradually reworked during these early years. This can be slowed by armoring the outlet bed. Consequently, during a second phase, sediment storage takes place upstream of the moraine, within the intramorainic plain. The frontal moraine constitutes a form of resistance, a stable base level. Sediment transfer decreases and the system therefore remains blocked until a flood is triggered, which dismantles the armoring and restores the upstream-downstream sedimentary continuum. During a third phase, sediment flow may then reach a maximum: sediments are first reworked from the bed of the flow channel, while the intramorainic plain constitutes a new sediment source that can feed downstream transfers (Cossart and Fort 2008a).

In summary, the peak of sediment export sequences typically occurs a few decades after glacial retreat and is not necessarily coupled with the rate of glacial retreat and the amount of meltwater available (Marren 2005; Cossart and Fort 2008a; Roussel *et al.* 2018).

5.3.2. *Paraglacial fluvial metamorphoses on a secular scale*

The river metamorphoses studied in the French Alps have become part of valley landscapes on a centennial scale under the double effect of climate change, particularly around the Little Ice Age (LIA), and then anthropogenic developments (Bravard 1989, 2010). River metamorphosis is a change of state in the balance of the river with its environment. There is a temporal correlation between these metamorphoses and climate change, initially at the time of the transition from the medieval climatic optimum to the LIA, and then at the time of the transition from the end of the LIA to contemporary global warming.

Initially (14th–19th Centuries), river metamorphosis is under climatic control with a morphological response of meandering to braiding. Before the 14th Century, the archives allow us to speak of a certain hydrological tranquility with meandering rivers and a unique channel. The human occupation of the major beds is attested by the ports and mills. The LIA is illustrated by large morphogenic floods, associated with an increase in rainfall and glacial emptying. The hydrosedimentary response corresponds to a rise in bed height by lateral inflow from the slopes in a context of high relief energy. Rivers have reacted to this climatic deterioration with floods of unusual magnitude and frequency. Torrents and rivers with torrential regimes are then characterized by considerable water and sedimentary inputs from destabilized slopes. As the sediment inflows exceeded the carrying capacity of the rivers, the riverbeds rose, sometimes by several meters, and above all they widened significantly to occupy an increasing part of the major riverbed. The valleys had narrow beds incised into their alluvial floor. The beds became wide and braided in which the waters divided.

This metamorphosis, driven by climate change, occurred to the great despair of the societies bordering the Isère valley, which were losing bridges, mills and farmland. This first river metamorphosis, marked by the transition from meandering to braiding, has been demonstrated in four sectors in south-eastern France. They are the Isère in the Grésivaudan valley (upstream from Grenoble), the Arve in the Cluses basin, the Rhône in the Dauphiné lowlands, and the Rhône downstream from the confluence with the Ain.

The LIA was therefore a period of abundant supply of sediment load through reactivation of slope processes, a period of strong sedimentary transits and strong alluvial storage in the main valleys. Disruptions in the balance occurred first in the headwaters of the catchment areas and on lower-order rivers (from the 14th Century onwards).

Then, there was a longer term alluvial progradation and the river metamorphosis on the higher ranked rivers appears only during the 18th Century. However, even if we can consider this first paraglacial fluvial metamorphosis under climatic control, the anthropic pressure had been able to lower the resistance thresholds of the environments.

The first fluvial metamorphosis led to a response from societies, hence a second fluvial metamorphosis marked by the transition from braiding to a single channel from the 19th Century onwards. Societies sought to eliminate the risk of land loss (erosion control), to eliminate the risk of flooding, to gain land on rivers, to exploit water economically (industrial, agricultural, electricity), to extract aggregates for building and public works, to extract dead wood (to facilitate flood runoff, the operation of mills, navigation). From then on, work upstream of the watersheds was set up in the 1860s through reforestation. On the riverbeds, the diking and correction of torrential beds aimed to stabilize the longitudinal profile and reduce the slope by building sill stairs and dams. In the areas of the alluvial fans, dikes were built to contain the sediments. The construction of hydroelectric reservoirs (white coal) resulted in the diversion of watercourses and the use of water for agriculture. The hydrological and sedimentological response was manifested by the reduction of sedimentary inputs into the major riverbed, the increase in the erosion capacity of the liquid flow, and the incision of riverbeds by 8 to 14 m was seen between 1950 and 1980, with incisions of up to 50 cm per year. Thus, a new fluvial metamorphosis took place, from braiding to a single channel.

Thus, in anthropized geosystems such as alpine hydrosystems, river metamorphoses are not under the sole control of hydroclimatic and geomorphological changes. They are increasingly constrained by anthropogenic interventions. The major problem lies in the artificialization of the river in relation to its environment and above all in the irreversibility of the metamorphosis. Moreover, this anthropization of river and coastal systems prevents the progradation of deltas, such as the Rhône, whereas unconstrained, non-anthropized paraglacial deltas are prograded elsewhere in the world.

5.4. Sedimentary stocks or the end of the paraglacial sedimentary cascade

5.4.1. *Temporary storage areas on a secular scale*

The construction of temporary storage areas on a secular scale is well documented in the basin heads of mountain areas, especially in the Alpine massif. The transfer of sediments is a function of the moraine/watercourse coupling and the disturbance caused by glacial retreat from the LIA (see section 5.3.1). The beginning of the retreat phase in the Alps dated from the end of the 19th Century and lasted until the first half of the 20th Century. The thickness of valley glaciers then decreased by several hundred meters. In connection with this, lowering of the base level – the glacier tongue – occurred on the juxtaposed and proglacial margins.

Despite this, sediment export was limited by blocking effects that persist upstream of the lateral and frontal moraines (see Figure 5.12). The paraglacial sedimentary reservoirs developed around these buildings jointly show sedimentary fattening during the first decades of deglaciation.

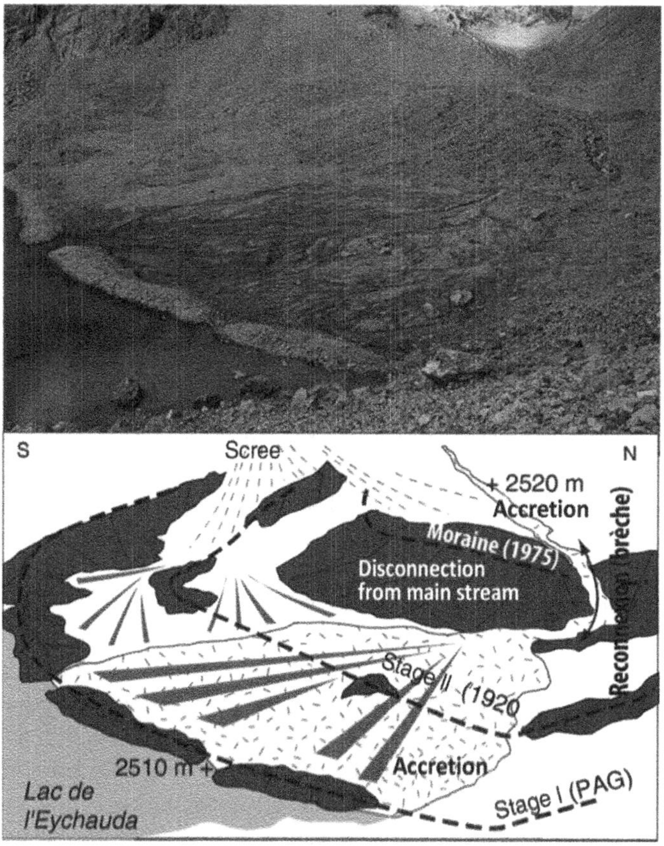

Figure 5.12. *Example of sedimentary filling on a margin undergoing post-LIA de-icing (Eychauda glacier, Hautes-Alpes, France, 2009). For a color version of this figure, see www.iste.co.uk/mercier/climate.zip*

COMMENT ON FIGURE 5.12.– *Moraine barriers remain effective dams on a secular scale, just as overburden lakes are sediment sinks. In the event of an upstream-downstream reconnection, incision of the drains leave moraine deposits in blind spots, where they are difficult to remobilize due to their disconnection (source: design and construction, É. Cossart).*

The obliteration of this double blocking effect often occurred in the middle of the 20th Century, but seemed essentially linked to an external, random event. The destocking was in fact favored by the occurrence of low frequency/high magnitude floods, such as the events of 1948 or 1957 in the Écrins massif. By widening breaches in the moraine buildings (lateral and frontal) and obliterating armoring phenomena, these floods promoted progressive reconnections between the paraglacial sedimentary reservoirs and the sedimentary cascade. The sedimentary export from the basin heads then became maximum, even if the variations in glacial volumes were less spectacular than at the beginning of the paraglacial phase. This injection of sediment was at the origin of a fluvial metamorphosis of rivers, perceptible throughout the Écrins massif between 1960 and 1980 (Cossart and Fort 2008a), while similar processes have been well demonstrated in the Arve valley, in the Chamonix by sector (Berthet 2016).

5.4.2. *Interglacial-scale temporary storage areas*

Areas in a paraglacial context are characterized by a high rate of sediment production. Sediment budgets calculated at the Holocene scale show that this volume was mainly provided by large mass movements following glacial retreat (Cossart and Fort 2008b; Cossart *et al.* 2017). The deposits associated with these mass movements are, however, difficult to dismantle through gullying and runoff processes.

In high altitude areas these deposits remain confined to the valley bottoms, creating dam phenomena that interrupt the sedimentary cascade and create large sediment traps (volume = 10^7 m^3). Well-documented in the Alpine region (see Figure 5.13), these sealing effects are still effective several millennia after their formation, as evidenced by their weak incision. This incision is controlled and reduced by the phenomenon of armoring of the riverbeds. This armoring is all the more effective as the power of the (still nascent) watercourses is low, making them incapable of remobilizing the large blocks that make up the deposit of mass movements. The river system is thus segmented, alternating sections of gorges through collapse deposits and wide plains backed by durable sediment reserves (see Figure 5.14).

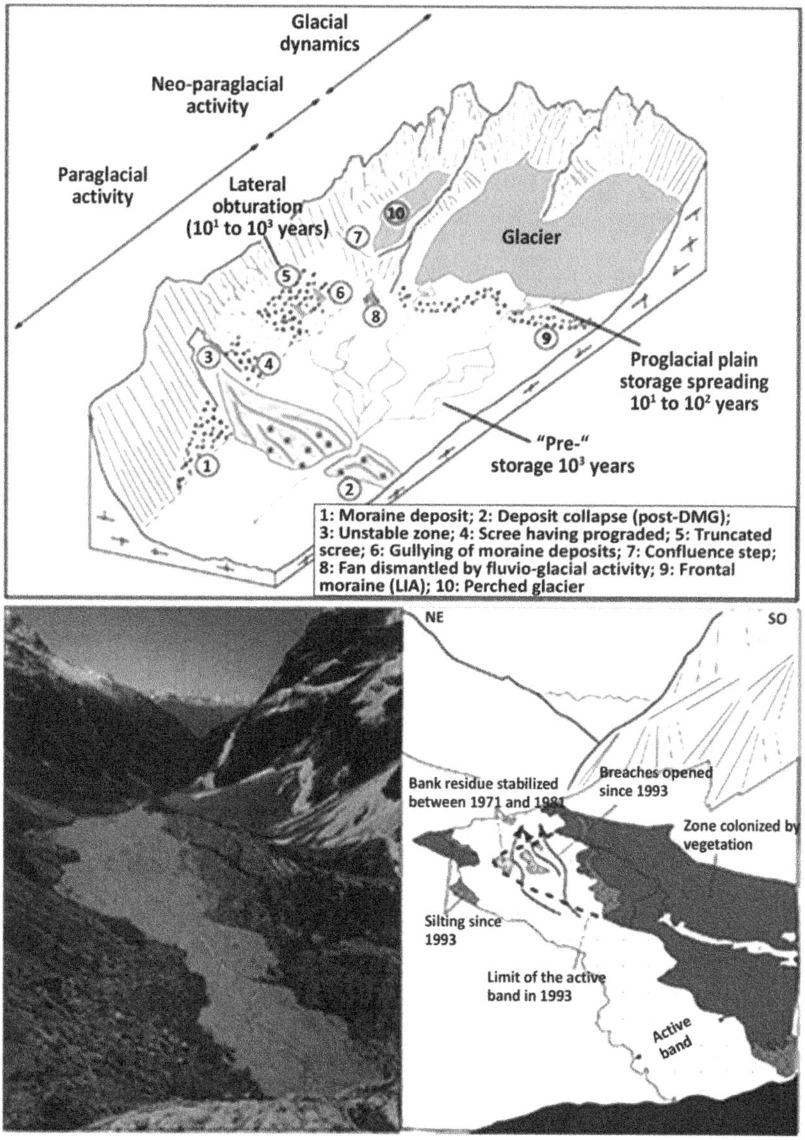

Figure 5.13. *Examples of sustainable storage phenomena in a paraglacial context. The mass movement deposits, like the Pré de Madame Carle deposit (Hautes-Alpes) (bottom), are effective dams that hinder sedimentary transfer on a millennial scale (source: design and construction, É. Cossart). For a color version of this figure, see www.iste.co.uk/mercier/climate.zip*

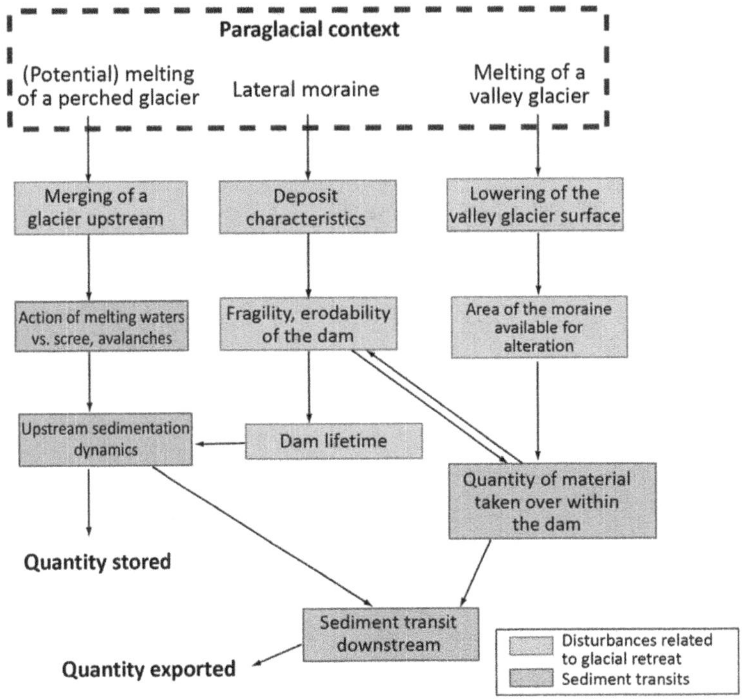

Figure 5.14. *Sedimentary budgets of upper watersheds affected by mass movements impeding sediment flow. The reservoir thus created causes an accretion from which sediment export remains very small in relation to the magnitude of the sedimentary volumes involved, brought by the slopes (source: modified from Cossart and Fort 2008b). For a color version of this figure, see www.iste.co.uk/mercier/climate.zip*

In the high latitudes, the dismantling of the slopes of fjords and other glacial troughs cannot contribute to the sedimentary cascade for two combined reasons. The landslides occur mostly in the lower parts of the valleys, in unconfined areas, creating a remoteness unfavorable to any subsequent remobilization by rivers. More particularly, in Iceland as in Svalbard, it has been shown that these deposits take place on fluvio-glacial terraces, without any possible direct connection with the watercourses likely to remobilize them (Cossart *et al.* 2013). The entire sequence at the Tardiglacial/Holocene transition is thus characterized by a fattening of the foot of the slopes.

The consequence of this operation is that only the glaciers seem to be able to evacuate the large quantities of debris located at the foot of the slope. This reinforces the hypothesis that glacial valleys in high latitudes, as well as high altitudes, undergo sedimentary ablation mainly in the post-glacial period, whereas during the glacial phase, glaciers simply evacuate the material accumulated beforehand (see, for example, arguments in Le Cœur 1997; Spagnolo and Clark 2009). This point of view converges notably with the work of Bentley and Dugmore (1998), or more recently of Coquin *et al.* (2015), who see postglacial landslides as the main agent of the widening of troughs and fjords, hitherto considered to be the result of glacial erosion alone.

5.4.3. *Final storage areas*

Within the paraglacial cascade, the sediments eventually end up in sinks where they are permanently stored. These accumulation zones are located in particular in lakes for the land areas of high latitudes, where they are numerous in connection with the recent history of deglaciation (areas of overglacial glaciation, disorganization of the drainage network, etc.). On the other hand, permanent sinks, such as the bottom of fjords and oceans, accumulate sediments carried by the calving of sea-front glaciers and from river inflows associated with the melting of land glaciers.

5.5. Conclusion

In terms of time, the term paraglacial refers primarily to a morphological sequence, an erosion crisis, which follows the preceding glacial sequence. This major temporal sequence in the evolution of environments during phases of climate warming is characterized by processes that are not glacial, but rather induced by the melting of the ice and involved in the cascading transfer of sediments. The two preeminent paraglacial processes are, on the one hand, post-glacial decohesion, which affects rock walls, and, on the other hand, runoff fed by the melting of glacier ice and dead ice, which remobilizes loose sediments. Thus, forms of ablation and accumulation associated with these dynamics are representative of these paraglacial dynamics. The large landslides resulting from post-glacial decohesion build coarse heterometric deposits that are difficult to remobilize later in time and space. On the other hand, the very active runoff during the paraglacial sequence is at the origin of the construction of gullying, incision and accumulation forms (at the origin of terraces, alluvial cones, sandar). The areas affected by these landscape metamorphoses are therefore numerous and have undergone a complex morphogenic system. The cascade systems approach is currently the most widely used to quantify sediment and energy transfers between source and storage areas. One of the consequences of this approach is that, around the general models established by Church and Ryder (1972) and then by Ballantyne (2002), many authors document the effects of local contexts showing the variability of paraglacial responses as a function of external variables (lithology, tectonics, climatic domain).

5.6. References

Ballantyne, C.K. (2002). Paraglacial geomorphology. *Quaternary Science Reviews*, 21(1), 1935–2017.

Ballantyne, C.K. and Benn, D.I. (1996). Paraglacial slope adjustment during recent deglaciation and its implications for slope evolution in formerly glaciated environments. In *Advances in Hillslope Processes*, Anderson, M.G., Brooks, S.M. (eds). John Wiley, New York.

Bentley, M. and Dugmore, A.J. (1998). Landslides and the rate of glacial trough formation in Iceland. *Quaternary Proceedings*, 6, 11–15.

Berthet, J. (2016). L'évolution géomorphologique des systèmes torrentiels proglaciaires dans la vallée de Chamonix-Mont Blanc. PhD thesis, Université Grenoble Alpes, Grenoble.

Bourriquen, M., Mercier, D., Baltzer, A., Fournier, J., Costa, S., Roussel, E. (2018). Paraglacial coasts responses to glacier retreat and associated shifts in river floodplains over decadal timescales (1966–2016), Kongsfjorden, Svalbard. *Land Degradation and Development*, 29(11), 4173–4185.

Bravard, J.-P. (1989). La métamorphose des rivières des Alpes françaises à la fin du Moyen Âge et à l'époque moderne. *Bulletin de la Société Géographique de Liège*, 25, 145–157.

Bravard, J.-P. (2010). Discontinuities in braided patterns: The River Rhône from Geneva to the Camargue delta before river training. *Geomorphology*, 117(3/4), 219–233.

Church, M. and Ryder, J.M. (1972). Paraglacial sedimentation: Consideration of fluvial processes conditioned by glaciation. *Geological Society of America Bulletin*, 83, 3059–3072.

Church, M. and Ryder, J.M. (1989). Sedimentology and clast fabric of subaerial debris flow facies in a glacially influenced alluvial fan – A discussion. *Sedimentary Geology*, 65, 195–196.

Coquin, J., Mercier, D., Bourgeois, O., Cossart, É., Decaulne, A. (2015). Gravitational spreading of mountain ridges coeval with Late Weichselian deglaciation: Impact on glacial landscapes in Tröllaskagi, northern Iceland. *Quaternary Science Reviews*, 107(1), 97–213.

Coquin, J., Mercier, D., Bourgeois, O., Feuillet, T., Decaulne, A. (2016). Is the gravitational spreading a precursor of the landslide of Stífluhólar (Skagafjörður, Northern Iceland)? *Géomorphologie : relief, processus, environnement*, 22(1), 9–24.

Cossart, É. (2005). Évolution géomorphologique du haut bassin durancien Alpes du Sud, France depuis la dernière glaciation. Contribution à la compréhension du fonctionnement du système paraglaciaire. PhD thesis, Université Paris-Diderot Paris 7, Paris.

Cossart, É. and Fort, M. (2008a). Sediment release and storage in early deglaciated areas: Towards an application of the exhaustion model from the case of Massif des Écrins (French Alps) since the Little Ice Age. *Norsk Geografisk Tidsskrift – Norwegian Journal of Geography*, 62, 115–131.

Cossart, É. and Fort, M. (2008b). Consequences of landslide dams on alpine river valleys: Examples and typology from the French Southern Alps. *Norsk Geografisk Tidsskrift – Norwegian Journal of Geography*, 62, 75–88.

Cossart, É., Mercier, D., Decaulne, A., Feuillet, T. (2013). An overview of the consequences of paraglacial landsliding on deglaciated mountain slopes: Typology, timing and contribution to cascading fluxes. *Quaternaire*, 24, 13–24.

Cossart, É., Mercier, D., Decaulne, A., Feuillet, T., Jónsson, H.P., Sæmundsson, Þ. (2014). Impacts of post-glacial rebound on landslide spatial distribution at a regional scale in northern Iceland (Skagafjörður). *Earth Surface Processes and Landforms*, 39(3), 336–350.

Cossart, É., Mercier, D., Coquin, J., Decaulne, A., Feuillet, T., Jónsson, H.P., Sæmundsson, Þ. (2017). Denudation rates during a postglacial sequence in Northern Iceland: Example of Laxárdalur valley in the Skagafjörður area. *Geografiska Annaler*, 99(3), 240–261.

Cossart, E., Viel, V., Lissak, C., Reulier, R., Fressard, M., Delahaye, D. (2018). How might sediment connectivity change in space and time? *Land Degradation and Development*, 29, 2595–2613.

Decaulne, A., Cossart, É., Mercier, D., Coquin, J., Feuillet, T., Jónsson, H.P. (2016). An early Holocene age for the Vatn landslide (Skagafjörður, central northern Iceland): Insights into the role of postglacial landsliding on slope development. *The Holocene*, 26(8), 1304–1318.

Jackson, L.E., MacDonald, G.M., Wilson, M.C. (1982). Paraglacial origin for terraced river sediments in Bow Valley, Alberta. *Canadian Journal of Earth Sciences*, 19, 2219–2231.

Knight, J. and Harrison, S. (2018). Transience in cascading paraglacial systems. *Land Degradation and Development*, 29, 1991–2001.

Le Cœur, C. (1999). Rythmes de dénudation tertiaire et quaternaire en Écosse occidentale. *Géomorphologie : relief, processus, environnement*, 5(4), 291–303.

Marren, P.M. (2005). Magnitude and frequency in proglacial rivers: A geomorphological and sedimentological perspective. *Earth-Science Reviews*, 70(3/4), 203–251.

Mercier, D. (2008). Paraglacial and paraperiglacial landsystems: Concepts, temporal scales and spatial distribution. *Géomorphologie : relief, processus, environnement*, 14(4), 223–234.

Mercier, D. (2011). *La géomorphologie paraglaciaire. Changements climatiques, fonte des glaciers et crises erosives associées*. Éditions universitaires européennes, Sarrebruck.

Mercier, D. and Laffly, D. (2005). Actual paraglacial progradation of the coastal zone in the Kongsfjorden area, western Spitsbergen (Svalbard). In *Cryospheric Systems: Glaciers and Permafrost*, Harris, C., Murton, J. (eds). Geological Society, special publication, London.

Mercier, D., Étienne, S., Sellier, D., André, M.-F. (2009). Paraglacial gullying of sediment-mantled slopes: A case study of Colletthøgda, Kongsfjorden area, West Spitsbergen (Svalbard). *Earth Surface Processes and Landforms*, 34, 1772–1789.

Mercier, D., Cossart, É., Decaulne, A., Feuillet, T., Jónsson, H.P., Sæmundsson, Þ. (2013). The Höfðahólar rock avalanche (Sturzström): Chronological constraint of paraglacial landsliding on an Icelandic hillslope. *The Holocene*, 23(3), 431–445.

Mercier, D., Coquin, J., Feuillet, T., Decaulne, A., Cossart, É., Jónsson, H.P., Sæmundsson, Þ. (2017). Are Icelandic rock-slope failures paraglacial? Age evaluation of seventeen rock-slope failures in the Skagafjörður area, based on geomorphological stacking, radiocarbon dating and tephrochronology. *Geomorphology*, 296, 45–58.

Moreau, M., Mercier, D., Laffly, D., Roussel, E. (2008). Impacts of recent paraglacial dynamics on plant colonization: A case study on Midtre Lovénbreen foreland, Spitsbergen (79° N). *Geomorphology*, 95(1/2), 48–60.

Peras, A., Decaulne, A., Cossart, É., Coquin, J., Mercier, D. (2016). Distribution and spatial analysis of rockslides failures in the Icelandic Westfjords: First results. *Géomorphologie : relief, processus, environnement*, 22(1), 25–35.

Roussel, E., Toumazet, J.-P., Marren, P.M., Cossart, É. (2016). Iceberg jam floods in Icelandic proglacial rivers: Testing the self-organized criticality hypothesis. *Géomorphologie : relief, processus, environnement*, 22(1), 37–49.

Roussel, E., Marren, P.M., Cossart, É., Toumazet, J.-P., Chenet, M., Grancher, D., Jomelli, V. (2018). Incision and aggradation in proglacial rivers: Post-Little Ice Age long-profile adjustments of southern Iceland outwash plains. *Land Degradation and Development*, 29, 3753–3771.

Spagnolo, M. and Clark, C.D. (2009). A geomorphological overview of glacial landforms on the Icelandic continental shelf. *Journal of Maps*, 5, 37–52.

Strzelecki, M., Long, A.J., Lloyd, J.M., Malecki, J., Zagorski, M. (2018). The role of rapid glacier retreat and landscape transformation in controlling the post-Little Ice Age evolution of paraglacial coasts in central Spitsbergen (Billefjorden, Svalbard). *Land Degradation and Development*, 29(6), 1962–1978.

6

Spatial Impacts of Climate Change on Periglacial Environments

Denis MERCIER[1] and Étienne COSSART[2]

[1] *Sorbonne University, Paris, France*

[2] *Jean Moulin University Lyon 3, France*

6.1. Introduction

Periglacial environments now represent 25% of the Earth's surface, of which 13 to 18% is permafrost, in the high latitudes of both hemispheres and in the mountains and highlands of the mid and low latitudes (Ballantyne 2018).

In the context of current climate change, knowledge of periglacial processes and the degradation of permafrost are challenges for a global understanding of our planet, the consequences of which do not solely affect periglacial areas. According to Ballantyne (2018, p. 371):

> "The battle to limit global change has barely begun, but periglacial environments are in the front line of a conflict that will affect humanity and ecosystems throughout the world."

6.1.1. *Definition of periglacial*

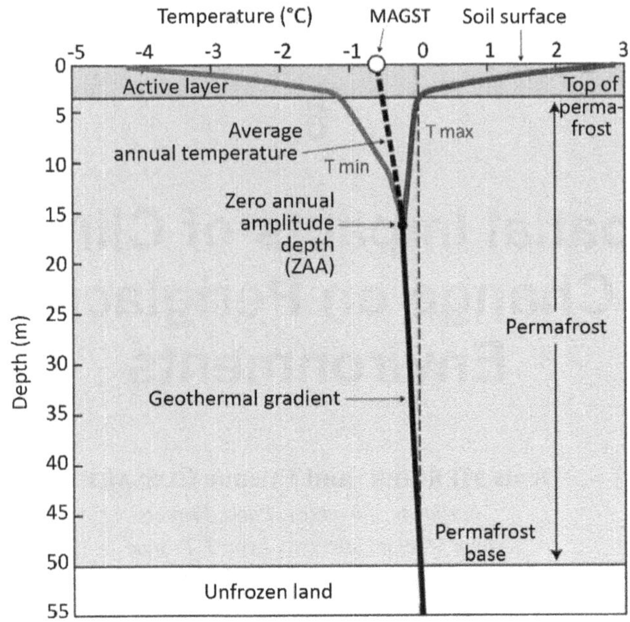

Figure 6.1. *Vertical profile of permafrost. The active layer (gray) thaws each summer and refreezes in winter when air temperatures become negative. The temperature of the permafrost (in blue) remains constantly ≤ 0°C. Dotted line represents the mean annual ground surface temperature (MAGST) (source: based on Ravanel 2010). For a color version of this figure, see www.iste.co.uk/mercier/climate.zip*

The term periglacial is composed of the Greek prefix *peri*, meaning around, and glacial derived from the Latin *glacies*, meaning ice. The term was proposed in 1909 by the Polish geologist Walery Łoziński. It refers to processes that are mainly driven by the alternation of freezing and thawing in rocks and soils (gelifraction, gellifluxion, geliturbation). The term also includes the shapes created by these same processes, such as cones and scree slopes, stone circles, pingos, palsas, etc. The term periglacial is more widely used to refer to cold environments in which these processes associated with the freeze-thawing of soils and rocks are predominant (French 2017; Ballantyne 2018). Although etymologically the term means "around glaciers", periglacial environments may be geographically hundreds or thousands of kilometers away from the nearest glacier, such as in Canada, Alaska, or northern Eurasia. These periglacial spaces are characterized by low temperatures and are often associated with areas of permafrost. The term permafrost refers to ground or rocky outcrops that have a negative temperature for at least two consecutive years. The presence of permafrost fundamentally influences periglacial processes and surface shaping by providing an impermeable barrier and high subsurface moisture during the melting season, allowing freeze-thaw processes to develop within the active layer that is typically 0.5 to 2 m thick (see Figure 6.1).

6.1.2. *Present and past spatial extent of periglacial environments*

An arbitrary limit of +3°C for mean annual air temperature is used to encompass cold environments related to seasonal surface frost. The most visible landscape boundary for periglacial environments is the tree line or timber line for both high latitudes and high altitudes in alpine environments. However, this limit should be taken with caution, as there are many anthropogenic interferences in the positioning of the tree line.

The most common classification of areas with permafrost distinguishes between areas with continuous permafrost which account for 90–100% of the area, areas with discontinuous permafrost (50–90%) and areas with sporadic permafrost for less than 50%. In the Northern Hemisphere, permafrost is present over 23 million km^2, or about 25% of the land area (see Figure 6.2). The thickness of permafrost varies greatly from one area to another. From just a few meters, it can reach up to 1,500 meters in the continental areas of Siberia, where it formed during the cold Pleistocene periods at the periphery of the great ice sheets. Along the coasts of the Arctic Basin, there is also submarine permafrost inherited from the Pleistocene cold periods, when the sea level was 120 meters lower than today, providing continental conditions for these now submerged areas (Overduin *et al.* 2019).

According to the ground surface temperature map of the northern hemisphere produced by scientists at the German Polar Institute (Obu *et al.* 2018), permafrost areas with mean annual ground temperatures below 0°C account for 15% of the land area (see Figure 6.3).

The mapping of permafrost extent in high-altitude areas is being continuously improved as a result of the development of geophysical prospecting methods, coupled with the systematic use of geostatistical modeling to extrapolate local observations. The French school has particularly improved knowledge of alpine terrain from field surveys (Bodin 2007; Perrier 2014; Magnin 2015; Marcer 2018), which can serve as a reference.

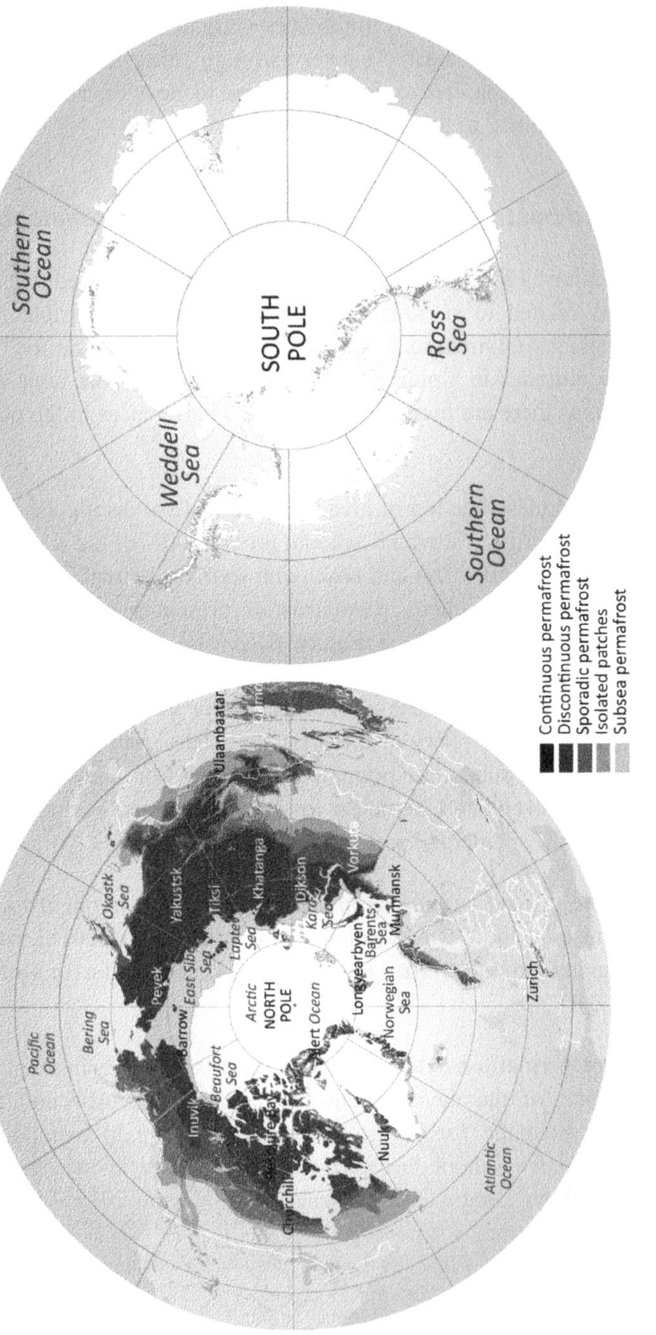

Figure 6.2. *Maps of permafrost in both hemispheres (source: Brown et al. 1997). For a color version of this figure, see www.iste.co.uk/mercier/climate.zip*

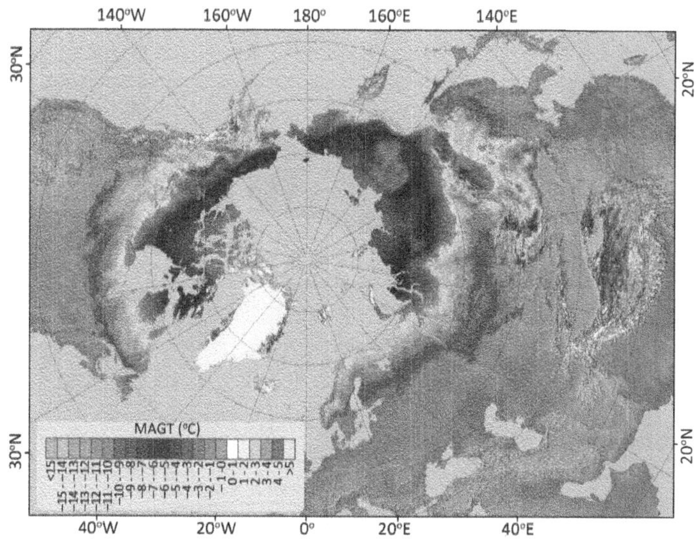

Figure 6.3. *Map of mean annual soil temperatures (2000–2016) in the Northern Hemisphere (source: Obu* et al. *2018). For a color version of this figure, see www.iste.co.uk/mercier/climate.zip*

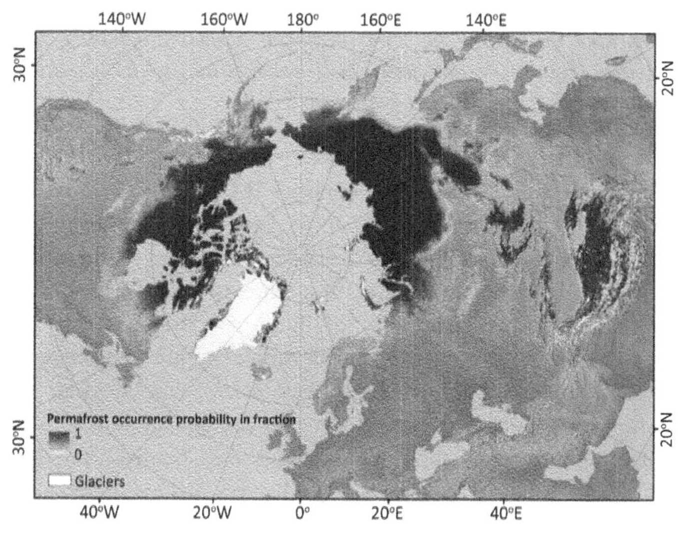

Figure 6.4. *Probability of occurrence of permafrost for the Northern Hemisphere derived from the mean annual ground temperature map (Figure 6.3) (source: Obu* et al. *2018). For a color version of this figure, see www.iste.co.uk/mercier/climate.zip*

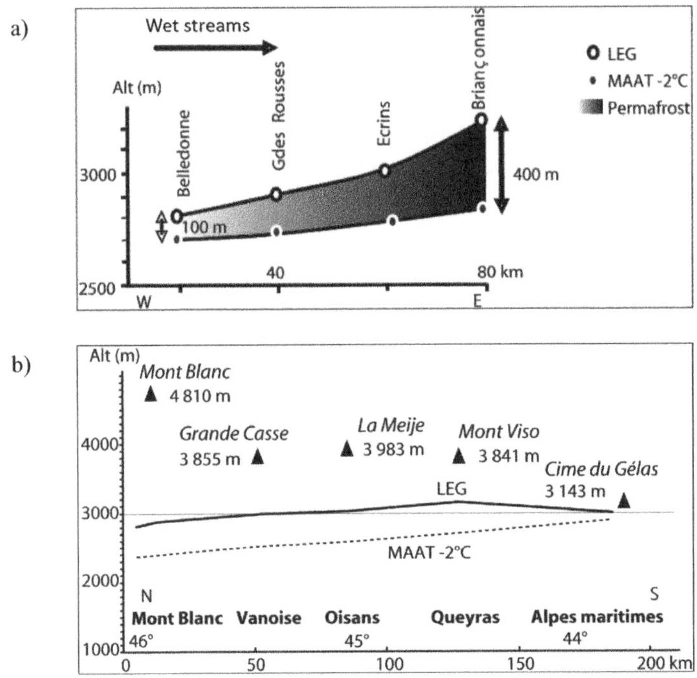

Figure 6.5. *Extension of the permafrost fringe in the French Alps. The fringe lies between the MAAT −2°C and the limit of glacial equilibrium (LGE), beyond which glaciers develop (source: conception and realization by É. Cossart, from the synthesis of Bodin 2007; Perrier 2014; Magnin 2015).*

In general, the elevation of the −2°C isotherm (MAAT −2°C) is used to delineate the discontinuous permafrost domain. According to this average marker (see Figure 6.5), permafrost appears from 2,500 to 2,600 meters in the northern sectors of the Alps (Mont Blanc, Vanoise), while this limit tends to rise to 2,800 meters towards the lower latitudes and the inner part of the Alps (Briançonnais, Ubaye). This general pattern should not obscure the great variability in the altitude of the lower permafrost limit. Indeed, all recent studies show that singular topoclimatic conditions (strong shading effects, concavity that can constitute cold air mass traps, etc.) can cause permafrost to appear at altitudes well below what the theoretical altitude of the −2°C isotherm would suggest. In very shaded areas, permafrost has been observed around 2,400 meters above sea level, even in the southern French Alps (Perrier 2014). Next to so-called "typical" permafrost (Delaloye 2004), where energy transfers between the atmosphere and the permafrost roof are typically vertical by conduction or convection, marginal (or "atypical") permafrost may coexist. This permafrost is associated with singular energy exchanges, mainly in the horizontal plane by advection (including air circulation in *openwork* material, or by water flows). Finally, the weight of legacies often influences the distribution of the current permafrost; its inertia helps to explain why the thermal conditions of these frozen soils still partly reflect those of the Little Ice Age.

6.2. Melting permafrost and paraperiglacial geomorphological crises

6.2.1. *Definition of paraperiglacial*

By analogy with paraglacial crises, it is possible to describe processes, forms, spaces and temporalities characterized by the disappearance of permafrost, particularly in the context of warming (Mercier 2008). The adjective paraperiglacial has thus been defined (Mercier 2008). These adjustments have been documented with considerable delay (in comparison with the paraglacial studies), probably due to the difficulty of reconstructing permafrost dynamics. The permafrost is not easily visible in the landscape and is also subject to particularly complex heat exchanges within the cryosphere (see Figure 6.6).

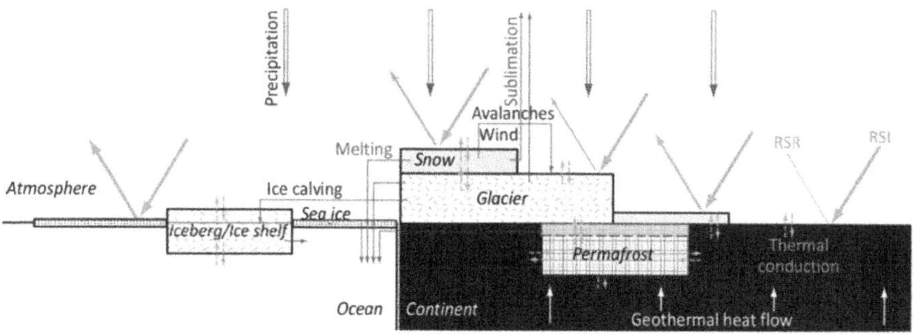

Figure 6.6. *All radiative exchanges affecting permafrost in a complex system within the cryosphere, driven by numerous feedbacks related to interactions with the ice field and snow cover in particular (source: based on Cossart 2014). For a color version of this figure, see www.iste.co.uk/mercier/climate.zip*

Geomorphologically, the term paraperiglacial refers to the Earth's surface processes, sedimentary accumulations, landforms, systems and landscapes that are directly conditioned by permafrost degradation (Mercier 2008). Thus, interglacial periods, such as the Holocene and the current period of global warming subsequent to the Little Ice Age (LIA) are sequences of paraperiglacial evolution. Melting permafrost in the face of new thermal imbalance, generates processes, stimulated in particular by the heating of ground ice as it approaches its melting point, or even the disappearance of this ice and the concomitant release of water into the substrate.

The past extent of permafrost, reconstructed using geomorphological indicators and other data, makes it possible to assess the extent of areas that have been, or are still being, affected by permafrost degradation. In the Pleistocene, the "Last Permafrost Maximum" (Vandenbergh *et al.* 2014) was characterized by the presence of permafrost over nearly one-third of the continental areas (see Figure 6.7) during a period dated between 25 and 17 ka. In detail, the remarkable extent, even at relatively low latitudes, of permafrost in Eurasia contrasted with a more limited extent in North America. This result was interpreted by a marked continentality in Eurasia, limiting the thickness of the snow cover and its insulating capacity between the atmosphere and the substrate. The construction of such a map, on a large spatial scale, for the LIA remains to be created, its realization is made particularly complex by the fact that permafrost had a high inertia. The presence and thermal state of permafrost in the LIA probably only partially reflected the climatic conditions of that period, which was too short to have created a state of equilibrium between permafrost, associated geomorphic forms, and climatic conditions.

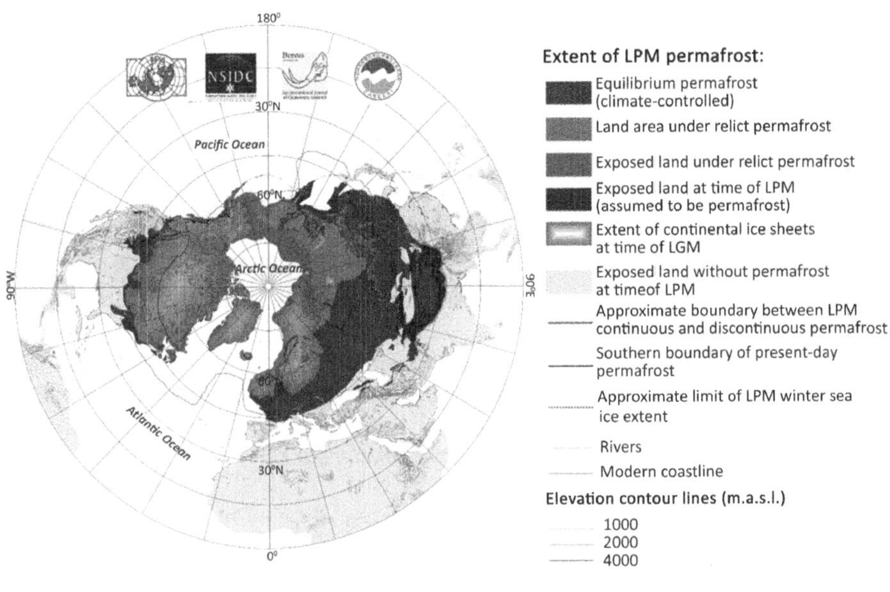

Figure 6.7. *Extension of permafrost during the Last Permafrost Maximum (LPM) in the Northern Hemisphere (source: Vandenbergh* et al. *2014). For a color version of this figure, see www.iste.co.uk/mercier/climate.zip*

6.2.2. *Paraperiglacial processes and forms*

The processes and forms associated with this paraperiglacial dynamics can be analyzed as a cascading paraperiglacial geosystem (Mercier 2008; see Figure 6.8).

From the source zones (wall, slope deposits, valley bottom deposits, shoreline deposits), periglacial processes (gelifraction, geliturbation, etc.) and azonal-related processes at work in these cold environments construct so-called "periglacial" patterns that constitute primary periglacial deposits (scree slopes, boulder fields, pingos, palms, etc.). In the sequence that follows, thermal conditions lead to the degradation of permafrost and the onset of periglacial processes (mass movements, runoff, debris flows, etc.) that remobilize primary deposits and constitute new forms of accumulation.

The processes at work are stimulated in particular by the injection of water into the substrate, either directly by the direct melting of permafrost ice or more indirectly by the melting of snow cover and the increase in liquid precipitation (classic in periods of warming). The substitution of liquid water for ice causes a loss of volume in the substrate and a settling phenomenon called "thermokarst". This process was described as early as the first half of the 20th Century in Eurasia (Ermolaev 1932) and North America (Muller 1947), in particular because of the profound damage it causes to infrastructure. Then, the increased imbibition of the substrate materials causes a destabilization of the substrate, which can adopt a plastic behavior, or even exceed a liquidity threshold. Finally, even if the ice does not reach its melting point, its heating is sufficient to give it a rheological behavior characterized by greater plasticity. In other words, while cold ice can be an excellent cement, ensuring the cohesion of the substrate, ice close to its melting point can be conducive to ground instabilities, particularly on steep slopes. These two permafrost degradation processes (disappearance of ice from the ground, heating of the ice from the ground) most often act jointly, on the same site, from the moment the radiative balance brings heat into the ground. This contribution is first materialized by a thickening of the active layer and the lowering of the permafrost roof. Gradually, in deeper areas, the thermal gradient adjusts to the heat input, which propagates and brings the remaining ice closer to the melting point.

New paraperiglacial deposits (alluvial cones, thermokarstic depressions, stratified deposits, etc.) are thus created and constitute a sedimentary stock from which the aeolian, fluvial, littoral and marine azonal-dynamics draw most of their contributions.

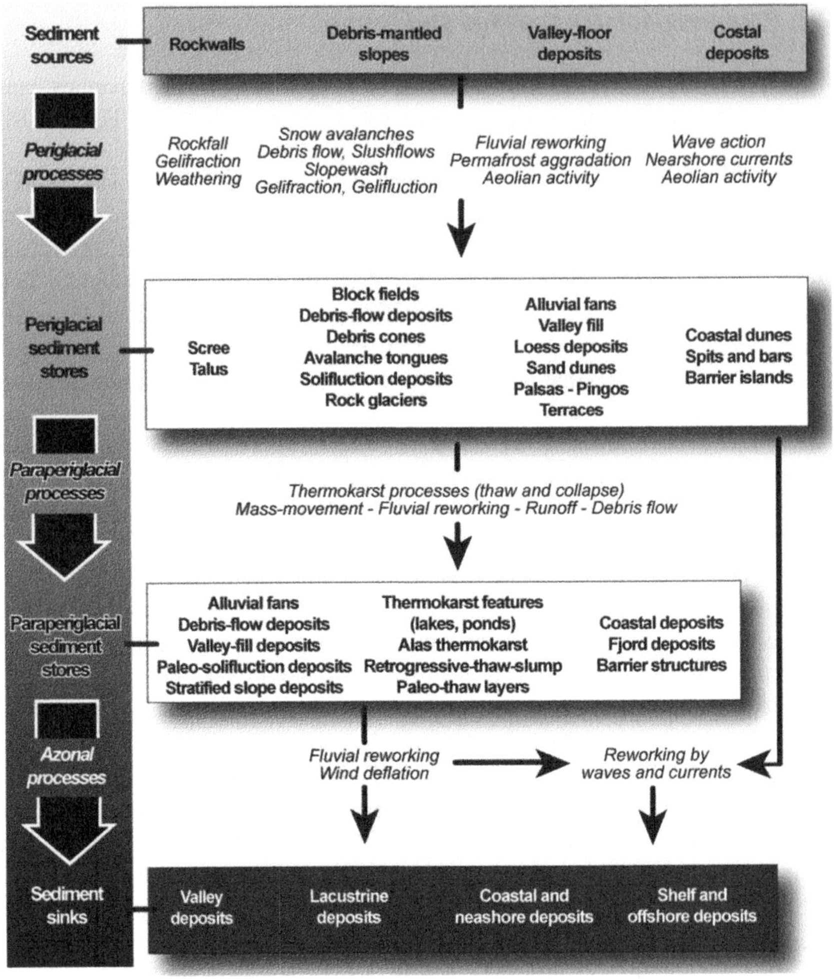

Figure 6.8. *Simplified diagram of cascade sediment transfer in a paraperiglacial context (source: based on Mercier 2008)*

In coastal paraperiglacial systems, the relationship between rock outcrops and the sea may be a function of the presence and nature of the inherited continental deposits that now encroach on the coastal realm. With the rise in sea level during the Holocene, erosion has resulted in the retreat of deposits that were carved into cliffs. The coarse sediments removed from the periglacial deposits line the base of the cliffs and form a bulwark against the onslaught of the sea. In this context, most coastlines on both sides of the North Atlantic are experiencing, or have experienced, conditions of paraperiglacial evolution similar to those of the Holocene and other interglacial periods, such as the Eemian. For example, the coastlines of France covered by periglacial deposits during the cold Pleistocene sequences, as in the Weichselian, can be considered as paraperiglacial coasts (Mercier 2008). The Normandy coastline of the Cotentin peninsula, with these head deposits cut into cliffs during the Flandrian transgression, is a good example. The same applies to the coastal platforms of Galicia (Blanco-Chao *et al.* 2007).

6.3. Periglacial coastal environments in high latitudes in the face of climate change

High latitude permafrost areas experience increased air and ground temperatures, resulting in an increase in active layer thickness. The responses of Arctic spaces to this contemporary warming depend in particular on the environments concerned.

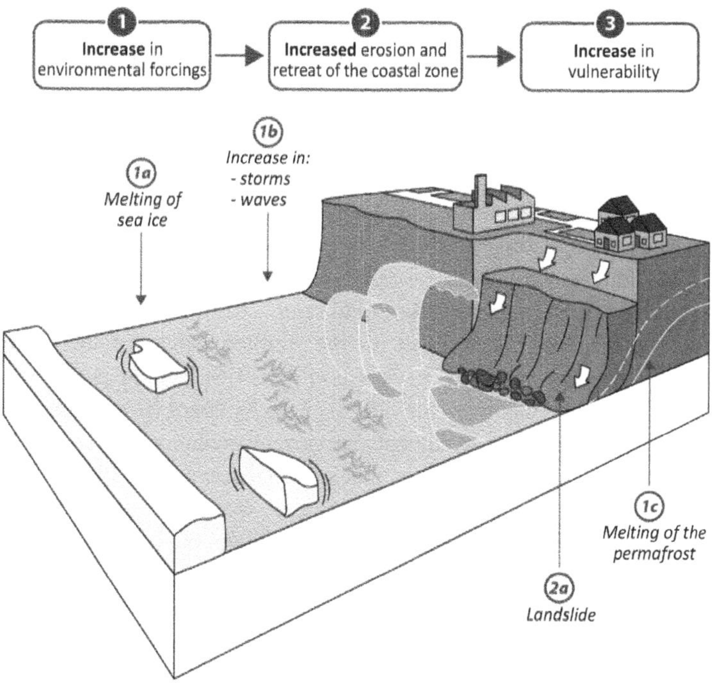

Figure 6.9. *Changing environmental forcing on the Arctic Basin coastlines and the implications for shoreline erosion and issues (source: design: D. Mercier; drawing: F. Bonnaud, Sorbonne University, 2019). For a color version of this figure, see www.iste.co.uk/mercier/climate.zip*

Figure 6.10. *Large active paraperiglacial landslides related to contemporary melting of permafrost on Herschel Island, Canada, on the southern shore of the Beaufort Sea (69° 36' N, 139° 04' W) (source: Alfred Wegener Institute, Potsdam, Germany). For a color version of this figure, see www.iste.co.uk/mercier/climate.zip*

The permafrost shorelines of the Arctic Basin represent 34% of the world's coastlines. The issues at stake in their evolution go beyond the zonal framework of high latitudes alone, since they release particulate organic carbon and methane each year in relation to permafrost melting (Fritz *et al.* 2017; Tanski *et al.* 2017). Many recent studies have been carried out on the southern shore of the Beaufort Sea (Jones *et al.* 2009; Obu *et al.* 2017; Coutre *et al.* 2018; Irrgang *et al.* 2017, 2019), along the Siberian coast (Günther *et al.* 2015) or the Svalbard (Bourriquen *et al.* 2018; Strzelecki *et al.* 2018). They document the current evolution of these sedimentary accumulations inherited from the last glacial sequence of the Pleistocene, whose environmental conditions during the Holocene were conducive to permafrost aggradation. Now, rising air and ground temperatures in the Arctic are leading to an increase in the thickness of the active layer and the disappearance of the fast ice that protects it from wave action (see Figures 6.9 and 6.10). For example, some coastlines are experiencing erosion rates of 25 to 30 m per year (Jones *et al.* 2009; Wobus *et al.* 2011). The average retreat rate for shorelines along the Yukon coast is on the order of 0.7 m per year (Irrgang *et al.* 2017) and on the order of 1.4 m per year along the northern coast of Alaska (Couture *et al.* 2018). However, some coastal areas are experiencing conversely progradation in relation to massive sediment input from ice-covered catchments such as in Spitsbergen (Strzelecki *et al.* 2018; Bourriquen *et al.* 2018) or landslide masses in relation to large landslides in soft coastlines (retrogressive thaw slump) such as along the Beaufort Sea coast (Obu *et al.* 2017).

Based on current erosion rates (1950–2011), work shows that for a 210 km stretch of the Beaufort Sea coastline between the U.S.-Canada border and the town of Tapqaq (Shingle Point), the infrastructure used by the local population is under serious threat (Irrgang *et al.* 2019). Two projections, one conservative (rate of retreat of 0.7 m per year) and one more pessimistic (rate of retreat of 2.2 m per year), show that between 850 and 2,660 ha could be eroded by 2100 resulting in a loss of 45% (S1) to 61% (S2) of the infrastructure.

6.4. Periglacial environments at high altitudes in the face of climate change

Now that the spatial extent of the permafrost is well known and geophysical surveys are increasing, knowledge is progressing on the delineation of the fringe affected by degradation. Pioneering work by Marcer (2018) and Marcer *et al.* (2019) shows that areas in a shaded position and located at an altitude of less than 2,800 meters, or in a position well exposed to incident solar radiation and located at an altitude of less than 3,200 meters, show symptoms of degradation (see Figure 6.11).

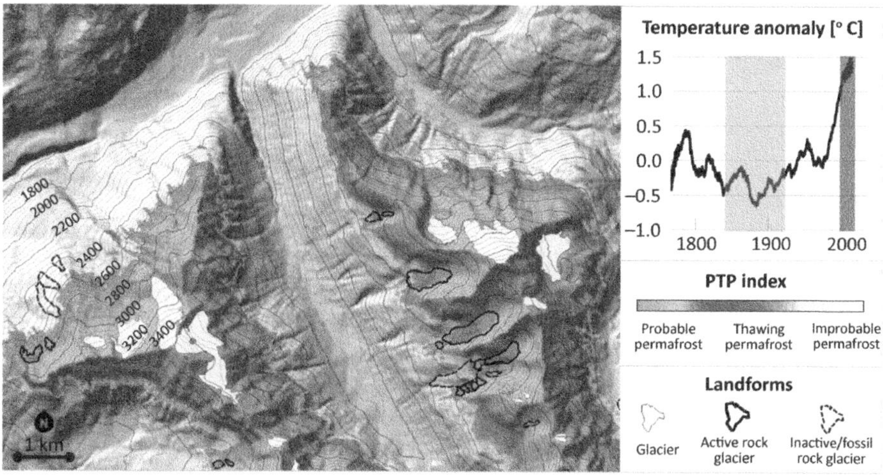

Figure 6.11. *Identification of permafrost fringe subject to degradation. Modeling carried out in the Bessans – Mont-Cenis sector (Haute Maurienne, Savoie, France) based on geomorphological and geophysical observations and geostatistical modeling (source: Marcer* et al. *2019). For a color version of this figure, see www.iste.co.uk/mercier/climate.zip*

6.4.1. *Gravity dynamics and permafrost wall degradation*

In rock walls, permafrost degradation is very directly materialized by the increase in gravity (see Figure 6.12). In these steeply sloping areas, the thickening of the active layer causes the disappearance of the cement that could have been created by ice within the rock material. Even if ice remains present at depth, its thermal state close to the melting point is not always sufficient to maintain the cohesion of the rock compartments. In addition to these two triggering parameters, there is a preparatory factor related to the progressive injection of liquid water into the bedrock, which lubricates the diaclases where the ice had previously ensured cohesion. This water can come from the melting of the permafrost, as well as from the increase in liquid precipitation in a context of global warming.

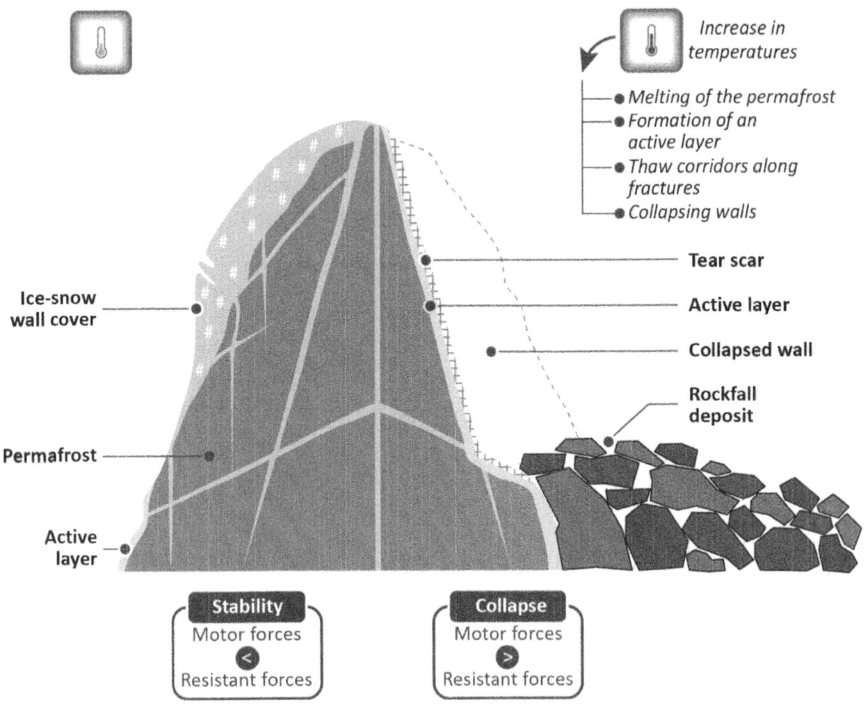

Figure 6.12. *Wall permafrost degradation causing gravity phenomena on high mountain slopes in the context of global warming (source: design D. Mercier, drawing F. Bonnaud, Sorbonne University; modified from Ravanel 2010). For a color version of this figure, see www.iste.co.uk/mercier/climate.zip*

These factors combine with preconditioning factors, which make high mountain walls structurally unstable: steep topography (>30–40°), geological material that is often tectonized and intensely fractured. It should be noted that in addition to these preconditioning factors, which play over a very long period of time (several thousand years), post-glacial decompression may also occur. In fact, the rock walls of the high mountain alpine cirques are affected by a stop-loss phenomenon, linked to the thinning of the tongues of ice since the last glacial maximum, such as since the LIA. Paraperiglacial dynamics are then relayed, or even exacerbated, by paraglacial dynamics.

The reconstitution of gravity events in the Chamonix valley, through investigations and the use of archival documents, is revealing of the current paraperiglacial dynamics (see Figure 6.13). The rise in temperatures observed at the end of the 1980s was materialized by an increase in the number of occurrences of hot summers, with a paroxysm that is reached after the scorching summer of 2003. Since then, rock falls have occurred with high frequency. In detail, it is sometimes difficult to understand whether a year is a good one for weather-related triggers of rockfalls, or because of multi-year cumulative effects (see Permafrost Warming Preparatory Factor). However, the chronicle shows a radically different functioning of the system from what was observed before the increase in temperatures (Ravanel and Deline 2015). Beyond the temporal aspect, the spatial dimension also shows the extent of the adjustments in progress. The altitudes at which the current collapses are taking place are singularly high (fringe 3,000 to 4,000 meters), assumed until recently to be sheltered from the consequences of current climate change.

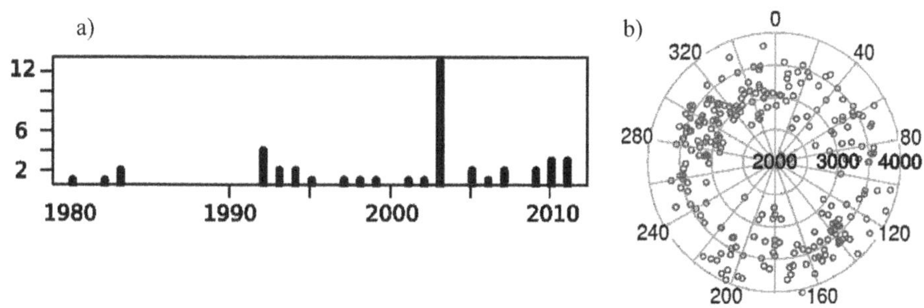

Figure 6.13. *(a) Annual number of collapses in the Aiguilles de Chamonix since 1980, established by photo comparison. (b) Altitude (in m) and orientation (in °) of all collapses recorded between 2007 and 2011 in the Mont Blanc massif (source: Ravanel and Deline 2015). For a color version of this figure, see www.iste.co.uk/mercier/climate.zip*

6.4.2. *Gravity dynamics and permafrost degradation in loose formations*

There is a growing number of case studies to reconstruct the geomorphologic evolution of high mountain surface formations from rock glaciers since the middle of the 20th Century, thus showing the symptoms of paraperiglacial degradation.

These symptoms already appear on aerial photographs from the 1950s: melt furrows (sometimes hectometer-wide) within the rock glaciers. They are sometimes accompanied by crevasses (a few meters long), suggesting the presence of ice at depth.

In several cases in the southern Alps, a network of thermokarstic lakes appeared in the 1970s within rock glaciers (Perrier 2014). In the well-tracked case of Lac Rouge (Clarée valley), a thermokarstic lake with a surface area of about 50 m^2 is, for example, well-observed in 1971 aerial photographs. It is still present and partially filled with water. A cross-section currently shows the presence of a massive ice cap (4 m high above the lake surface) under a thin layer of surface debris (20 to 40 cm) (see Figure 6.14). Since 2008, the area of the thermokarst lake has been steadily increasing, reflecting not only topographic subsidence but also accelerated creep of the entire rock glacier. Downstream, an 85 m long landslide occurred on the western part of the front of the rock glacier: it can be interpreted as the culmination, the paroxysm of a progressive destabilization in relation to an increase in the water content of the material constituting the rock glacier. This paroxysmal phenomenon has also been observed in the Bérard rock glacier (Bodin *et al.* 2015). Here again, the first signs of destabilization date back to the early 2000s, showing the progressive nature of this preparatory evolution, before the catastrophic displacement suddenly set in (two major phases during the summer of 2006 mobilized more than 0.25 Mm^3 of ice and debris).

The acceleration of rock glacier creep under warming conditions is now well known (Delaloye *et al.* 2012; Schoeneich *et al.* 2014), and the potential consequences on the catastrophic destabilization of these rock glaciers have been well identified (Marcer 2018). In addition to the classic role of topography, the increase in air temperature in the 1990s, successive heat waves (the summers of 2003 and 2006 in particular) seem to have prepared for the occurrence of these destabilizations. More generally, these preparatory factors durably create favorable conditions for the triggering of these hazards, even if they finally occur in relation to the hydro-nivo-meteorological conditions of the weeks preceding the triggering.

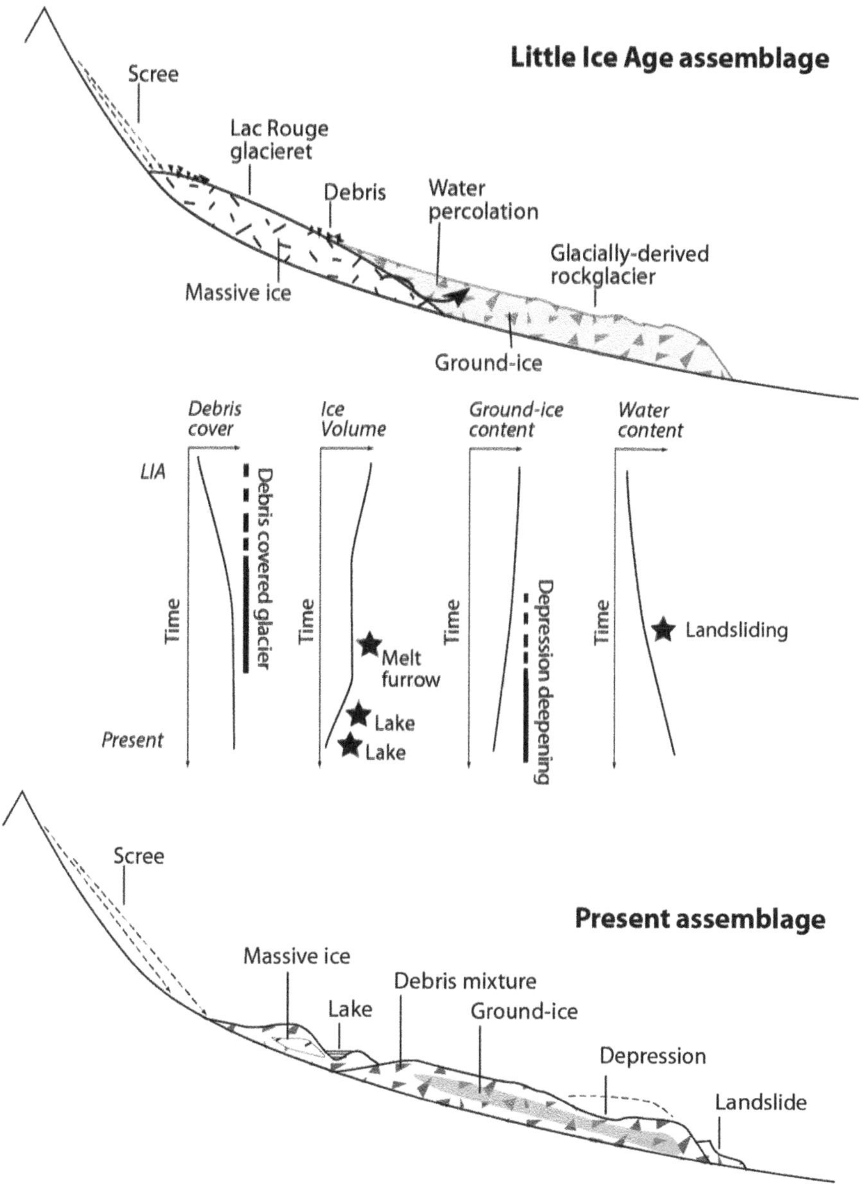

Figure 6.14. *Post-LIA evolution of a rock glacier in a paraperiglacial context (source: design and construction by É. Cossart)*

6.4.3. *The impact of global warming on high-mountain practices*

Gravity dynamics increase with global warming and the melting of permafrost walls; they may even affect outdoor tourism activities, particularly mountaineering (Mourey 2019) or the practice of downhill skiing (Duvillard *et al.* 2015).

In terms of mountaineering, the dangerousness of the routes is increasing, as are the technical difficulties of certain routes, which may exclude the less experienced category of climbers who wish to access high mountain refuges. In summer, especially in hot periods, some routes become too dangerous and it is preferable to shift the use of the routes to spring, autumn or even winter. The example of the normal ascent route to Mont Blanc illustrates this danger exacerbated by the ongoing global warming (Mourey 2019). Every year around 17,000 people attempt to reach the summit of the Alps at 4,809 m. Between 1990 and 2017, the High Mountain Gendarmerie Platoon (PGHM) intervened 347 times, involving 387 people, between the Tête Rousse refuge and the Goûter refuge. The accidents took place on the Goûter couloir at an altitude of 3,270 m and on the rocky ridge leading to the Goûter needle at 3,863 m. They were extremely serious: 102 people died, 230 were injured and 55 were unharmed. Around 82% of the victims were men and 18% women, of 37 different nationalities. Around 84% of the victims were not supervised by professionals. The reasons for the accidents were mainly unscrewing, rock falls, overuse of the access road and the insufficient technical and physical level of the candidates for the Alpine summit. The frequency of rock falls – linked to gneiss fracturing, the call of the void, the degradation of permafrost and the rapid melting of snow – increases between 1100 and 1330 hours, at the very time when the use of this sector is at its highest, particularly on the descent, when the climbers are the most tired (Mourey 2019).

In terms of the practice of alpine skiing, it is more the physical infrastructure (ski lifts, shelters) that are affected by paraperiglacial hazards. Indeed, gravity deformations, when they affect loose formations or rock faces, are likely to damage these infrastructures, with considerable economic damage, even if the companies operating the ski areas remain discreet (Duvillard 2019). In a pioneering and sensitive study, a first census of high-mountain infrastructures was carried out in the French Alps. This census is based on the currently exposed infrastructure, whose location is confronted with data on the potential distribution of permafrost (see Figure 6.15). A total of 1,769 infrastructures were identified in areas with probable permafrost and thus susceptible to ground deformation (Duvillard *et al.* 2015). In addition to this census, which shows the extent of the threat within the French Alps, several renowned resorts (Val Thorens, Les Deux Alpes) are faced with hazards that are sufficiently significant to require recourse to particularly costly geotechnical solutions.

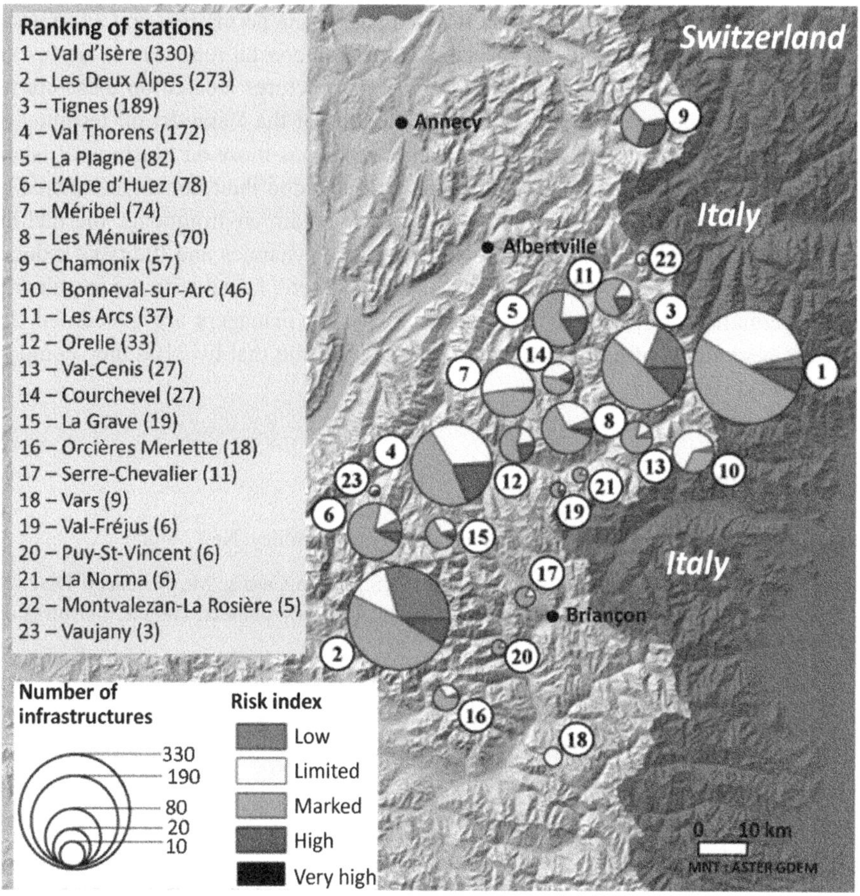

Figure 6.15. *Distribution of infrastructure at risk in stations of the French Alps (source: based on Duvillard* et al. *2015). For a color version of this figure, see www.iste.co.uk/mercier/climate.zip*

6.5. Conclusion

Periglacial areas are subject to paraperiglacial erosion crises caused by the melting of the permafrost, which are geomorphologically materialized by large scale ground deformations, affecting loose formations such as rocky substrates, slope areas as well as flat areas. Faced with this generalized process, the melting permafrost calls into question the stability of buildings built on foundations embedded in the frozen ground. When destroyed in this way, gas or oil pipeline networks are at risk of destabilization. On the Arctic coasts, which have the originality of being at the interface between a submarine permafrost and a terrestrial permafrost, the disappearance of the latter generates coastal retreat rates of several tens of meters per year, endangering certain infrastructures located on these coasts. This is notably the case for part of the installations of the Varandei oil terminal on the shores of the Barents Sea. Some inhabitants, such as those on Shishmaref Island off the coast of Alaska, even decided in 2006 to abandon their homes threatened by the accelerated retreat of the coastline. In high mountain environments, the melting of the permafrost of the walls increases the gravity dynamics and the risks induced for mountaineers, hikers and for the infrastructures there. Faced with the inertia with which permafrost responds to environmental forces, managers are concerned that only the beginnings of hazards are currently visible, and that by 2100 they could be on a dreaded scale.

6.6. References

Ballantyne, C.K. (2018). *Periglacial Geomorphology*. John Wiley, New York.

Blanco-Chao, R., Perez-Alberti, A., Trenhaile, A.S., Costa-Casais, M., Valcacerl-Diaz, M. (2007). Shore platform abrasion in a para-periglacial environment, Galicia, northwestern Spain. *Geomorphology*, 83, 136–151.

Bodin, X. (2007). Géodynamique du pergélisol de montagne : fonctionnement, distribution et évolution récente. L'exemple du massif du Combeynot (Hautes Alpes). PhD thesis, Université Paris-Diderot, Paris.

Bodin, X., Schoeneich, P., Deline, P., Ravanel, L., Magnin, F., Krysiecki, J.-M., Echelard, T. (2015). Le permafrost de montagne et les processus géomorphologiques associés : évolutions récentes dans les Alpes françaises. *Journal of Alpine Research*, 103–2.

Bourriquen, M., Mercier, D., Baltzer, A., Fournier, J., Costa, S., Roussel, E. (2018). Paraglacial coasts responses to glacier retreat and associated shifts in river floodplains over decadal timescales (1966–2016), Kongsfjorden, Svalbard. *Land Degradation and Development*, 29(11), 4173–4185.

Brown, J.O.J., Ferrians, O.J., Heginbottom, J.A., Melnikov, E.S. (1997). Circum-Arctic map of permafrost and ground-ice conditions. U.S. Geological Survey in Cooperation with the Circum-Pacific Council for Energy and Mineral Resources, Washington, DC. Circum-Pacific Map Series CP-45, scale 1:10,000,000, 1 sheet.

Cossart, É. (2014). Des sources sédimentaires à l'exutoire : un problème de connectivité ? Réflexions sur le fonctionnement géomorphologique des bassins montagnards. HDR, Université Blaise-Pascal, Clermont-Ferrand.

Couture, N.J., Irrgang, A., Pollard, W., Lantuit, H., Fritz, M. (2018). Coastal erosion of permafrost soils along the Yukon Coastal Plain and fluxes of organic carbon to the Canadian Beaufort Sea. *Journal of Geophysical Research: Biogeosciences*, 123, 406–422.

Delaloye, R. (2004). Contribution à l'étude du pergélisol de montagne en zone marginale. PhD thesis, Université de Fribourg, Fribourg.

Delaloye, R., Morard, S., Barboux, C., Abbet, D., Gruber, V., Riedo, M., Gachet, S. (2012). Rapidly moving rock glaciers in Mattertal. In *Jahrestagung der Schweizerischen Geomorphologischen Gesellschaft*, 21–31.

Duvillard, P.-A. (2019). Déstabilisation des terrains-supports d'infrastructures en contexte de réchauffement climatique dans la haute montagne alpine française. PhD thesis, Université Grenoble Alpes, Saint-Martin-d'Hères.

Duvillard, P.-A., Ravanel, L., Deline, P. (2015). Évaluation du risque de déstabilisation des infrastructures de haute montagne engendré par le réchauffement climatique dans les Alpes françaises. *Journal of Alpine Research*, 103–2.

Ermolaev, M. (1932). Geologitcheskiy i geomorfologitcheskiy otcherk ostrova B. Lakhovskog. *Trudy Sovĕta po izutcheniyou proizvodit. sil., ser. Yakutskay*a, 8.

French, H. (2017). *The Periglacial Environments*, 4th edition. John Wiley, New York.

Fritz, M., Vonk, J.E., Lantuit, H. (2017). Collapsing Arctic coastlines. *Nature Climate Change*, 7, 6–7.

Günther, F., Overduin, P.P., Yakshima, I.A., Opell, T., Baranskaya, A.V., Grigoriev, M.N. (2015). Observing Muostakh disappear: Permafrost thaw subsidence and erosion of a ground-ice-rich island in response to arctic summer warming and sea ice reduction. *The Cryosphere*, 9, 151–178.

Irrgang, A., Lantuit, H., Manson, G.K., Günther, F., Grosse, G., Overduin, P.P. (2017). Quantification of shoreline movements along the Yukon Territory mainland coast between 1951 and 2011. *PANGAEA* [Online]. Available at: https://doi.org/10.1594/PANGAEA.874343.

Irrgang, A., Lantuit, H., Gordon, R.R., Piskor, A., Manson, G.K. (2019). Impacts of past and future coastal changes on the Yukon coast-threats for cultural sites, infrastructure, and travel routes. *Arctic Science*, 5(2), 107–126.

Jones, B.M., Arp, C.D., Jorgenson, M.T., Hinkel, K.M., Schmutz, J.A., Flint, P.L. (2009). Increase in the rate and uniformity of coastline erosion in Arctic Alaska. *Geophysical Research Letters*, 36, L03503.

Magnin, F. (2015). Distribution et caractérisation du permafrost des parois du massif du Mont-Blanc : une approche combinant monitoring, modélisation et géophysique. PhD thesis, Université Grenoble Alpes, Saint-Martin-d'Hères.

Marcer, M. (2018). Déstabilisation des glaciers rocheux dans les Alpes françaises : une évaluation à l'échelle régionale et locale. PhD thesis, Université Grenoble Alpes, Saint-Martin-d'Hères.

Marcer, M., Serrano, C., Brenning, A., Bodin, X., Goetz, J., Schoeneich, P. (2019). Evaluating the destabilization susceptibility of active rock glaciers in the French Alps. *The Cryosphere*, 13, 141–155.

Mercier, D. (2008). Paraglacial and paraperiglacial landsystems: Concepts, temporal scales and spatial distribution. *Géomorphologie : relief, processus, environnement*, 14(4), 223–234.

Mourey, J. (2019). L'alpinisme à l'épreuve du changement climatique. Évolution géomorphologique des itinéraires, impacts sur la pratique estivale et outils d'aide à la décision dans le massif du Mont Blanc. PhD thesis, Université de Savoie, Chambéry.

Muller, S.W. (1947). *Permafrost of Permanently Frozen Ground and Related Engineering Problems*. J.W. Edwards, Ann Arbor.

Obu, J., Lantuit, H., Grosse, G., Günther, F., Sachs, T., Helm, V., Fritz, M. (2017). Coastal erosion and mass wasting along the Canadian Beaufort Sea based on annual airborne LiDAR elevation data. *Geomorphology*, 293, 331–346.

Obu, J., Westermann, S., Kääb, A., Bartsch, A. (2018). Ground Temperature Map, 2000–2016, Northern Hemisphere Permafrost [Online]. Available at: https://pangaea.figshare.com/articles/Ground_Temperature_Map_2000-2016_Northern_Hemisphere_Permafrost/11057513/1.

Overduin, P.P., Schneider von Deimling, T., Miesner, F., Grigoriev, M.N., Ruppel, C.D., Vasiliev, A. (2019). Submarine permafrost map in the Arctic modeled using 1-D transient heat flux (SuPerMAP). *Journal of Geophysical Research: Oceans*, 124(6), 3490–3507.

Perrier, R. (2014). Suivi local et régional du pergélisol dans le cadre du changement climatique contemporain : application aux vallées de la Clarée et de l'Ubaye (Alpes du Sud, France). PhD thesis, Université Paris-Diderot, Paris.

Ravanel, L. (2010). Caractérisation, facteurs et dynamiques des écroulements rocheux dans les parois à permafrost du massif du Mont-Blanc. PhD thesis, Université de Savoie, Chambéry.

Ravanel, L. and Deline, P. (2015). Rockfall hazard in the Mont Blanc massif increased by the current atmospheric warming. In *Engineering Geology for Society and Territory*, Lollino, G. *et al.* (eds). Springer, Berlin, 2, 435–428.

Schoeneich, P., Bodin, X., Echelard, T., Kaufmann, V., Lieb, G.K. (2014). Velocity changes of rock glaciers and induced hazards. In *Engineering Geology for Society and Territory*, Lollino, G., Giordan, D., Crosta, G.B., Corominas, J., Azzam, R., Wasowski, J., Sciarra, N. (eds). Springer, Berlin.

Strzelecki, M.C., Long, A.J., Lloyd, J.M., Malecki, J., Zagorski, P., Pawlowski, L., Jasloski, M. (2018). The role of rapid glacier retreat and landscape transformation in controlling the post-Little Ice Age evolution of paraglacial coasts in central Spitsbergen (Billefjorden, Svalbard). *Land Degradation and Development*, 29(6), 1962–1978.

Tanski, G., Lantuit, H., Ruttor, S., Knoblauch, C., Radosavljevic, B., Strauss, J., Wolter, J., Irrgang, A.M., Ramage, J., Fritz, M. (2017). Transformation of terrestrial organic matter along thermokarst-affected permafrost coasts in the Arctic. *Science of the Total Environment*, 581/582, 434–447.

Vandenbergh, J., French, H.M., Gorbunov, A., Marchenko, S., Velichko, A.A., Jin, H., Cui, Z., Zhang, T., Wan, X. (2014). The last permafrost maximum (LPM) map of the Northern Hemisphere. *Boreas*, 43, 652–666.

Wobus, C., Anderson, R., Overeem, I., Maell, N., Clow, G., Urban, F. (2011). Thermal erosion of a permafrost coastline: Improving process-based models using time-lapse photography. *Arctic, Antarctic, and Alpine Research*, 43(3), 474–484.

7

The Impacts of Climate Change on the Hydrological Dynamics of High Latitude Periglacial Environments

Emmanuèle GAUTIER
University of Paris 1 Panthéon-Sorbonne, France

7.1. Periglacial regions strongly affected by recent climate change

The response of high latitude periglacial environments to climate change is very particular for two main reasons: (1) the rise in temperature is more pronounced than in other bioclimatic zones, and (2) the presence of permanently frozen ground (or permafrost).

7.1.1. *Much warmer winters*

The latest IPCC report (IPCC 2019) clearly states that for the period 2014 to 2018, average temperatures in the Arctic regions reached a level unseen since 1900. The long climate series in Russia provides a good account of the warming. At the Yakutsk station in eastern Siberia, for example, the average annual temperature was –10.3°C until the 1980s and has reached –7.8°C since 2000 (see Figure 7.1). Winter temperatures from January to March show the largest increase. In Yakutsk, these winter temperatures, which averaged –34°C until 1980, have been –30.2°C since 2000; this winter increase of more than 3°C is common to all high latitude regions.

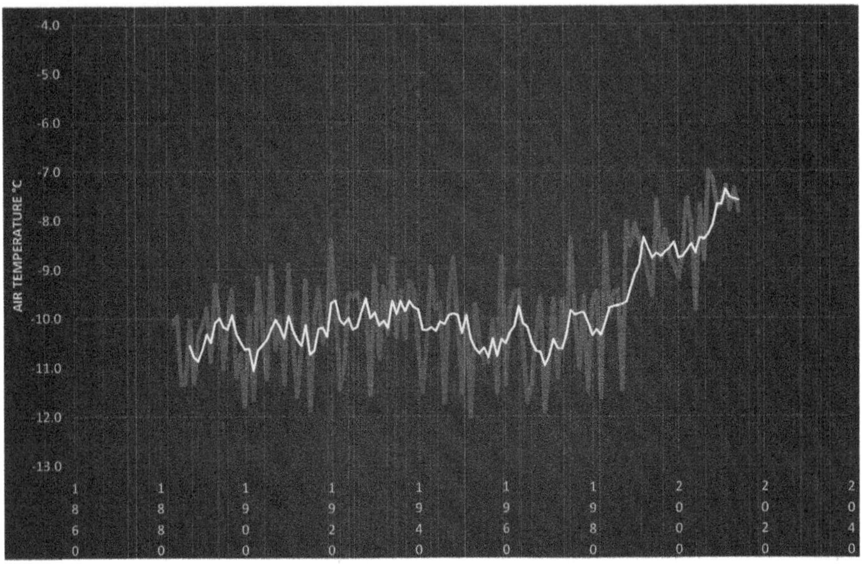

Figure 7.1. *Annual mean air temperatures at Yakutsk from 1880 to 2017 (Eastern Siberia, 62° 02' N; 129° 42' E, NOAA data). In blue: annual averages, in yellow: the five-year moving average. For a color version of this figure, see www.iste.co.uk/mercier/climate.zip*

7.1.2. *Permafrost and its sensitivity to air temperatures*

Permafrost covers about 22–23 million km² in the Northern Hemisphere. It is thickest and most continuous in Siberia, due to the limited spread of the ice sheet during the cold Pleistocene stages, unlike in North America (Brown *et al.* 2002). This allowed the cold wave to penetrate the ground, explaining the formation of permafrost up to 1,200 m thick in northern and eastern Siberia. The rise in atmospheric temperatures, particularly in winter, has led to a significant increase in permafrost temperatures, which for the decade 2007–2016 averaged 0.39 ± 0.10°C for all Arctic regions. Permafrost temperatures have thus risen by more than 2°C on average since the middle of the 20th century. The latest IPCC report (IPCC 2019) estimates that about 20% of this permafrost is highly vulnerable in the short term and will be gone by 2100.

While permafrost is sensitive to average air temperatures, it is also influenced by winter snow depth and summer precipitation (Park *et al.* 2016). While atmospheric humidity has increased in the Arctic, precipitation does not show a clear overall trend. The amount of snowfall in winter is increasing in northern Europe and southern Siberia. It is also noted that the duration of the snow cover has been greatly reduced, with a loss of 0.7 to 3.9 days per decade.

As a result, the snowpack melts much earlier in the spring. However, in eastern and northern Siberia, the duration of snow cover appears to be increasing (Bulygina *et al.* 2009). Very wet conditions have occurred more frequently since 1998, with storms affecting Siberia resulting in more abundant late summer and early winter precipitation (Iijima *et al.* 2016). Some years with excess precipitation have seen record flows, such as in 2007 in Siberian rivers (Shiklomanov and Lammers 2009; Gautier *et al.* 2018).

It has been shown that heat transfer between the atmosphere and the ground is more efficient in winter (on frozen ground) than in summer on moist ground. In addition, thicker snow inhibits heat transfer. The active layer above the permafrost thaws a few tens of centimeters, or more than a meter, in the summer and also plays an important role in permafrost temperatures. The thickening of the active layer, the sudden increase in its temperature and that of the permafrost below it, are closely related to the increase in soil moisture (Fedorov *et al.* 2014; Iijima *et al.* 2016). Thus, several factors contribute to the degradation of permafrost: on the one hand, warmer and snowy winters, and on the other hand, hot and humid summers promote the thickening of the active layer. The interactions between permafrost, water in the active layer and the atmosphere are therefore complex.

NOTE.– Frost-related forms in the soil develop in the permafrost of original models. Lowering temperatures in the frozen ground causes thermal shrinkage, creating block fracturing and deformation (Tricart and Cailleux 1967; Pissart 1970). The most original forms are, among others, ice wedges. The cracks created by thermal shrinkage fill with water in the summer when the ice thaws. These cracks freeze during the winter: an ice wedge forms and gradually enlarges during the following winters. The ice wedges then form large tundra polygons, the center of which is lower than the sides formed by the ice wedges. Other frost-related forms exist: in soil composed of fine sediments, slow freezing also causes segregating ice to form. Ice lenses are then formed, which, as they grow, create mounds (pingo, palsars) (Tricart and Cailleux 1967; Pissart 1970; Ballantyne 2018).

7.2. The influence of permafrost on hydrological functioning

Permafrost determines a very particular hydrological functioning. In areas where permafrost is thick and continuous, it can be considered to act as an impermeable barrier blocking water infiltration (Brown *et al.* 2002; Woo 2012). The contribution of groundwater to surface flows is then minimal, particularly in colder areas with thick and continuous permafrost. This specificity explains why, in contrast to more temperate zones, river flows are very reduced or even non-existent in winter. The active layer contributes to surface and subsurface flows. The active layer then functions as a small perched aquifer, resting on permanently frozen ground, especially if the latter is ice-rich (see Figure 7.2). In areas with discontinuous or thin permafrost, the groundwater share of stream flow increases.

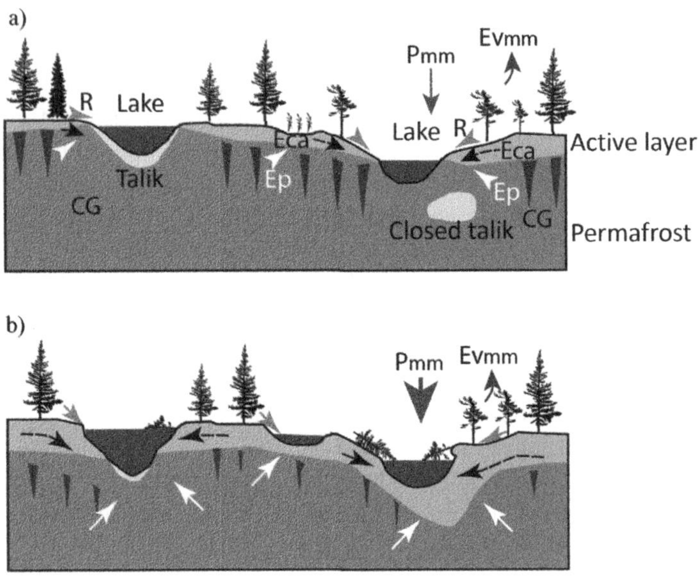

Figure 7.2. *Surface hydrologic functioning of permafrost zones: (a) undisturbed; (b) disturbed by climate change. For a color version of this figure, see www.iste.co.uk/mercier/climate.zip*

COMMENT ON FIGURE 7.2.– *Pmm: precipitation; Evmm: evapo-transpiration and evaporation; R: surface runoff; Eca: water in the active layer; CG: ice wedges; Ep: permafrost thawed water. The thickness of the active layer is exaggerated for readability (source: adapted from Walvoord and Kurylyk 2016).*

The presence of talik (unfrozen water in permafrost) can contribute to feeding ponds, lakes and streams if it is "open", that is, hydrologically connected to surface waters. Talik can also be closed (or isolated) within the permafrost. Talik can form and persist even in winter below a watercourse (due to the energy required for phase change). The presence of "open" water (lake, channels) at the surface is a determining factor in the existence of talik.

For example, in the Mackenzie Delta in Canada, talik occurs under channels and numerous lakes with bottom temperatures in the range of 5°C to 6°C (Burn and Kokelj 2009). Finally, the upwelling of water, very rich in dissolved substances through permafrost, can also promote the presence of talik. This upwelling of thermal water that freezes at the surface can lead to the accumulation of layers of ice on the surface ("*aufeis*" in German, "*naled*" in Russian).

7.2.1. *Numerous wetlands in periglacial environments*

The specificity of surface hydrological functioning partly explains the extent of wetlands in periglacial regions. Contrary to what might have been expected in these predominantly dry environments in terms of precipitation, the periglacial regions are very rich in surface water and the presence of numerous wet depressions, ranging in size from a pond to a very large lake, is a dominant feature of the landscapes. Water from the active layer, snowmelt or summer rainfall infiltrates little (or not at all) and circulates poorly. Ponds or lakes have little or no connection to each other and supply relatively little water to the network of large rivers. Similarly, on floodplains, the abandoned channels that are poorly (or not at all) connected to the main channel remain in water until the following winter.

The gently undulating plateau landforms, often interspersed with glacial deposits, and the unfinished post-glacial reorganization of the river system also account for poor drainage in permafrost regions. Finally, the presence of ice polygons in the permafrost also promotes the development of these wetlands.

The two main types of wetlands appear in the satellite image in Figure 7.3: (a) the lakes and wetlands on the plateau; and (b) the abandoned channels in the Vilyuy floodplain. On the plateau, the vegetation is of the taiga type; the clear halo around the lakes is related to the absence of trees on the banks and in the wetlands. In the north-western corner, flows the Vilyuy River (left bank tributary of the Lena), within the southern bank, aeolian dunes resulting from the reworking of the river sands (very white sands on the picture). The lakes on the plateau are not connected to the rivers.

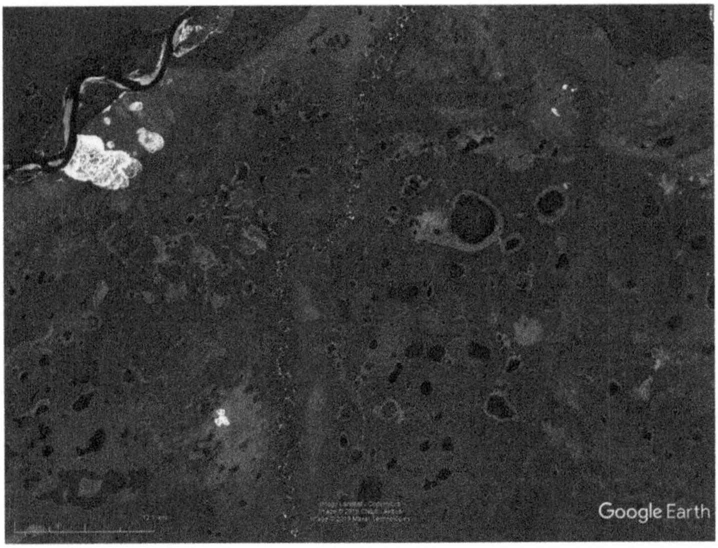

Figure 7.3. *Thermokarst depressions in Eastern Siberia (central from the main lake: 63° 51' 2.32" N – 123° 54' 54.47" E) (source: Google Earth). For a color version of this figure, see www.iste.co.uk/mercier/climate.zip*

7.2.2. The knock-on effects of climate change on slope hydrology

The contact between the water in these wetlands and the permafrost leads to a special process: thermokarst. This process is active in areas with thick, ice-rich permafrost. This particular permafrost, called "yedoma" in Russia, is more than 50% ice, either in the form of large ice wedges or massive ice layers. The presence of unfrozen water in the summer allows heat to penetrate, causing the permafrost to thaw, especially along the ice wedges. This causes the ground to subside and gullies to form. The banks of lakes in particular gradually retreat, sometimes to the point of forming large, partly dried-up alases (Soloviev 1973). The alas (Yakut term) are large depressions of variable size (from less than 1 km^2 to sometimes more than 100 km^2 over 5 to 20 m depth) such as those shown in Figure 7.3; they are created by the thawing of thick ice-rich permafrost. This thermokarst process occurred during the Holocene, due to the rise in temperatures in Arctic areas.

However, the current climate change explains a sudden acceleration of thermokarstic mechanisms. Large networks of ice wedges are thawing, creating polygonal ground collapse (see Figures 7.2 and 7.4). This leads to the individualization of pyramidal blocks on the banks of lakes and rivers and sometimes their collapse, the so-called "baydjarakhs" (see Figure 7.5).

Figure 7.4. *Lake development related to the degradation of large permafrost ice wedges (Yakutia, north of Pokrovsk, May 2013). Polygonal soils bulge and depressions fill with water. As the wetland area expands, the spruce forest disappears. The level before collapse is shown by the arrow at the top right of the photograph (source: E. Gautier). For a color version of this figure, see www.iste.co.uk/mercier/climate.zip*

Figure 7.5. *Baydjarakhs on the shore of a lake related to the degradation of large ice wedges in permafrost in Yakutia, Syrdak (source: photograph by F. Costard). For a color version of this figure, see www.iste.co.uk/mercier/climate.zip*

The recent evolution of the lakes is a good example of the effects of climate change. Using remote sensing, Zakharova *et al.* (2018) analyzed changes in lakes in parts of Yakutia. The water level has risen by 60 cm since 2006 causing the lakes to expand, although a regression appears to have been occurring since 2011. A little further north, Fedorov *et al.* (2014) show the very rapid evolution of some thermokarstic lakes, monitored by the Permafrost Institute team in Yakutsk since the 1990s: some have deepened by more than 3 m and widened by more than 50 m. The permafrost thawed at a rate three times faster after 2004; as a result, the contribution of permafrost degradation to the lakes is five times greater. During the abnormally wet period from 2004 to 2008, surface runoff and subsurface runoff in the active layer accounted for one-third of the water input into the lakes, whereas in previous periods they accounted for only 4 to 6%. Permafrost is particularly vulnerable in south-facing shorelines, creating lake asymmetry (Séjourné *et al.* 2015).

As a result of permafrost degradation, water transfers from the slopes to the river beds are increasing significantly. As they expand, lakes can become coalescent or even connect hydrologically to a river. In the latter case, the connection leads to an acceleration and increase in water flows in rivers.

NOTE.– Feedbacks between warming, wetland development and permafrost degradation. The expansion of wetlands due to permafrost degradation is accompanied by a reduction in forest cover: taiga trees are partly dependent on the presence of permafrost and flooded trees perish. This transformation of the landscape in turn causes accelerated degradation of the frozen ground: according to Kurylyk *et al.* (2016) wetlands absorb more energy than forests; this additional surface and subsurface energy accelerates the degradation of permafrost. This is confirmed by permafrost temperature measurements in Yakutia, where the ground temperature at a depth of 10 m is –3°C on average under forest, compared to –2°C under wet grassland (Fedorov *et al.* 2014).

7.3. The response of Arctic fluvial hydrosystems to ongoing climate change

Rivers in Arctic environments show an original hydrological functioning, largely related to the absence (or near-absence) of groundwater input. These rivers have one thing in common – their behavior is very immoderate: a long winter low-water season interrupted by a spring break-up flood (May–June) followed by a fairly rapid recession. Several rivers flowing through the subarctic and arctic zones are among the longest in the world: Mackenzie, Ob, Yenissei and Lena rivers are over 4,000 km long, Yukon and Kolyma rivers are over 3,000 km and 2,000 km long, respectively (see Figure 7.6). As a result, their meridian course (except for the Yukon River, which flows westward) causes them to drain in their middle and lower valleys, regions with colder climates than the regions where they originate. The Yukon River incorporates the combined influences of high mountain areas (glacial and snow influences); the same is true of some Siberian tributaries that drain the mountains (Chersky Mountains, Verkhoyansk, etc.). They will therefore respond differently to climatic disturbances. Northern Asia (Siberia), Alaska and northern Canada show similarities in their recent evolution (e.g. the increase in winter flows), but also strong differences (the same worsening of maximum flows is not found in Siberia and America).

	Total annual total volume sold, average annual throughput	Size of the catchment area (106 km²), specific discharge	Ratio between maximum monthly flow and minimum monthly flow	Share of permafrost in the watershed
Yenissei River Igarka (67° 43 N; 8° 48 E)	635–673 km^3 18,700 m^3 s^{-1}	2,576 7.26 L s^{-1} km^{-2}	13	33% of basin with continuous permafrost 55% of basin with discontinuous permafrost
Lena River Kusur (70° 68 N; 127° 39 E)	543–580 km^3 17,200 m^3 s^{-1}	2,47 6.9 L s^{-1} km^{-2}	35	77% of basin with continuous permafrost 23% of basin with discontinuous permafrost
Ob River Salekhard (66° 63 N; 66° 60 E)	401–427 km^3 12,700 m^3 s^{-1}	2,95 4.3 L s^{-1} km^{-2}	10	2% of basin with continuous permafrost 24% of basin with discontinuous permafrost
Mackenzie River Arctic Red River (67° 45 N; 133° 74 W)	316–325 km^3 9,190 m^3 s^{-1}	1,8 5.5 L s^{-1} km^{-2}	6,4	16% of basin with continuous permafrost 66% of basin with discontinuous permafrost
Yukon River Pilot Station (61° 93 N; 162° 88 W)	208 km^3 6,400 m^3 s^{-1}	0,857 7.5 L s^{-1} km^{-2}	13	23% of basin with continuous permafrost 77% of basin with discontinuous permafrost
Kolyma River Srednekolymsk (67° 47 N; 153° 59 E)	124–136 km^3 3,300 m^3 s^{-1}	0,65 5.1 L s^{-1} km^{-2}	86	Continuous permafrost throughout basin

Table 7.1. *Main rivers in Arctic regions (source: Serreze et al. 2002; Magritsky* et al. *2018*[1]*)*

1 http://www.r-arcticnet.sr.unh.edu; https://arcticgreatrivers.org/rivers/.

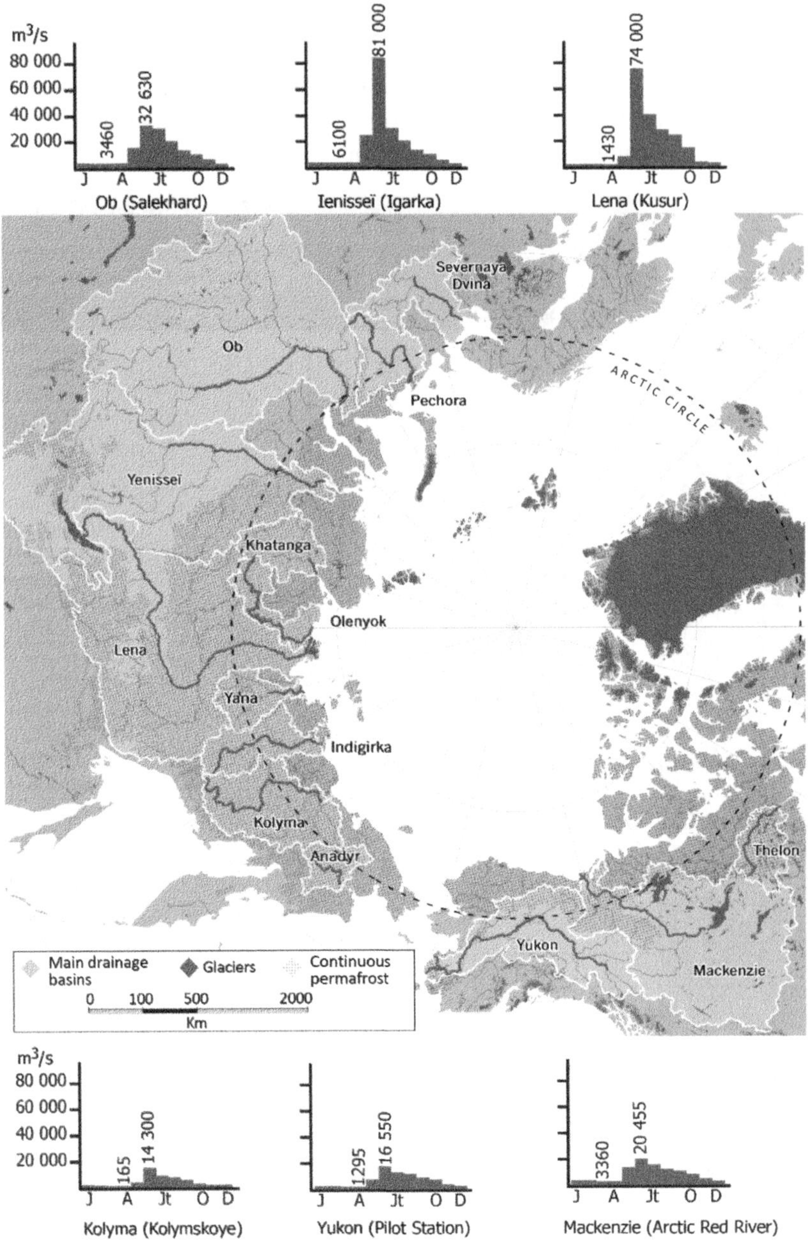

Figure 7.6. *Map of the main Arctic rivers with minimum and maximum monthly discharges (source: design E. Gautier; realization J. Cavero, 2019). For a color version of this figure, see www.iste.co.uk/mercier/climate.zip*

7.3.1. *River ice*

Rivers in the periglacial region freeze for several months in winter: Siberian rivers remain frozen for more than 200 days a year, Lena and Kolyma rivers remain under a shell of ice for more than 240–250 days a year. Similarly, the Mackenzie and Yukon rivers are ice-bound from October to May. The duration of this river ice varies with latitude.

Figure 7.7. *Lena in eastern Siberia in winter (source: photograph E. Gautier, April 2013). For a color version of this figure, see www.iste.co.uk/mercier/climate.zip*

Frazil (ice needles forming in the water and on the bottom of the bed) appears in early to mid-October. Then a few days later, ice "patties" drift out of the ice and finally form a compact and very immobile ice shell at the end of October (see Figure 7.7). River ice depends mainly on air temperatures, but also on the presence and thickness of snow, which has an insulating effect (Park *et al.* 2016). It has shown a clear trend towards thinning and a reduction in its duration over the last 20 years or so. Shiklomanov and Lammers (2014) estimate that the duration of this ice shell has decreased by seven days on the Severnaya Dvina, Lena and Yenissei rivers and 20 days on the Ob River since the beginning of the 21st century. The average thickness of the ice has lost more than 20 cm on the Lena River in the Yakutsk region, the capital of the Sakha Republic and a strategic location in Siberia, since the 1960s. The maximum thickness has also been greatly reduced (–73 cm, for example, on the Lena River, according to Shiklomanov and Lammers (2014). The reduction in thickness and duration is primarily related to air temperatures. The increase in winter discharge can also explain the reduction in ice cover. The earlier melting of the snow cover in spring also explains the earlier disappearance of ice.

During the rising flood that leads to breakup, river ice will play a key role: the water lifts the ice and ice jams form at the head of the islands and at the entrance to the secondary channels (see Figure 7.8 and Costard *et al.* 2014). These ice jams can reach more than 5 m in height. In some years, because of thick ice, or an early arrival of the upstream flood, the ice is slow to break up and drift downstream. The water then rises rapidly and floods the plain. This is what happened in May 2001 on the Lena River: two thirds of the city of Lensk was destroyed by the flood that rose 16 meters, blocked by the ice shell. Twenty-five thousand people were evacuated and more than 5,000 houses destroyed (Kichigina 2013). Even the bombers were unable to break through the ice.

Figure 7.8. *Lena River in eastern Siberia – ice jam. The flood of May 2010 reached a historical level with more than 56,000 $m^3 s^{-1}$ in Tabaga. The water level rose by more than 8 m in one night, the break-up was rapid and particularly damaging to the alluvial forest (source: photograph F. Costard, May 19, 2010). For a color version of this figure, see www.iste.co.uk/mercier/climate.zip*

The ice, as the flood rises, breaks up in a few days and drifts. This period has a very strong effect on the riparian vegetation (alluvial vegetation) acting like a bulldozer, the ice breaks up tree trunks on the banks and heads of islands (see Figure 7.9).

NOTE.– River ice has a definite economic interest. Rivers, when the thickness of the ice allows it, become passable routes in winter; it is easier to drive than on wet roads with muskeg in summer. The danger comes from the increasingly frequent "hot" (positive temperature) days in winter and, of course, (see section 7.1.1), the increasingly warmer winters. This often-pure river ice can be cut into large cubes and stored in cellars dug into the permafrost. In this way, people have access to drinking water all year round.

Figure 7.9. *Lena River in eastern Siberia – the consequences of the May 2010 ice breakup on alluvial vegetation. Note tree trunks broken cleanly by ice pans during breakup. In the background, willows, lying and partly buried under sandy deposits, are piled up and have made new stems (source: photograph E. Gautier, July 2010). For a color version of this figure, see www.iste.co.uk/mercier/climate.zip*

7.3.2. *Increasing winter low water levels*

The low-water period therefore corresponds to winter, with much lower water discharges than in the high-water period (see Table 7.1). Flows decrease from October to May and low flows occur in late winter: in March on the Mackenzie, Ob and Yenisei rivers and in April on other rivers draining colder regions. The weighting coefficients relating the discharge of the highest month to that of the lowest month clearly express the influence of continuous permafrost on flows (see Table 7.1): the higher the coefficient, the more indigent the low-water discharges are due to the influence of permafrost and, therefore, to low groundwater inputs (Yang *et al.* 2002; Gautier *et al.* 2018). The lower values of this coefficient, around 10, correspond to rivers (Mackenzie, Ob) draining the least "frozen" regions. On the Mackenzie River, the large lakes in the eastern part of the basin support low water levels, resulting in a relatively low ratio. Where continuous permafrost occupies most of the basin, the ratio exceeds 30; on the Kolyma River, whose basin is entirely "frozen", mean monthly discharge exceeds 14,000 $m^3 s^{-1}$ in June, compared with 165 $m^3 s^{-1}$ in April, a "record" ratio of 86.

The increase in this ratio between maximum and minimum water discharge clearly expresses the degradation of permafrost and its effect on increasing flows (Yang *et al.* 2002; YE, 2009; Gautier *et al.* 2018). In eastern Siberia, the Lena River at the Tabaga station (upstream of Yakutsk) clearly shows that the low values of this ratio (around 15–20) observed from the 1990s onwards are linked to a strong increase in low water levels in winter; they correspond to clearly warmer years (see Figure 7.10). However, it is the winter months that are most affected by warming. The year 2006 with a ratio of 40 is an exception, which is linked to a very high flow in May and not to very low water. Until the 1990s, this ratio varied between 25 and 45.

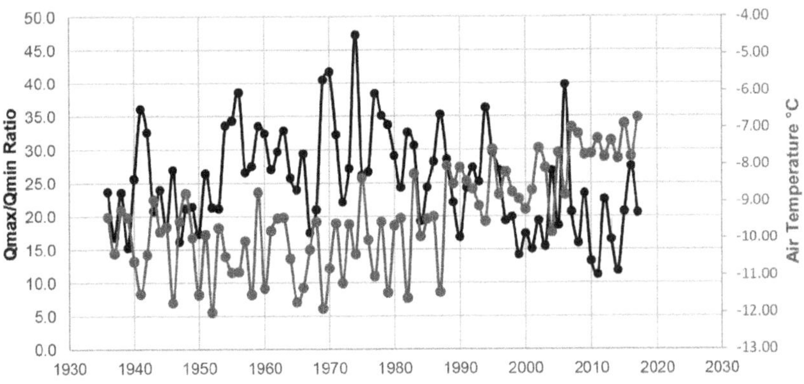

Figure 7.10. *The ratio between maximum and minimum water discharge rate (black curve) on the Lena River in eastern Siberia (Tabaga station, upstream of Yakutsk) and mean air temperatures at Yakutsk (blue curve) from 1936 to 2017 (source: adapted from Yang* et al. *2002; Ye et al. 2009; Gautier* et al. *2018). For a color version of this figure, see www.iste.co.uk/mercier/climate.zip*

Winter low water discharges have therefore increased significantly on most rivers in Arctic and sub-Arctic areas, as shown in numerous publications (Brabets and Walvoord 2009; Déry *et al.* 2009; Ye *et al.* 2009; Tan *et al.* 2011; Bring and Destouni 2013; Magritsky *et al.* 2018, etc.). On the middle Lena (Tabaga station), winter discharge increased by 22 to 32%, with the highest value recorded at the end of winter, during the April low-water period. Downstream, at the Kusur station, winter discharge increased by 45% due to the combined effect of the winter change and the dams on the Vilyuy River (a left-bank tributary, equipped with a large dam, which flows into the Lena River downstream of Yakutsk). Such increases in winter flows are recorded on all Siberian and northern European Russian rivers (Magritsky *et al.* 2018). On the Arctic and sub-Arctic rivers of Sweden, the same observations are made: increased winter discharge and earlier floods in connection with thickening of an earlier melting snow cover and degradation of permafrost (Matti *et al.* 2016).

In North America, the Mackenzie and Alaska rivers (including Yukon River) also show a marked increase in winter discharge, followed by a corresponding decrease in spring and summer flows (Bennett *et al.* 2015). This is related to milder and snowier winters. More specifically, (Brabets and Walvoord 2009) demonstrate a complex evolution since the mid-20th Century in the Yukon Basin. The increase in winter flows is related to greater input from underground water and subsurface flows. In April, there has also been an increase in discharge, associated with earlier increases in atmospheric temperatures, triggering earlier snowmelt in the basin and breakup of ice on the river. These changes appear to be driven by the Pacific Decadal Oscillation, which was positive (warmer) over the 1975–2005 period.

7.3.3. *Spring flooding and breakup*

The winter low-water season comes to an abrupt end with the arrival of the flood. This flood is triggered by rising temperatures in the spring, which cause the snow to melt fairly quickly. On the other hand, milder temperatures also melt the ice cover on the rivers. The active layer on the floodplain and on the slopes thaws more slowly during the summer, so that the snowmelt water will infiltrate very little and thus concentrate in the river beds. Large rivers, especially those flowing northward, flood in April in their upstream reaches (March for the Mackenzie River). This flood wave propagates downstream and arrives in more northerly areas where the river is still frozen. The water lifts the ice, breaks it up and carries it away, and the ice breaks up (Costard *et al.* 2014). Piles of ice blocks may accumulate at the front of islands and at the entrance to the secondary channels (see Figure 7.8). In some cases, the ice cover "resists" and the water level rises rapidly backwards, causing catastrophic flooding upstream, as seen above with the Lensk case in May 2001. During breakup, water discharge increases 10-fold (sometimes 50-fold) from one day to the next. On the Lena at Tabaga, in May 2010, the flow, which was about 1,300 $m^3 s^{-1}$ at the beginning of the month, rose to 9,500 $m^3 s^{-1}$ on May 17, reaching 51,600 $m^3 s^{-1}$ on May 20. During the night of May 19 to 20, the level rose by 8 m. This was the second highest flood recorded since the Russians began carrying out measurements on the Lena River, the first very large flood having been recorded in 1955 (52,700 $m^3 s^{-1}$).

Breakup flooding takes several weeks to progress downstream: on average, it takes eight weeks on the Mackenzie River; on the Lena River, it takes four to five weeks to reach the middle valley from the Pre-Baikal mountains, and another two to three weeks to reach the lower valley, which, as on the Mackenzie River, is about two months of delay between upstream and downstream. On rivers that follow a meridian route, breakup follows the spring progression, whereas on west-east oriented rivers, breakup occurs almost simultaneously throughout the system.

Breakup dates have been changing rapidly over the last 20 or 30 years. Generally speaking, the evolution is towards earlier flooding. In the Mackenzie Basin, flood discharge has advanced by one day per decade since 1970. For example, the peak flood advanced by five days between 1973 and 2011 on the Mackenzie River (Yang *et al.* 2015). Rivers in Alaska also show earlier floods (Bennett *et al.* 2015). On the Lena River, the onset of breakup, which generally occurred around May 22 in Yakutia, advanced by five days (and in some years by 10 days) between May 15 and 17, or even earlier (Gautier *et al.* 2018) (see Figure 7.11).

The flood peak, which may occur immediately after breakup or a few days later, seems to be disorganized. Flood peak on the Lena River generally occurred around May 30, an average of about eight days after the start of the rise in water level. Since the mid-1990s, this peak has been recorded either earlier than before or much later!

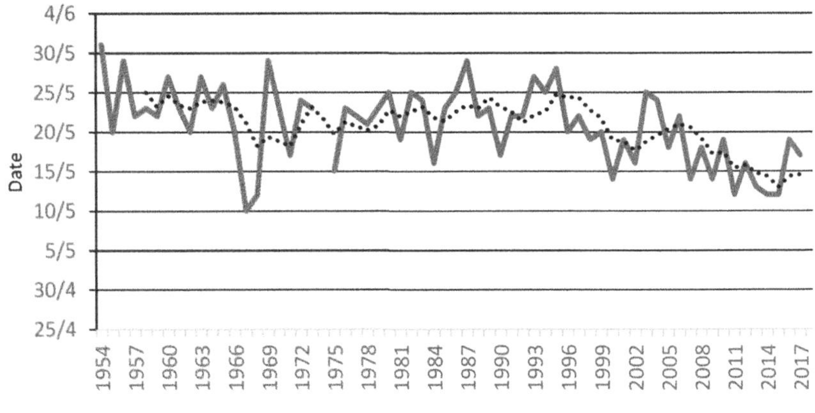

Figure 7.11. *Date of onset of breakup on the Lena at Tabaga in 1954 to 2017. In blue: date; dotted line: five-year moving average (source: adapted from Gautier* et al. *2018; Roshydromet data). For a color version of this figure, see www.iste.co.uk/mercier/climate.zip*

7.3.4. *The rapid evolution of water discharge*

One of the characteristics of Arctic rivers is the very high interannual variability in their flows (making it difficult to define changes in a statistically reliable way). High waters and floods commonly vary by a factor of three. However, interannual variability appears to have increased markedly since the early 21st Century. Recent trends in peak flows are contrasted and two patterns appear to be emerging, one on the American side and one on the Asian side.

In the American Arctic, there is a general trend of decreasing maximum flows (Bennett *et al.* 2015). This decrease in maximum discharge is noted over the Mackenzie River and is mainly related to earlier snowmelt (Yang *et al.* 2015). Over the Yukon, however, there does not appear to be a trend for the highest flows. Because the Yukon "mixes" several influences (glacial melt, snowmelt and active layer thaw), it is difficult to distinguish clear drivers of flood development in the Yukon.

Figure 7.12. *Flooding of the plain in May 2012 (Yakutia, upstream of Yakutsk) (source: photograph E. Gautier). For a color version of this figure, see www.iste.co.uk/mercier/climate.zip*

In Arctic Asia, very high floods have occurred more frequently since the beginning of the 21st Century (Shiklomanov and Lammers 2009; Gautier *et al.* 2018). On the middle Lena (Tabaga), 23 floods above 40,000 m^3 s^{-1} (corresponding to a height of more than 9 m in Yakutsk) have been measured since 1936: 39% of them have occurred since the beginning of the 21st Century. Not only are these large floods more frequent, but their duration is also increasing. The years 2005–2008 were particularly wet on the Siberian side, resulting in very high average discharge and floods. For example, in June 2006, the Lena River recorded a monthly average flow that was double the average value, and in 2007, a surplus of 30% was observed. In Siberia, 2007 was the year of records, since the rivers brought 25% more water to the Arctic seas (Shiklomanov and Lammers 2009). This exceptional hydrology should be seen in the context of warmer (+2°C) and wetter (+100 mm) years. It has since become apparent that this "exceptional" situation is becoming frequent. In both the American and Asian Arctic, we note the emergence of a new atmospheric circulation pattern that visibly increases hydrological variability (Shiklomanov and Lammers 2009; Overland and Wang 2010; Inoue *et al.* 2012; Iijima *et al.* 2016). Recent atmospheric change is causing the advection of warm, moist air into the Siberian Arctic, resulting in hot, humid summers (and early snowy winters).

Indeed, (Iijima *et al.* 2016) clearly identify a break in the years 2004–2005 with a more intense cyclonic activity that leads to an increase in humidity in the active layer in parallel with its thickening. Summer rains create stronger and longer floods. Since the beginning of the 21st Century, these climatic events have thus led to peaks of secondary summer floods (July to September): previously anecdotal, these summer floods have become commonplace. Figure 7.13 shows the recent years in which summer floods have occurred in Lena River; these "multipeak" years show a sharp increase in their frequency. In 2008, the August flood in the middle Lena valley was greater than the breakup flood (Gautier *et al.* 2018). In 2016, the flood in early August was almost equivalent to that in June. The effect of these summer floods also visibly altered the mean discharge in September (see Figure 7.14).

NOTE.– Consequences of changes in flows and seasonality of floods.

Several factors contribute to changing river sediment dynamics in permafrost regions. Longer and earlier floods and increased summer flooding contribute to the degradation of permafrost along river banks. During flooding, water erodes the bank both through a thermal process (penetration of the water heat wave into the frozen ground) and through mechanical erosion.

A recent study (Costard *et al.* 2014) has shown that bank erosion depends both on the duration of the flood (more than on its intensity) and on the date of the flood. In fact, summer floods, which have warmer water than spring floods, promote greater erosion of channel banks. This increasing erosion destabilizes the river bed.

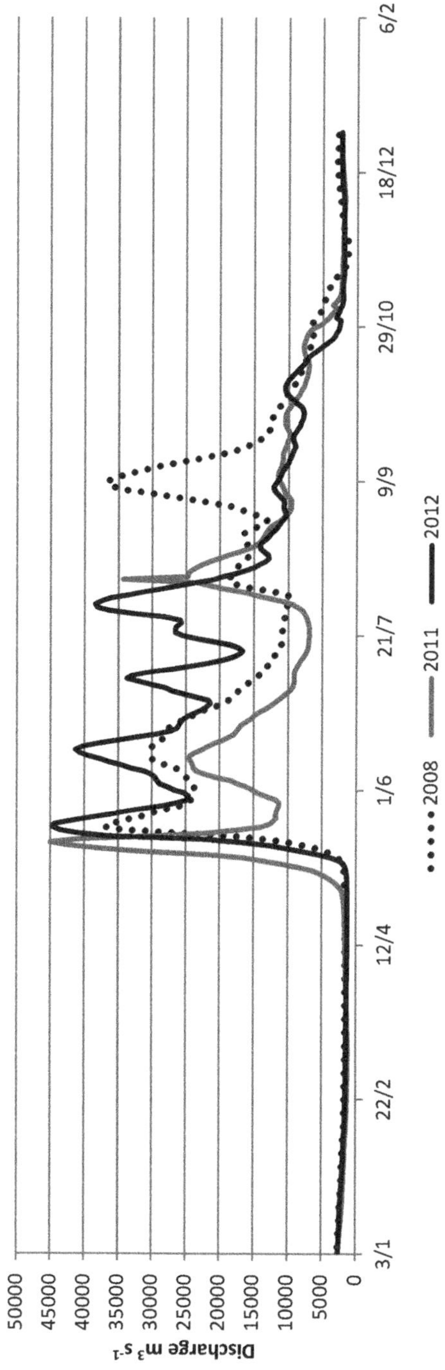

Figure 7.13. *Example of summer floods on the Lena at Tabaga (Roshydromet data). For a color version of this figure, see www.iste.co.uk/mercier/climate.zip*

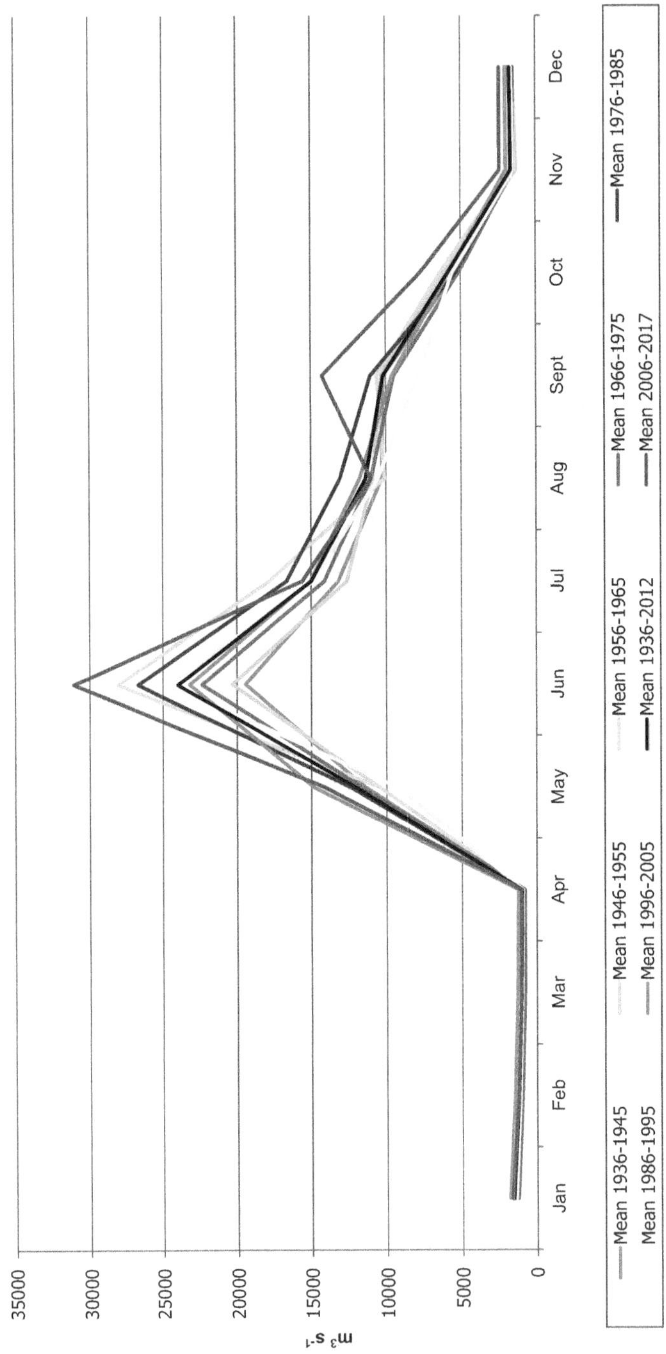

Figure 7.14. *Average monthly flows per decade from Lena to Tabaga station (1936–2017; Roshydromet data). For a color version of this figure, see www.iste.co.uk/mercier/climate.zip*

7.4. Conclusion

The increase in the water discharge of the great Arctic rivers is large and sudden; it is found in all rivers. Average annual flows have increased by 8–15% on average, which is significant in such a short period of time. Winter flows are increasing everywhere, while spring and summer flows are increasing strongly in northern Asia and northern Europe, and are stable or decreasing in the Americas. This exceptional growth reflects the vulnerability of the permafrost that occupies Arctic Russia over thicker and larger areas than elsewhere.

The hydrological consequences of permafrost degradation are very heavy for local populations. The increase in active layer moisture and precipitation (snow and/or rain) leads to an expansion of lakes and an accelerated transfer of water from slopes to river beds. On the riverbanks, the populations are confronted with two problems: the increase in the frequency of floods, as well as their unpredictability. Breakups, which used to be regulated almost like music paper, may occur earlier or much later. The increased erosion of river banks also implies greater instability of the river beds. Riverbeds play an important economic role, and navigation is at the forefront of this. The safety of navigation depends on water levels. A recent study (Scheepers *et al.* 2018) raises this issue by demonstrating the expected degradation of navigation conditions on the Mackenzie River by 2080, due to the increase in the duration of low water. On the Lena, where goods and people have only a short period of time to navigate, navigation services are obliged to replace the signs marking the navigable channel and to remove sand from many port sites due to the increased instability of the channels.

7.5. References

Ballantyne, C.K. (2018). *Periglacial Geomorphology*. John Wiley, New York.

Bennett, K.E., Cannon, A.J., Hinzman, L. (2015). Historical trends and extremes in boreal Alaska river basin. *Journal of Hydrology*, 527, 590–607.

Brabets, T.P., Walvoord, M.A. (2009). Trends in streamflow in the Yukon River Basin from 1944 to 2005 and the influence of the Pacific Decadal Oscillation. *Journal of Hydrology*, 371, 108–119.

Bring, A. and Destouni, G. (2013). Hydro-climatic changes and their monitoring in the Arctic: Observation-model comparisons and prioritization options for monitoring development. *Journal of Hydrology*, 492, 273–280.

Brown, J., Ferrians, O., Heginbottom, J.A., Melnikov, E. (2002). Circum-Arctic map of perma-frost and ground ice conditions, Version 2. *National Snow and Ice Data Centre* [Online]. Available at: https://nsidc.org/data/GGD318.

Bulygina, O.N., Razuvaev, V.N., Korshunova, N.N. (2009). Changes in snow cover over northern Eurasia in the last few decades. *Environmental Research Letter*, 4, 45026.

Burn, C.R. and Kokelj, S.V. (2009). The environment and permafrost of the Mackenzie Delta area. *Permafrost and Periglacial Processes*, 20, 83–105.

Costard, F., Gautier, E., Fedorov, A., Konstantinov, P., Dupeyrat, L. (2014). An assessment of the erosion potential of the fluvial thermal process during ice breakups of the Lena River (Siberia). *Permafrost and Periglacial Processes*, 25, 162–171.

Déry, S.J., Hernandez-Henriquez, M.A., Burford, J.E., Wood, E.F. (2009). Observational evidence of an intensifying hydrological cycle in northern Canada. *Geophysical Research Letter*, 36 L13402.

Fedorov, A.N., Ivanova, R.N., Park, H., Hiyama, T., Iijima, Y. (2014). Recent air temperature changes in the permafrost landscapes of northeastern Eurasia. *Polar Science*, 8, 114–128.

Gautier, E., Dépret, T., Costard, F., Virmoux, C., Fedorov, A., Grancher, D., Konstantinov, P., Brunstein, D. (2018). Going with the flow: Hydrologic response of middle Lena River (Siberia) to the climate variability and change. *Journal of Hydrology*, 557, 475–488.

Iijima, Y., Nakamura, T., Park, H., Tachibana, Y., Fedorov, A.N. (2016). Enhancement of Arctic storm activity in relation to permafrost degradation in eastern Siberia. *International Journal of Climatology*, 36(13), 4265–4275.

Inoue, J., Hori, M.E., Takaya, K. (2012). The role of Barents Sea ice in the winter time cyclone track and emergence of a warm-Arctic cold-Siberian anomaly. *Journal of Climatology*, 25, 2561–2568.

IPCC (2019). Special report on the ocean and cryosphere in a changing climate. Report, Polar Regions, Geneva [Online]. Available at: https://www.ipcc.ch/srocc.

Kichigina, N. (2013). Hydroclimatic change and analysis of floods in large river basins of southern East Siberia. *Hydrological Processes*, 27(15), 2144–2152.

Kurylyk, B.L., Hayashi, M., Quinton, W.L., McKenzie, J.M., Voss, C.I. (2016). Influence of vertical and lateral heat transfer on permafrost thaw, peatland landscape transition and groundwater flow. *Water Resources Research*, 52, 1286–1305.

Magritsky, D.V., Frolova, N.L., Evstigneev, V.M., Povalishnikova, E.S., Kireeva, M.B., Pakhomova, O.M. (2018). Long-term changes of river water inflow into the seas of the Russian Arctic sector. *Polarforschung*, 87(2), 177–194.

Matti, B., Dahlke, H., Lyon, S.W. (2016). On the variability of cold region flooding. *Journal of Hydrology*, 534, 669–679.

Overland, J.E. and Wang, M. (2010). Large-scale atmospheric circulation changes are associated with the recent loss of Arctic sea ice. *Tellus*, 62A, 1–9.

Park, H., Yoshikawa, Y., Oshima, K., Kim, Y., Ngo-Duc, T., Kimball, J.S., Yang, D. (2016). Quantification of warming climate-induced changes in terrestrial Arctic river ice thickness and phenology. *Journal of Climate*, 29, 1733–1754.

Pissart, A. (1970). Les phénomènes physiques essentiels liés au gel, les structures périglaciaires qui en résultent et leur signification climatique. *Annales de la Société Géologique de Belgique*, 93, 7–49.

Scheepers, H., Wang, J., Gan, T.Y., Kuo, C.C. (2018). The impact of climate change on inland waterway transport: Effects of low water levels on the Mackenzie River. *Journal of Hydrology*, 566, 285–298.

Séjourné, A., Costard, F., Fedorov, A., Gargani, J., Skorve, J., Massé, M., Mège, D. (2015). Evolution of the banks of thermokarst lakes in Central Yakutia (Central Siberia) due to retrogressive thaw slump activity controlled by insolation. *Geomorphology*, 241, 31–40.

Serreze, M.C., Bromwich, D.H., Clark, M.P., Etringer, A.J., Zhang, T., Lammers, R. (2002). Large-scale hydro-climatology of the terrestrial Arctic drainage system. *Journal of Geophysical Research*, 108, 8160.

Shiklomanov, A.I. and Lammers, R.B. (2009). Record Russian river discharge in 2007 and the limits of analysis. *Environmental Research Letter*, 112, 1–14.

Shiklomanov, A.I. and Lammers, R.B. (2014). River ice response to a warming Arctic-recent evidence from Russia. *Environmental Research Letter*, 9, 035008.

Soloviev, P.A. (1973). Thermokarst phenomena and landforms due to frost heaving in central Yakutia. *Biuletyn Peryglacjalny*, 23, 135–155.

Tan, A., Adam, J.C., Lettenmaier, D.P. (2011). Change in spring snowmelt timing in Eurasian Arctic rivers. *Journal of Geophysical Research*, 116, 1–12.

Tricart, J. and Cailleux, A. (1967). *Traité de géomorphologie, tome II. Le modelé des régions périglaciaires*. SEDES, Paris.

Walvoord, M.A. and Kurylyk, B.L. (2016). Hydrologic impact of thawing permafrost – A review. *Vadose Zone Journal*, 15(6), 1016.

Woo, M.K. (2012). *Permafrost Hydrology*. Springer, Berlin.

Yang, D., Kane, D.L., Hinzman, L.D., Zhang, X., Zhang, T., Ye, H. (2002). Siberian Lena River hydrologic regime and recent change. *Journal of Geophysical Research*, 107, 4694.

Yang, D., Shi, X., Marsh, P. (2015). Variability and extreme of Mackenzie daily discharge during 1973–2011. *Quaternary International*, 380/381, 159–168.

Ye, B., Yang, D., Zhang, Z., Kane, D.L. (2009). Variation of hydrological regime with permafrost coverage over Lena Basin in Siberia. *Journal of Geophysical Research*, 114, D07102.

Zakharova, E.A., Alexei, V., Kouraev, A.V., Guillaso, S., Garestier, F.R., Desyatkin, A., Desyatkin, R. (2018). Recent dynamics of hydro-ecosystems in thermokarst depressions in Central Siberia from satellite and *in situ* observations: Importance for agriculture and human life. *Science of The Total Environment*, 615, 1290–1304.

8

The Impacts of Climate Change on Watercourses in Temperate Environments

Gilles DROGUE
University of Lorraine, Metz, France

8.1. What is at stake?

8.1.1. *Spatial dynamics of climate zoning and river regimes*

On a global scale, temperate environments are typically defined by climate zoning. The Köppen-Geiger[1] climate classification (see Figure 8.1), for example, places temperate environments in the temperate zone (letter C or D of the major climate types) where the influence of the seasonal swing of the meteorological equator, land-sea relationships, continentality or relief leads to the following major regional climates:

– the temperate oceanic climate to the west characterized by westerly winds and disturbed currents in all seasons (Cfb type climate);

– the temperate Mediterranean climate (Csa and Csb), to the west, which in the summer experiences the rise of subtropical high pressures with that of the meteorological equator; in the cold season, fall-winter, high pressures move towards

1 A climate, according to this classification, is identified by a code of two or three letters (the 1st letter designates the main types of climate, the 2nd letter designates the dry season letter, the 3rd letter corresponds to the thermal letter).

the equator which leaves the Mediterranean regions under the influence of disturbed currents from the west;

– the temperate continental climate (type D climate and its variants), at a distance from the oceans or sheltered by a mountain range, is dominated in winter by the presence of a thermal anticyclone, which leaves little room for humidity; disturbances from the west can nevertheless reach these regions by passing over them and bringing a little snow; in summer, the continent is more open to oceanic influences and experiences its maximum rainfall;

– the humid climate of the east of the continents (also called humid subtropical, Cfa-type climate) which takes advantage of tropical summers; in winter, the atmosphere is very continental up to the coast, which can experience very snowy cold fronts (cold waves, etc.).

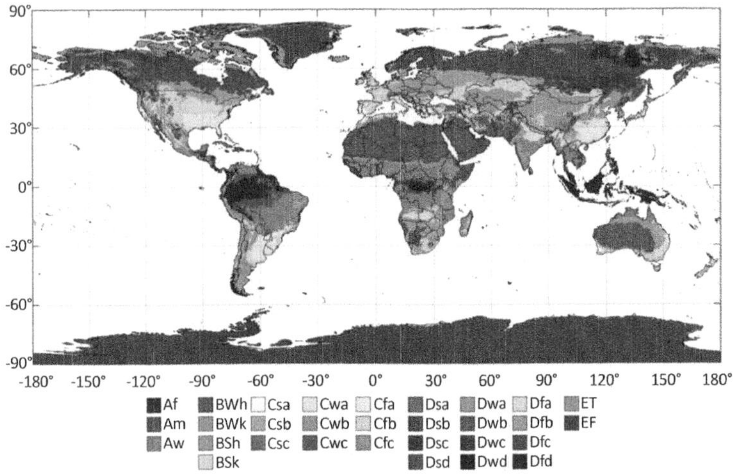

Figure 8.1. *The global climate mosaic according to the Köppen-Geiger classification. Map based on recorded climate data for the period 1980–2016 (source: Beck* et al. *2018). For a color version of this figure, see www.iste.co.uk/mercier/climate.zip*

The climatic factor being a predominant aspect in the hydrological regime[2] of rivers[3], it largely reflects the regional climate map. This map changes over time and is

2 Cyclical variations in river flow under the influence of climate, altitude, land use, etc.

3 (Pardé 1933) differentiates between the simple hydrological regime (e.g. the rainfall and evaporation regime of the Seine in Paris), the mixed regime (e.g. the snow-pluvial regime of the upper Mississippi) and the complex regime (e.g. that of the Rhône at Chancy downstream of Lake Geneva) where the influences of several tributaries are combined.

therefore a marker of climate change. In the northern hemisphere, for example, Spinoni *et al.* (2014) have observed a marked decline in the tundra climate (ET) in favor of the "humid" taiga type temperate climate (Dwc) since the Second World War. A change in the current spatial extent of temperate zone hydrological regimes is therefore to be expected in a context of rapidly changing global climate map (Beck *et al.* 2018).

8.1.2. *Watercourses: resource, vector and living environment*

In addition to their regime, watercourses in temperate environments are characterized by their function as a resource, a vector and an environment:

– in metropolitan France, for example, in an average year, the runoff of internal origin of the six major river basins[4], combining surface runoff via rivers and underground flow, amounts to 193 km^3/year, that is 3,200 m^3/inhab/year (Andreassian and Margat 2014). This amount, which reflects an abundant resource of the basins in relation to their population, is used for different purposes, the largest of which is energy (see Figure 8.2). In regards to all sample from Figure 8.2, 70.4% accounts for continental surface water (rivers and water bodies);

– water flow through rivers are also a preferred vector for elements transported in solution (especially pollutants) or in suspension, such as sediments carried in the bed of a river, for example. Thus, the solid flow of the Seine at Poses, upstream of Rouen, is estimated at 700,000 t/year (Meybeck 2001); by transporting suspended matter, flowing water also helps to shape geomorphological patterns, as shown by the Badlands or Lapiez landscapes, for example;

– running water is a living environment for many plant and animal organisms, which are very sensitive to sudden or gradual variations in river flows: severe low-water periods, which generally result in a rise in water temperature and a drop in oxygenation, provide more precarious living conditions for fish fauna, hence the need to define an ecological flow (Lamouroux *et al.* 2016).

Assessing the impacts of sudden or gradual climate change on watercourses in temperate environments therefore requires quantitative and qualitative knowledge of the flows that pass through them. This knowledge is useful and necessary to understand the link between the temporal variability of hydrology and that of climate-induced explanatory variables (precipitation, evapotranspiration, etc.).

4 Artois-Picardie, Rhin-Meuse, Seine-Normandie, Loire-Bretagne, Adour-Garonne, Rhône-Méditerranée-Corse.

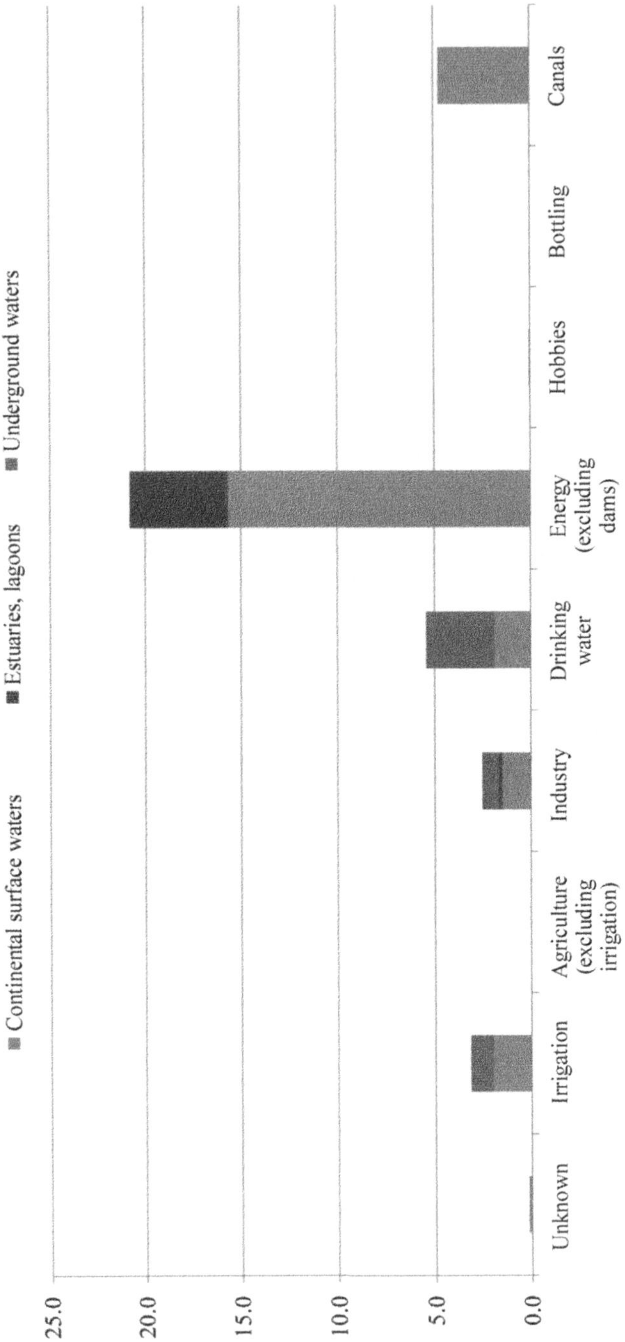

Figure 8.2. *Breakdown of the total volume of abstractions (in billions of m^3) according to the different uses declared and the type of water removed (mainland France and overseas territories). Data 2016 (source: Chataigner and Michon 2019). For a color version of this figure, see www.iste.co.uk/mercier/climate.zip*

8.1.3. *The (dis)equilibrium between precipitation, evapotranspiration and flow in temperate environments*

To illustrate this point, we refer to an original method of representation, which gives a synthetic description of the water balances of the watersheds (see Figure 8.3). This mode of representation is an adimensional space that is constructed from the comparison of the adimensional indices of aridity (P/E: ratio of mean annual rainfall to the mean annual PET[5] of the watershed) and yield or flow coefficient (Q/P: ratio of mean annual flow to the mean annual rainfall of the watershed).

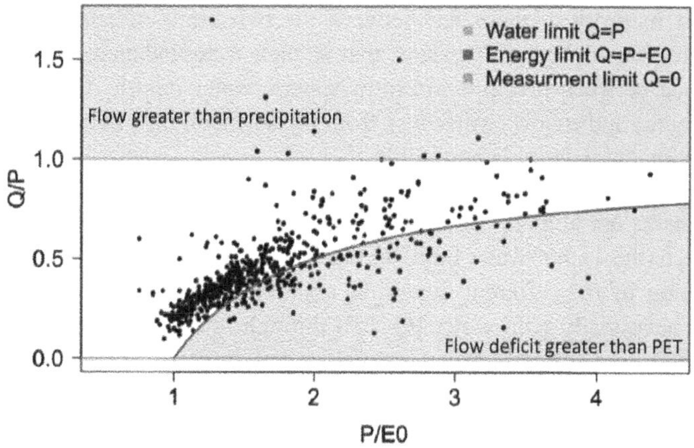

Figure 8.3. *Application of the dimensionless representation Q/P = f(P/PET) to a sample of 609 basins located in metropolitan France. On the graph: E0 = PET (source: Lebecherel 2015). For a color version of this figure, see www.iste.co.uk/mercier/climate.zip*

For basins where underground flows are negligible on the interannual water balance, the placement of a basin in this space is constrained by two limits of realism of the water balance (Lemoine 2008; Coron 2013):

– the constraint of inferiority of the flows compared to the precipitation on average (i.e. limit Q = P or y = 1): the actual evapotranspiration (AET) represents only a very small fraction of the PET and consequently the flow is equal to the rain;

– the constraint of inferiority of the flow deficit (P – Q), which can be assimilated to the AET compared to the PET on average (i.e. limit P – Q = PET, i.e.

5 Potential evapotranspiration (PET) of the watershed or evaporative capacity of the watershed (in the form of energy) when water availability is not a limiting factor.

y = 1 – 1/x): the AET reaches the potential rate (PET), the quantity of water remaining for the flow being then the difference P – PET.

Figure 8.3 illustrates the placement of 609 French catchments representing a wide spectrum of temperate climates (continental, oceanic, Mediterranean, including mountain climate since medium mountain rivers are included in the sample), lithologies (sedimentary rocks, magma, etc.), land use patterns (agricultural areas, grassland areas, forest areas) and whose runoff is not influenced by anthropogenic activities (Lebecherel 2015).

The vast majority of them fall between the two limits of realism of the water balance. They are conservative basins, that is, their water balance is the result of the usual balance between precipitation (atmospheric water input), AET (water loss to the atmosphere) and runoff (water loss from the surface), the underground influence on the inter-annual balance is negligible.

Some basins are atypical in terms of flow: they exceed the realistic limit of the water balance (this is the "water limit" in Figure 8.3). In addition to the hypothesis of a bad estimation of the different quantities observed, these situations correspond to basins that necessarily gain water (Q > P) through inter-basin exchanges that feed rivers underground. Other basins are below the limit of realism of the energy balance (this is the "energy limit" in Figure 8.3) which means that Q < P – PET: a basin cannot lose more water to the atmosphere than there is energy to evaporate it (Lebecherel 2015). These watersheds are non-conservative, they do not just "lose" water through evapotranspiration and in-stream flow, they "leak" to a regional aquifer (Lebecherel 2015).

How does the dimensionless space Q/P = f (P/PET) shed light on our problem? By describing the genesis of river flow in temperate environments as a complex response of catchments to precipitation and evapotranspiration flows, it reflects the fact that hydroclimatic analysis of the impacts of climate change on rivers must adapt to a very wide range of hydrological situations. Among these are rivers, albeit a minority, which can be described as "hydrological monsters"[6] because of their unexpected or seemingly insoluble behavior. This can defeat current hydrological models that attempt to replicate their behavior. Understanding and explaining the reasons for hydrological monstrosities is therefore fundamental, as it contributes to making hydrological models more robust and therefore more efficient in predicting the effects of climate change on rivers.

---

6 This neologism was the subject of a scientific workshop organized by Cemagref entitled "*La cour des miracles de l'hydrologie*" (Paris, June 18–20, 2008).

In addition to the challenge of hydrodiversity, there is also the challenge of identifying issues related to the temporal variations in hydrological phenomena under the effect of climate, whether between the past and the present or between the present and the future.

8.1.4. *The study of past climate impacts*

The issues raised by the retrospective analysis of river flows under climate change are varied. They are associated with quantification and methodological issues. A brief inventory can be drawn up on the basis of the work of Drogue (2013) and Perrin (2015):

– what is the "natural" variability of the systems studied over time? Answering such a question requires quantifying the characteristics of past hydrological phenomena (floods, low water levels, recession) and their long-term evolution;

– are the changes currently observed (or expected in the future) within the range of natural variability in the systems? The challenge here is to assess the statistical significance of the changes at the time scales considered;

– what have been the causes and consequences of past changes? This question focuses on understanding the dynamics of systems and finding the drivers of change.

The first two questions focus on detecting correlations between temporal variability in hydrology and climate explanatory variables; but correlations are not causal. It is therefore necessary to search for the causes of the changes identified in order to be able to attribute an origin to them, which is the subject of the third issue listed.

8.1.5. *The study of future climate impacts*

In prospective analysis, these are the questions raised by the analysis of variations in river flows under the effect of climate cover issues of quantification, modeling and adaptation to change (Drogue 2013; Perrin 2015):

– What will be the evolution of river flows? Answering such a question requires quantifying likely future changes.

– What will be the consequences on human activities (resource, risk)? The challenge here is to quantify the future vulnerability to the risk of water shortage or excess water and to compare water resources with needs.

– Will the estimated changes require adaptation strategies to be defined? This question raises the question of whether or not it is necessary to develop strategies for the evolution of uses, demand, management or development in the long term, the objective of which would be to compensate for the impact of climate change on rivers (through the construction of reservoir dams, for example).

8.1.6. *Summary*

8.1.6.1. *Scientific assessment of stream flow*

In order to meet the above-mentioned challenges, it is imperative to have long chronicles of good quality flow. Between the past and the present, they serve to answer the question of whether there is a climate indicator in the already observable hydrological changes. Between the present and the future, they are used to make flow projections from hydrological models, "calibrated" to current flows and "forced" by climate scenarios.

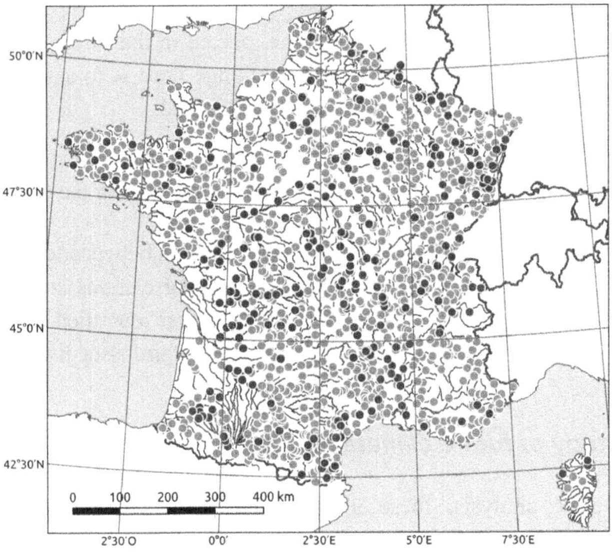

Figure 8.4. *Observation of daily flows. For a color version of this figure, see www.iste.co.uk/mercier/climate.zip*

COMMENT ON FIGURE 8.4.– *In blue, stations observing daily flows from the French Hydro Bank considered influenced and available over the 1990–2016 period. In green, observation station of daily flows considered as not influenced by man over the same period. In red, observation station of daily flows considered as long, that is, available over at least 70 years (source: Banque Hydro française, realization D. François).*

Unfortunately, the measurement of river flow is relatively recent in most countries in temperate zones. In France, for example, long chronicles of continuous flow (100 to 150 years) are rare and concern only a limited number of rivers (see Figure 8.4). The French hydrometric measurement network has only really developed since the 1960s[7], which means that the available flow data are not very old, unlike the climate data, which are older.

On the other hand, in order to be able to detect the influence of the climate on the already observable hydrological changes, the influence of non-climatic factors must be limited as much as possible, such as uses (withdrawals, various hydraulic developments, etc.), metrology (relocation of stations, change of instrumentation, changes in the stage-discharge relationship, etc.) or land use (urbanization, land clearing, etc.). This further reduces the sample of candidate rivers (see Figure 8.4) unless an attempt is made to "naturalize" the flows by adding to the observed flows influenced by hydrological alterations due to human activities (Terrier 2016). A smart way to protect the long flow records useful for climate change studies is to establish a reference hydrometric network for low-flow climate monitoring (Giuntoli *et al.* 2012) (see Figure 8.5).

Figure 8.5. *Flow observation station on the Moselle at Rupt-sur-Moselle (Vosges department, France); this station is part of the reference network for the climatic monitoring of low water levels (source: Vigicrues – DREAL GRAND EST). For a color version of this figure, see www.iste.co.uk/mercier/climate.zip*

7 Nearly 85% of the hydrological series stored in the HYDRO central database began in the 1960s (Legros *et al.* 2015).

8.1.6.2. *Scientific evaluation of modeling tools*

If the retrospective analysis uses flow observations, it is not possible to use them to analyze changes in river flow between the present and the future. To find out where, when and how much climate-induced hydrological changes will occur, the hydrologist then usually uses simulations or projections from climate modeling tools (ocean-atmosphere general circulation models) and watershed-scale water cycle modeling tools (hydrological models).

But to move from the IPCC's global expertise on global climate change to the assessment of impacts on rivers, many obstacles to diagnosis must be overcome. In particular, the diagnosis is altered by the "congenital" uncertainty of the hydrological impact model. Two major elements contribute to this: the imperfection of the model, which is only a simplified representation of the water cycle, and its lack of climatic robustness, in other words, its inability to maintain its qualities under unknown climatic conditions.

As pointed out by Coron (2013), predicting river flow in a context of global warming over the next few decades is a new demand. Historically, hydrological modeling has not been designed with climate robustness in mind, but rather as a tool for understanding flow processes within a watershed or for filling gaps in flow records. To compensate for the lack of robustness of hydrological models, it is necessary to continue the research work already undertaken on "crash tests" in water cycle modeling in order to more firmly establish the causes of this "pathology" (Coron 2013).

Finally, the variety of stakes and problems related to the treatment of climate change in hydrology shows that it can only take place conditionally with the help of a strong interaction between all water stakeholders (hydrologists, users, management services, elected officials, climate and water cycle modelers, experimentalists, etc.).

8.2. Hydrological changes already "observable"

8.2.1. *The case of metropolitan France*

8.2.1.1. *Multi-decadal fluctuations*

As the very long flow records from the French Hydro Bank are tainted by anthropogenic influences (changes in instrumentation, pumping, dams, etc.), a hydrometeorological reconstruction is necessary to describe the multi-decadal variability of flows and try to correlate it with that of climate. An example of a hydrometeorological reconstruction is given in Figure 8.6 for the mean annual flow of the Seine at Poses. A strong interannual variability exists on the flows of the Seine at Poses, with deviations from the mean that can reach several hundred cubic meters per second, for a mean of 475 m^3/s (Bonnet 2018).

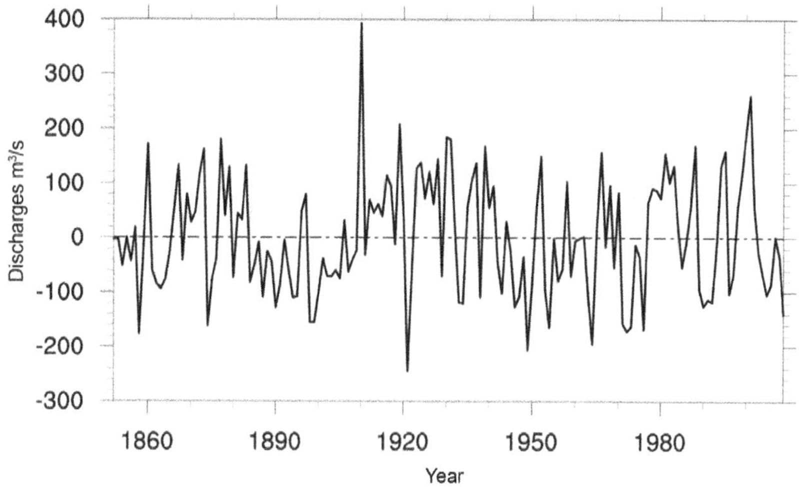

Figure 8.6. *Flow deviations from the interannual average calculated over the available period. River: the Seine at Poses. Flows are reconstructed from knowledge of precipitation, temperature and the state of the atmosphere (source: Bonnet 2018)*

Multi-decadal variability is also apparent in the inter-annual variability of the flows of the Seine River in Paris as shown in Figure 8.7. Both the observations and the hydrometeorological reconstruction show strong multi-decadal variations over the 20th Century, with a strong increase in mean flow in the 1920s (positive phase) followed by a decrease until the 1960s (negative phase).

Starting in the 1960s, a shorter, decadal variability also appears to be present in flows. These variations in mean flow are actually the result of seasonal variations. On the Seine in Paris, these annual variations are strongest in spring and winter and seem to significantly modulate secular trends (Dieppois *et al.* 2016).

According to Boé and Habets (2014) this multi-decadal variability can also be observed in the flows of other French rivers such as the Loire or the Rhône. It is at its maximum in spring, with significant variations in flows of up to 40% of average flows (Boé and Habets 2014).

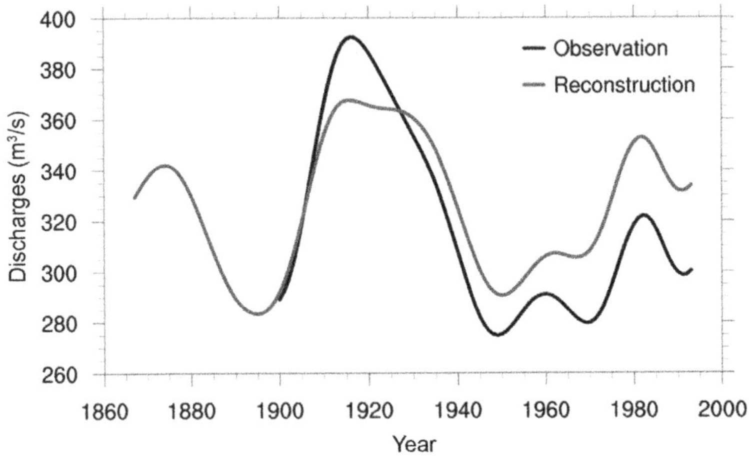

Figure 8.7. *Annual flows of the Seine in Paris (Austerlitz). In black the observations, in red a hydrometeorological reconstruction elaborated from different hypotheses. The curves are smoothed by a low-pass filter (source: Bonnet 2018). For a color version of this figure, see www.iste.co.uk/mercier/climate.zip*

8.2.1.2. *Attribution of these changes*

In spring, the main mode of internal climate variability influencing Europe on multi-decadal time scales is the Multi-decadal Atlantic Variability (MAV), also known as the Atlantic Multi-decadal Oscillation (AMO) (Deser *et al.* 2010).

Studies by Sutton and Dong (2012) and Boé and Habets (2014) suggest that in spring, a positive MAV anomaly (i.e. higher-than-average recorded sea surface temperatures in the northern Atlantic Ocean) is associated with positive pressure anomalies between central and eastern Europe, located between negative pressure anomalies in the eastern Atlantic and northeastern Europe. This type of atmospheric circulation corresponds to an increase in southerly flows over France, that is an increase in warm, dry air, generally associated with negative precipitation anomalies over France, and, consequently, a decrease in flow rates. These hypotheses are supported by the work of Bonnet (2018) which explored a period of about 50 years longer than that investigated by the above-mentioned studies.

Since the multi-decadal variations in summer and autumn flows are largely influenced, via groundwater and soil moisture, by spring precipitation, it is likely that the multi-decadal variability of summer and autumn flows also depends in part on the MAV (Bonnet 2018).

8.2.2. *Continental trends: Western Europe*

8.2.2.1. *Changes detected*

As long flow series are rare in Europe, we focus here on spatial coverage. Using an unpublished sample of 441 small, weakly influenced watersheds in 15 European countries (Stahl *et al.* 2010), we were able to detect the following significant trends over an observational window from 1962 to 2004 (Stahl *et al.* 2010):

– a decrease in annual runoff in southern and eastern Europe and generally an increase elsewhere;

– an increase in flow during the winter months in most basins and a marked decrease in April that extends across Europe until August.

Although locally things are more complex, the general trend is towards an increase in the minimum monthly flow for rivers of the lowland type (Baltic Sea region) and a decrease in the minimum monthly flow for rivers of the rainfall/evaporation type (Germany, Spain, France, United Kingdom).

The trend analysis carried out on a low-water indicator, the average minimum flow over seven consecutive days (known in France as the VCN7, or minimum consecutive volume for seven days) reveals a sharp fall, particularly in central Germany and the United Kingdom. The opposite trend is observed in the Alps, Switzerland and the western part of Austria, where the trend in this index is positive. Finally, for the majority of the rivers analyzed, the date of occurrence of VNC7 is getting earlier and earlier.

These trends confirm the observations made at national and regional scales and are consistent with the spatial organization that emerges from simulations of the effect of future climate change on rivers in Europe.

8.2.2.2. *Link to climate change*

In the case of France and, more broadly, Western European countries, the hydrological trends detected by Stahl *et al.* (2010) on low water flows are to be interpreted in the light of the 1987–1988 event or "climate shift" (Brulebois *et al.* 2015). This abrupt climate change resulted in a sudden increase in mean annual temperature with an amplitude of +1°C in a context of globally stationary precipitation.

The 1987–1988 event was detected in Western Europe from instrumental data at different spatial scales and is the result of marked seasonal warming in winter, spring and summer. As this period was not marked by a change in instrumentation, it cannot be explained by an artificial source. Rather, it would result from a combination of nested factors (from Laat and Crok 2013): an increase in greenhouse gas (GHG) concentration and a reduction in aerosols would have been superimposed on the internal climate variability dictated by the MAV described above. The effect of warming on low-level flow water would be direct (increased AET) and indirect (increased water demand) (Brulebois *et al.* 2015).

8.3. Hydrological projections

8.3.1. *For French rivers*

The recent evolutions detected by Stahl *et al.* (2010) and Brulebois *et al.* (2015) on French rivers are in agreement with the expected effects of climate change predicted by climate models. These have been used as a basis for many prospective studies on French rivers (Carroget *et al.* 2017; Thirel *et al.* 2019). The thesis of Dayon (2015) constitutes the most recent knowledge base on hydrological projections in metropolitan France for the coming decades. It is based on the 5th IPCC report (CMIP5[8] experiment), which uses representative profiles of changes in greenhouse gas concentrations[9] to simulate the evolution of global warming over the 21st Century and beyond, and on 12 climate simulations from the European "Ensemble" project. The simulations of Dayon (2015), which are a reference in the scientific community, show that, over the next few decades, the changes in annual flows will be more intense in the Loire, Garonne and Rhône rivers than the maximum changes observed during the 20th Century.

8 Fifth Coupled Model Intercomparison Project of the World Climate Research Programme that formed the basis for the IPCC 5th Progress Report. This project provides a common experimental framework for meteorological institutes around the world.

9 Radiative Concentration Pathway or RCP.

By mid-century, winter changes are expected to be moderate. For all RCP scenarios, flows increase in the Alpine catchments as well as very slightly in the Pyrenees.

In contrast, in summer, flow changes are negative across the country due to a general decrease in precipitation and intense warming. Projections for the distant future (2070–2100) show that without a rapid stabilization of GHG emissions (scenarios RCP4.5 and RCP6.0), low-flow rates will decrease over the whole country. Even under the RCP2.6 scenario ("optimistic" GHG mitigation scenario), low-flow rates are likely to decrease throughout the southern half of the country. The strongest changes are expected in the south of the country, particularly in the southwest. On the Garonne, for example, with the RCP8.5 scenario ("pessimistic" scenario of increase in GHG emissions), low water flows will, on average over all climate scenarios, decrease by more than 50%. Flood flows change little on average overall. For all scenarios, they do not change or increase slightly in the north of the country, whereas they generally decrease in the south of the country. Dayon (2015) finally concludes to a strengthening of the spatial contrast between the north and the south of France.

8.3.2. *For continental Europe*

Hydrological projections at this spatial scale are broadly similar to those of Dayon (2015). For example, Roudier *et al.* (2015) simulate the evolution of river flows by feeding three very different pan-European hydrological models with a set of 11 climate models from the Euro-cordex experiment. The hydrological impacts focus on a global warming of +2°C compared to the pre-industrial period, whatever the time horizon reached in the climate simulations.

In this context, flood intensity is expected to increase south of 60°, except in some regions (Bulgaria, Poland, southern Spain) where the upward trend in floods is not significant (see Figure 8.8). The sign of the trend is particularly robust in Romania, Ukraine, Germany, France, and northern Spain. The results for low-water levels (see Figure 8.9) are less robust, especially over the duration of the low-water level, for which the results are rather scattered in some regions. The severity and duration of low water levels are expected to increase in Spain, France and Italy, Greece, the Balkans and southern England and Ireland. The study by Roudier *et al.* (2015) also shows that it is in France, Spain, Portugal, Ireland, Greece and Albania that the impact of a global warming of +2°C will be the most extreme on floods and low flows.

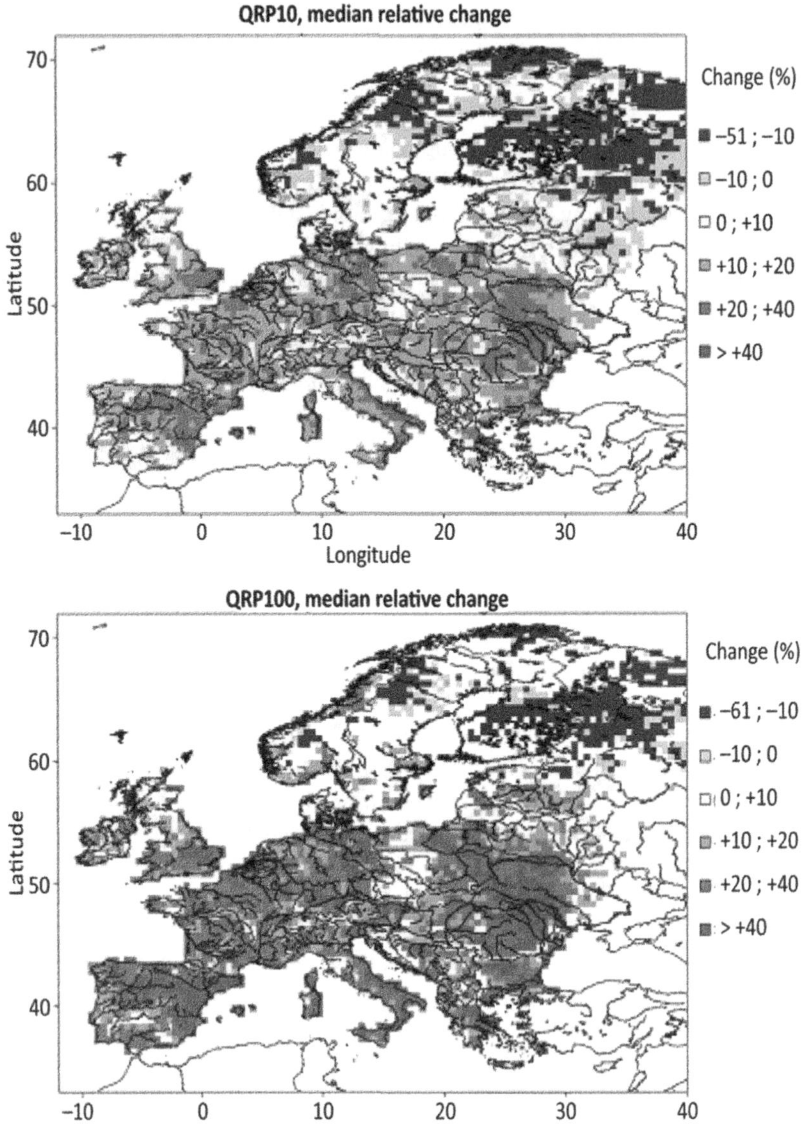

Figure 8.8. *Top: overall median (33 climate simulations) of the relative changes (%) in the annual maximum flood discharge that occurs on average once every 10 years. Bottom: overall median (11 climate simulations) of the relative change (%) in the annual maximum flood discharge that occurs on average once every 100 years. Only significant changes are shown in color (source: Roudier* et al. *2015). For a color version of this figure, see www.iste.co.uk/mercier/climate.zip*

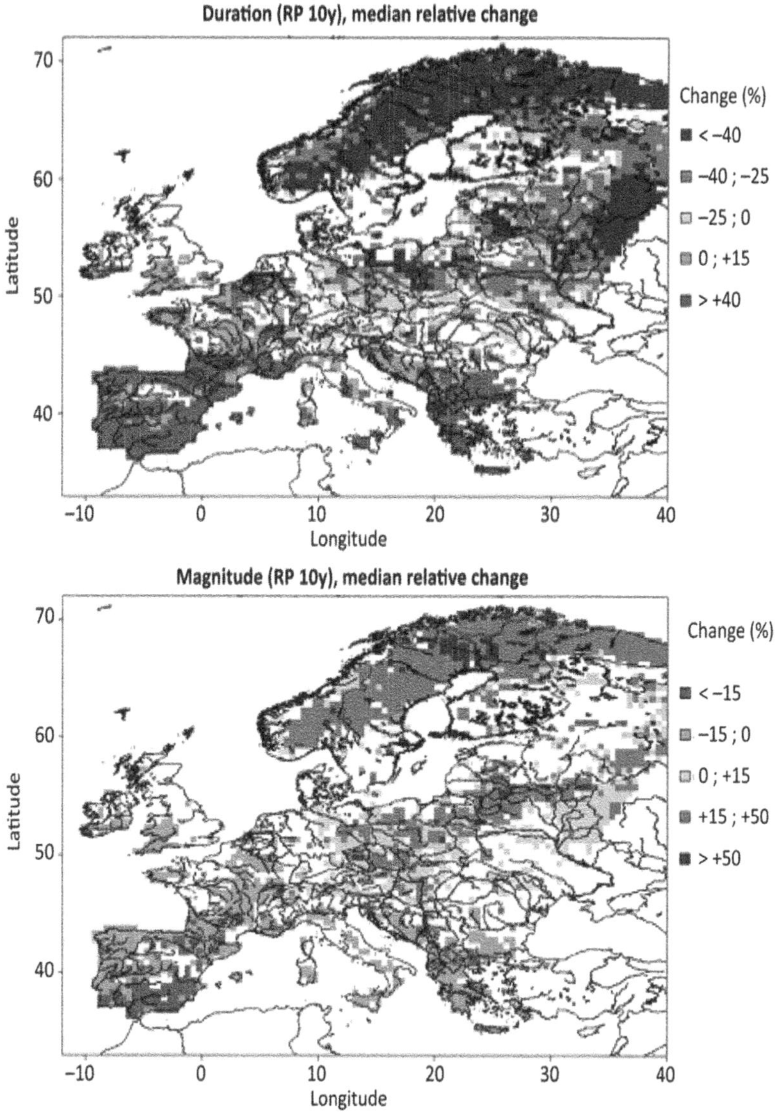

Figure 8.9. *Top: overall median (22 climate simulations) of the relative change (%) in the duration of low flow (here the 20th percentile of the classified flow curve) that occurs on average once every 10 years. Bottom: overall median (22 climate simulations) of the relative change (%) in the value of the low flow (20th percentile of the graded flow curve in this case) that occurs on average once every 10 years. Only significant changes are shown in color (source: Roudier* et al. *2015). For a color version of this figure, see www.iste.co.uk/mercier/climate.zip*

8.4. Conclusion

This chapter on the major issues and most recent knowledge related to the impacts of climate change on rivers in temperate environments shows that, over the recent past, there is a clear causality between natural climate fluctuations (expressed on interannual to multi-decadal scales) and river hydrology. In the temperate zone of the Northern Hemisphere, the impact of contemporary global warming on river flows is already visible, particularly during low-water periods, and the chain of causes and effects on river regimes is beginning to be better understood: "natural" rivers respond to climate change according to their "climate elasticity" determined by the watershed filter and the climatic context.

In France, the climate transition we are currently experiencing is beginning to affect mountain rivers (with a decrease in the buffer role of snow and glaciers) and to amplify the severity of low water levels without modifying the annual abundance of rivers and the intensity of floods. This climatic transition will lead us, in successive stages, to a real thermal shock (+3 to +4°C on annual average) in a few decades. By that time, our world is likely to be climatically extreme, with an aggravation of the hydrological contrasts between the northern and southern regions in the temperate zone of the Northern Hemisphere (it is where river flows are the weakest that they are most likely to decrease on average), and an amplified variability between years, between seasons, this increased variability making flows less exploitable (Andreassian and Margat 2014). In this context, the risk of water crises is well established, in Europe and more particularly in the Mediterranean basin (Viollet *et al.* 2016). The human factor will be particularly important in the capacity of modern societies to face water tensions and to make choices, and it will be necessary to implement adaptation policies aimed at compensating for the negative effects of climate change (Fernandez *et al.* 2014). Finally, it should be noted that if the geography of changes and deadlines are beginning to become clearer in temperate environments, it is thanks to progress in numerical modeling of the climate and the water cycle on different spatial scales (global, continental, regional, local) and to a growing synergy between models and field data collected in the framework of monitoring and forecasting river flows.

8.5. References

Andreassian, V. and Margat, J. (2014). *Allons-nous manquer d'eau ?* Le Pommier, Paris.

Beck, H.E., Zimmermann, N.E., McVicar, T.R., Vergopolan, N., Berg, A., Wood, E.F. (2018). Present and future Köppen-Geiger climate classification maps at 1-km resolution. *Scientific Data*, 5.

Boé, J. and Habets, F. (2014). Multi-decadal river flow variations in France. *Hydrology and Earth System Sciences*, 18(2), 691–708.

Bonnet, R. (2018). Variations du cycle hydrologique continental en France des années 1850 à aujourd'hui. PhD thesis, Université de Toulouse, Toulouse.

Brulebois, E., Castel, T., Richard, Y., Chateau-Smith, C., Amiotte-Suchet, P. (2015). Hydrological response to an abrupt shift in surface air temperature over France in 1987/88. *Journal of Hydrology*, 531, 892–890.

Carroget, A., Perrin, C., Sauquet, E., Vidal, J.P., Chazot, S., Chauvot, M., Rouchy, N. (2017). Explore 2070 : quelle utilisation d'un exercice prospectif sur les impacts des changements climatiques à l'échelle nationale pour définir des stratégies d'adaptation ? *Sciences eaux et territoires*, 22, 4–11.

Chataigner, J. and Michon, J. (2019). Prélèvements quantitatifs sur la ressource en eau (données 2016). Agence française pour la biodiversité, bulletin no. 5 [Online]. Available at: http://ode.creativ3.com/spip.php?article592.

Coron, L. (2013). Les modèles hydrologiques conceptuels sont-ils robustes face à un climat en évolution ? Diagnostic sur un échantillon de bassins-versants français et australiens. PhD thesis, AgroParisTech, Paris.

Dayon, G. (2015). Évolution du cycle hydrologique continental en France au cours des prochaines décennies. PhD thesis, Université de Toulouse, Toulouse.

Deser, C., Alexander, M.A., Xie, S.-P., Phillips, A.S. (2010). Sea surface temperature variability: Patterns and mechanisms. *Annual Review of Marine Science*, 2, 115–143.

Dieppois, B., Lawler, D.M., Slonosky, V., Massei, N., Bigot, S., Fournier, M., Durand, A. (2016). Multidecadal climate variability over northern France during the past 500 years and its relation to large-scale atmospheric circulation. *International Journal of Climatology*, 36, 4679–4696.

Drogue, G. (2013). Études hydro-climatologiques régionales. Applications à l'évolution du climat et aux écoulements de rivière dans un espace transfrontalier. HDR thesis, Université Paris-Diderot, Paris.

Fernandez, S., Martin, M.A., Troy, B., Verdier, J., Viollet, P.L. (2014). Prospective et tensions sur l'eau. Des crises de l'eau en 2050 ? Summary note, SHF/AFEID /Académie de l'eau, Paris.

Giuntoli, I., Maugis, P., Renard, B. (2012). *Évolutions observées dans les débits des rivières en France. Sélection d'un réseau de référence et analyse de l'evolution temporelle des régimes des 40 dernières années*. ONEMA, Paris.

de Laat, A.T.J. and Crok, M.A (2013). Late 20th Century European climate shift: Fingerprint of regional brightening? *Atmospheric and Climate Sciences*, 3, 291–300.

Lamouroux, N., Augeard, B., Baran, P., Capra, H., Le Coarer, Y., Girard, V., Gouraud, V., Navarro, L., Prost, O., Sagnes, P., Sauquet, E., Tissot, L. (2016). Débits écologiques : la place des modèles d'habitat hydraulique dans une démarche intégrée. *Hydroécologie appliquée*, 20, 1–27.

Lebecherel, L. (2015). Sensibilité des calculs hydrologiques à la densité des réseaux de mesure hydrométrique et pluviométrique. PhD thesis, AgroParisTech, Paris.

Le Moine, N. (2008). Le bassin-versant de surface vu par le souterrain : une voie d'amélioration des performances et du réalisme des modèles pluie-débit ? PhD thesis, Université Pierre et Marie Curie, Paris.

Legros, C., Sauquet, E., Lang, M., Achard, A.-L., Leblois, E., Biton, B. (2015). Les annuaires hydrologiques de la Société hydrotechnique de France : une source d'information patrimoniale pour la connaissance de l'hydrologie en France. *La Houille Blanche*, 4, 66–77.

Meybeck, M. (2001). *Flux et temps de séjour des matières en suspension*. PIREN-Seine, Paris.

Pardé, M. (1933). *Fleuves et rivières*. Armand Colin, Paris.

Perrin, C. (2015). De l'expertise globale sur l'évolution du climat par le GIEC à l'évaluation des impacts sur les cours d'eau en France. In *Cycle "chaud et froid sur le climat"*. Espace Vasarely, Antony.

Roudier, P., Andersson, J.C.M., Donnelly, C., Feyen, L., Greuell, W., Ludwig, F. (2015). Projections of future floods and hydrological droughts in Europe under a +2°C global warming. *Climatic Change*, 135(2).

Spinoni, J., Vogt, J., Naumann, G., Carrao, H., Barbosa, P. (2014). Towards identifying areas at climatological risk of desertification using the Köppen–Geiger classification and FAO aridity index. *International Journal of Climatology*, 35, 2210–2222.

Stahl, K., Hisdal, H., Hannaford, J., Tallaksen, L.M., van Lanen, H.A.J., Sauquet, E., Demuth, S., Fendekova, M., Jodar, J. (2010). Streamflow trends in Europe: Evidence from a dataset of near-natural catchments. *Hydrology and Earth System Sciences*, 14, 2367–2382.

Sutton, R.T. and Dong, B. (2012). Atlantic Ocean influence on a shift in European climate in the 1990s. *Nature Geoscience*, 5(11), 788.

Terrier, M. (2016). Évaluation des procédures de naturalisation pour la reconstitution de débits sur le bassin-versant de la Seine. Report, Université Nice Sophia Antipolis, Nice.

Thirel, G., Gerlinger, K., Perrin, C., Drogue, G., Renard, B., Wagner, J.-P. (2019). Quels futurs possibles pour les débits des affluents français du Rhin (Moselle, Sarre, Ill) ? *La Houille Blanche*, 5/6, 140–149.

Viollet, P.L., Verdier, J., Martin, M.A. (2017). Les tensions sur l'eau et les crises associées en Europe et dans le bassin méditerranéen d'ici 2050. Risques et mesures d'adaptation envisageables, Version 1 [Online]. Available at: https://www.shf-hydro.org/maj/phototheque/photos/pdf/synthesetensionseau2017v1.pdf.

9

Spatial Impacts of Melting Central Asian Glaciers: Towards a "Water War"?

Alain CARIOU
Sorbonne University, Paris, France

9.1. Societies and economies dependent on the cryosphere

9.1.1. *The possibility of water scarcity and "water war"?*

The glaciers and snow cover of the world's large mountain ranges play a crucial economic role, since it is estimated that meltwater provides drinking and irrigation water for about 1/6 of the world's population (Francou and Vincent 2007), as in the densely populated plains below the Himalayas and the Andes. However, no region in the world appears to be as dependent on the cryosphere for its development as Central Asia[1] (Kaser *et al.* 2010). Indeed, the region's unique climate and topography means that 70 to 80% of the flow of the major rivers, which feed the Central Asian plains, is irrigated by melting of snow and glaciers.

Consequently, the accelerated melting of the mountain cryosphere as a result of global warming will profoundly change regional hydrology, suggesting that the region will face severe water scarcity in the very short term. The situation is all the more

1 The concept of Central Asia refers here to the five post-Soviet republics (Kazakhstan, Kyrgyzstan, Tajikistan, Turkmenistan and Uzbekistan) and the Uighur Autonomous Region of Xinjiang (Western China). The choice of such a regional unit is justified by the fact that all these territories are linked by transboundary rivers and river basins.

worrying since the impacts of climate change are combined with a largely degraded water situation. Indeed, since the 1950s, titanic hydro-agricultural programs, deployed by the Soviet (Uzbekistan, Kazakhstan, Turkmenistan) and Chinese (Xinjiang) powers to improve millions of hectares of desert land through irrigation, have caused serious environmental problems, one of the most emblematic of which is the disappearance of the fourth largest lake in the world by surface area, the Aral Sea. The development model based on irrigated agriculture, particularly for cotton cultivation, has led to overexploitation and wastage of water resources, which has led to competition and tension for water between the States for various uses such as irrigation water, hydroelectricity or drinking water. The sharing of this resource is all the more complex as the watersheds of the large regional rivers are transboundary, like the Amu Darya basin, which is shared by six countries (Afghanistan, Tajikistan, Kyrgyzstan, Uzbekistan, Turkmenistan, and Iran).

In this context where geopolitical tensions over the sharing of such a resource and global warming are juxtaposed, water is becoming a major issue, so much so that some media Cassandras do not hesitate to predict an imminent shortage that will lead to "water wars" between riparian countries. What is the real situation? The situation is much more complex than it seems. We still know very little about Central Asia's glaciers, so it is difficult to assess the effects of climate change on the region's relatively well-endowed water resources, contrary to popular belief. Thus, much more than scarcity, it is the change in the hydrological regime of rivers that is likely to have a profound impact on the socio-economic balance of the entire region.

9.1.2. *"Water tower" mountains for arid depressions*

Located in the heart of the most massive continent on the planet, Central Asia, stretched from the Caspian Sea to Xinjiang, is arid because of its distance from oceanic moisture sources. Continentality imposes its mark on an environment where dryness shapes landscapes dominated by the immensity of steppes and deserts. Although the region is part of the great arid diagonal stretching from the Sahara to the deserts of China, it is not without water: powerful rivers find their source here, crossing it to end their course in great continental depressions: the Caspian, Aral, Balkhash, Dzungarian and Tarim basins. This paradox is due to the fortunate contrast of the relief (see Figure 9.1). Indeed, Central Asia is structured by a succession of large dry depressions – where rainfall is less than 200 mm/year – surrounded by the Pamir and Tien Shan mountain ranges, where rain is much more prevalent, since the continental characteristics of the climate change with altitude by an increase in cloud cover and rainfall.

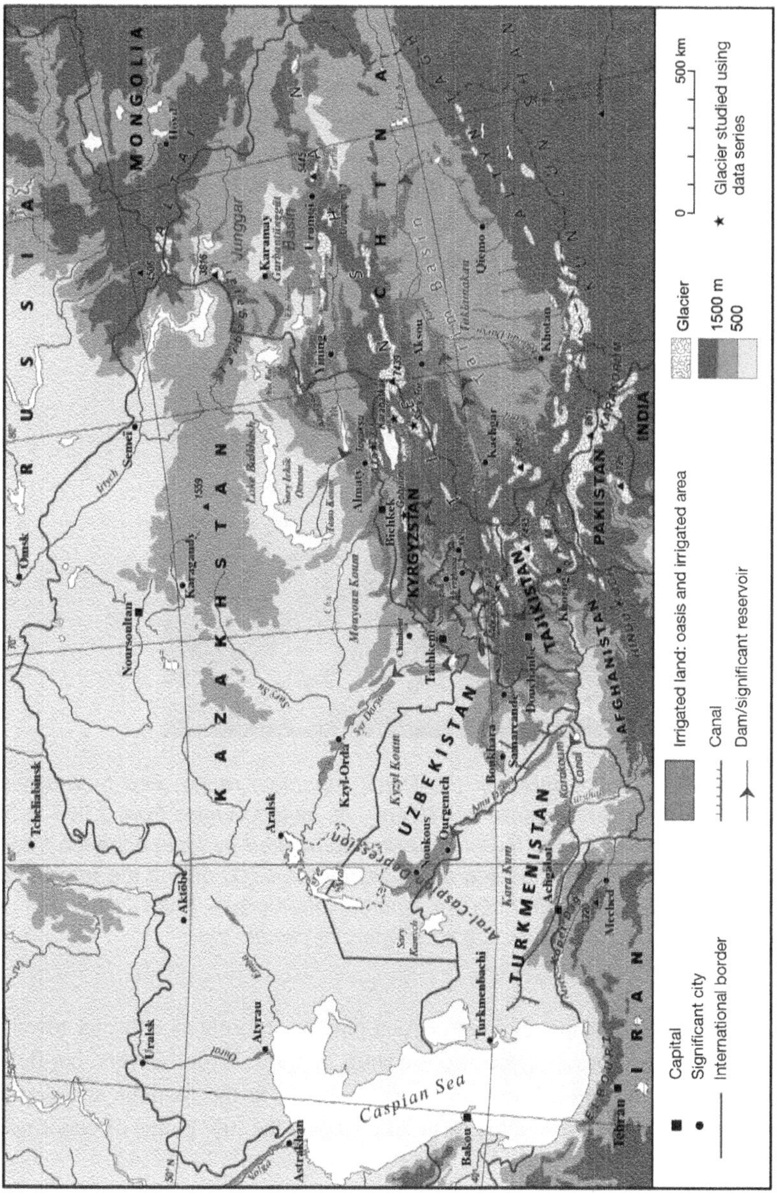

Figure 9.1. *Arid depressions fed by "water tower" mountains (source: Cariou 2019). For a color version of this figure, see www.iste.co.uk/mercier/climate.zip*

Belonging to the Alpine-Himalayan orogenic system, the high plateaus (4,000 to 5,000 m) and the massifs that sometimes exceed 7,000 meters in altitude[2] are covered with a snow cover and ice forming an imposing set of mountain glaciers. For example, the Tien Shan mountain range, which stretches from east to west over 2,500 km and is 250 to 350 km wide, had 7,590 glaciers in 2009. These covered a total area of approximately 12,949 km^2 and a volume of approximately 1,840 km^3. There are 11,834 km^2 of glaciers in the Pamir, a massif to which the Fedchenko glacier belongs, which, with a length of 72 km, is considered to be the longest glacier in the world outside the polar regions (Lambrecht *et al.* 2014; see Figure 9.2). While the glaciers in the western part of the Tien Shan and Pamir mountain ranges are "winter-accumulating", due to concentrated precipitation between December and May, the eastern part of these same massifs has "summer-accumulating" glaciers, as precipitation falls from June to September.

Figure 9.2. *The Fedchenko glacier in the Pamir mountain range. Fed by multiple tributaries, the glacier covers an area of 579 km^2. Its altitude varies from 2,900 m at the terminal tongue to 6,300 m in the high basins dominated by Independence Peak (6,940 m). Its equilibrium line is located at an altitude of 4,700 m and the maximum thickness of the glacier reaches 1,000 m (source: photograph Knith 2011). For a color version of this figure, see www.iste.co.uk/mercier/climate.zip*

Consequently, the mountains of Central Asia play a vital role as a "water tower" for the densely populated oases of the arid lowlands (Cariou 2015a). Rivers that flow down the slopes escape from the mountains to bring life to the foothills and desert plains before dying in the desert sands or in lake basins, the low points of endorheic

2 In the Pamir, the Kongur peaks at 7,649 m (Xinjiang) and the Ismail Samani peak (ex-communist peak), located in Tajikistan, at 7,495 m. The highest peak of the Tien Shan is the Jengish Chokusu at 7,439 m (Kyrgyzstan).

depressions[3]. Thus are born the long powerful rivers of the Tarim, the Ili, the Syr Darya and the Amu Darya. The latter has an average flow of 79.3 km^3 per year, which is quite comparable to the large exogeneous rivers of the Middle East such as the Tigris and the Nile, which carry 50 and 83 km^3 per year respectively through the deserts. In Central Asia, more than 70 million people depend on water from major mountain systems where the hydrological contribution of the cryosphere to river flow is fundamental. It generally represents between 70 and 80% of the average annual flow. In the Amu Darya basin, snowmelt accounts for 45% compared to 25% for glacier runoff (19 km^3). In the neighboring Syr Darya basin, the contribution of snowmelt represents 70% for 9% of water of glacial origin (3.4 km^3). This difference can be explained by an ice-covered surface of about 2% for the Amu Darya basin against 0.15% for the Syr Darya basin (Savoskul and Smakhtin 2013). In the Tarim basin, glacier runoff and snowmelt represents 41% and 39% of the total flow respectively. In spite of the regime disturbances caused by climate change, the contribution of glacier runoff to Central Asian rivers varies from 5–40% in the plains and reaches 70% in the upper basins, as shown by the hydrograph of the Sokh River, a left tributary of the Syr Darya River born in the Pamir Alai Mountains (Kemmerikh 1972; see Figure 9.3).

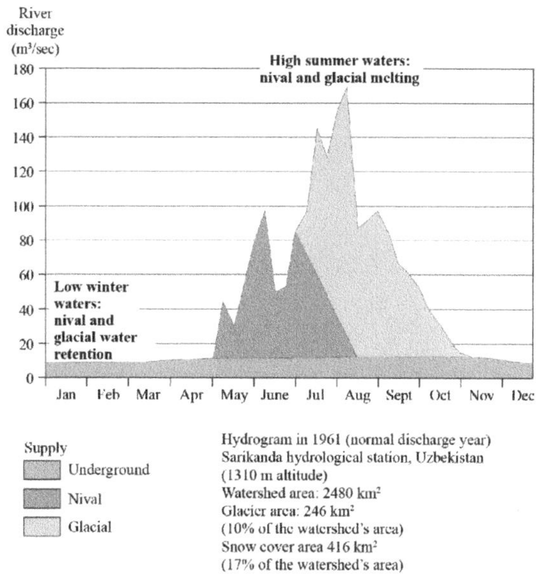

Figure 9.3. *Hydrological regime recharge patterns of the Sokh River (source: based on Kemmerikh 1972). For a color version of this figure, see www.iste.co.uk/mercier/climate.zip*

3 Endorheism: regions whose rivers do not reach the sea but end their course in continental depressions.

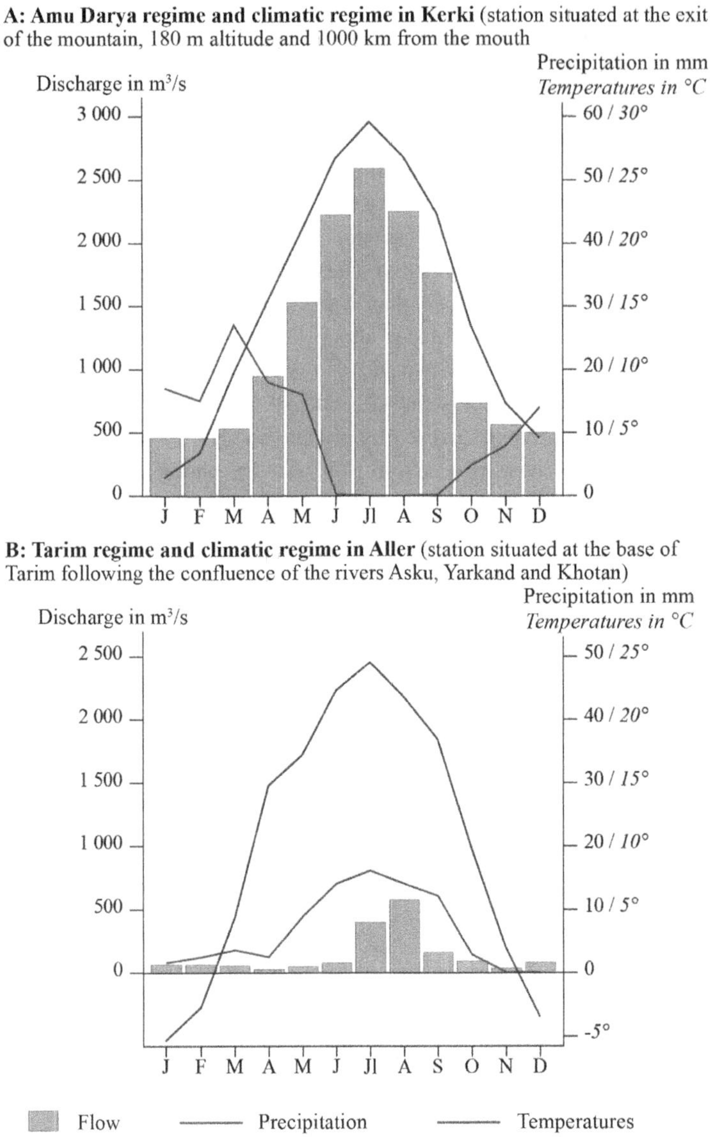

Figure 9.4. *Comparison of climatic and hydrological regimes in the Amu Darya and Tarim basin (source: Cariou 2019). For a color version of this figure, see www.iste.co.uk/mercier/climate.zip*

The hydrological regime of the rivers is generally synchronous with the summer drought on the plains, which is also the warm period of the growing season, which is crucial for the development of oasis agriculture. Conversely, low-water occurs in winter, the cold period of the agricultural off-season. In the Aral-Caspian basin, the nivo-glacial regime, with abundant water from late spring to mid-summer, is in perfect harmony with the growing season when drought is relentless (see Figure 9.4a).

Before the regulation of flows by dams, it was at the most favorable time that the high waters loaded with silt came to invigorate the crops, so that the Amu Darya delta could be described as "Central Asian Egypt". On the other hand, hydrological conditions are less favorable for agriculture in the Tarim desert basin. Due to mass melting of snow and ice in summer, combined with convective rainfall, summer high water (July to September) accounts for 75% of the total annual runoff (see Figure 9.4b). This concentration of summer flows is critical for seeding and spring vegetative startup. Between March and May, crop water requirements account for 30 to 40% of the total annual volume required for irrigation, while river flows during these three months provide only 10 to 20% of the total annual flow. Thus, contrary to the catastrophist opinion widely disseminated by the media, water is not scarce in Central Asia, unlike the Middle East and North Africa, the regions of the world with the poorest water resources. Thanks to the mountains, the per capita availability of renewable water is globally sufficient, as shown by the ranking of the various states according to the Falkenmark index, which measures relative abundance or scarcity of water by a ratio comparing the resources of a territory to its population. This indicator shows us that in 2017, all the Central Asian territories are above the shortage threshold set at 1,000 m^3/capita/year (2,740 L/day) and are for the most part well supplied with water since they are in the range of 2,000 to 6,000 m^3/capita/year. In 2017, the per capita renewable water resource is 5,955 m^3 in Kazakhstan, 4,302 m^3 in Turkmenistan, 4,257 m^3 in Kyrgyzstan and 2,456 m^3 in Tajikistan[4]. In this table, only Uzbekistan is an exception since, with 1,531 m^3 per inhabitant, the country is in a situation of water stress, the threshold of which is set at 1,700 m^3. This means that there is a chronic shortage of water with adverse effects on human health, economic development and general well-being (Falkenmark 1989). This overall situation of relative water abundance does not, however, prevent tensions over the resource, mainly due to poor governance and the choice of an inappropriate development model for the region (Cariou 2015b).

4 See the Aquastat, 2017 database, Food and Agriculture Organization of the United Nations (FAO).

9.1.3. *Tensions between riparian and rival states*

Water in Central Asia is the source of tensions related to both its sharing and its competition of uses. The main problem concerns sharing at the international level. In the transnational Aral basin, the current dispute between the upstream states (Kyrgyzstan and Tajikistan) and the downstream states (Uzbekistan, Turkmenistan and Kazakhstan) is a legacy of water management formalized in the 1970s by the Soviet federal government. In the catchment area, then politically unified by the USSR and considered as a single economic entity, priority was given to agricultural water use in the dry lowlands, which explains the large volumes allocated to Uzbekistan, Turkmenistan and Kazakhstan. Soviet Central Asia was the only region, together with the Caucasus, to offer warm lands to the USSR. Thanks to this climatic privilege and the diversion of surface water that was "unnecessarily" lost in endorheic depressions, more than 8 million hectares of desert land were transformed into irrigated areas, especially for subtropical crops such as cotton and rice. As a result, the downstream countries that withdraw the most water are paradoxically the least endowed, which places them in a situation of dependence on the upstream mountain countries where the resource is formed (see Figure 9.5).

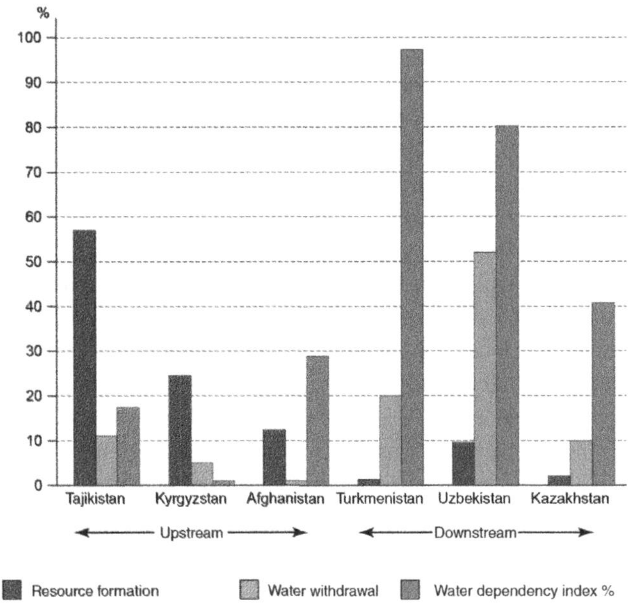

Figure 9.5. *Water resource formation and withdrawals in the Aral basin states (%) (source: based on data from Dukhovny and Sokolov 2003; Micklin 2002; Aquastat FAO 2015). For a color version of this figure, see www.iste.co.uk/mercier/climate.zip*

For this reason, Kyrgyzstan and Tajikistan have been equipped with dams to help regulate and store water for downstream irrigated areas (see Figure 9.6). In order to optimize agricultural production, dam management has favored summer releases of water in correlation with the agricultural season and storage during the rest of the year to protect against seasonal and interannual variability in runoff. This regulation system made hydroelectric production take a back seat, so that the winter power supply could not keep up with consumption demand at its peak during that season. In order to compensate for this imbalance, the Soviet government implemented a compromise based on water-energy barter. In exchange for the delivered water, upstream countries received hydrocarbons produced by Uzbekistan, Turkmenistan and Kazakhstan. It should be noted that in this water allocation designed on the scale of Soviet Central Asia, Afghanistan, which contributes 12.4% to the water supply of the Aral basin, was ignored in the sharing. It must be said that the country, long forced into political instability and civil wars, has not been able to use or demand more water.

Figure 9.6. *The Kirov Dam on the Talas River (Kyrgyzstan). Emblematic of the Soviet period, the dam was mainly built (1965–1975) to develop cotton cultivation. With a storage capacity of 0.55 km^3, it allows the irrigation of 220,000 hectares in Kyrgyzstan and 80,000 hectares in Kazakhstan (source: photograph by A. Cariou, 2014). For a color version of this figure, see www.iste.co.uk/mercier/climate.zip*

With the disintegration of the Soviet Union in 1991, this integrated system collapsed because each State redefined its priorities unilaterally without considering integrated river basin management. In order to reduce their energy dependence and to develop their hydropower potential, the authorities of Tajikistan and Kyrgyzstan multiplied dam construction projects (Rogoun, Kambarata I and II, Dashtidjum) and modified the water management of existing dams, which caused tensions with the downstream countries. This is the case in Kyrgyzstan with the Toktogul Dam

(19.5 km^3 of storage capacity), where a shift in regulation from an "irrigation regime" to a "hydroelectric regime" is gradually taking place (see Figure 9.7). The same is true in Tajikistan with the Nurek Dam where the volume used during winter to produce hydropower has increased from 2 km^3 during the Soviet period to 4.2 km^3 from the 2000s.

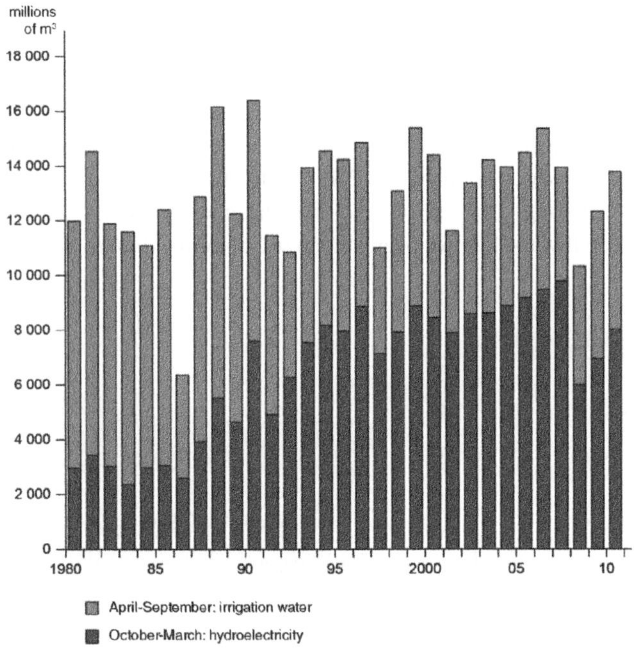

Figure 9.7. *Management of water releases from the Toktogul Dam in Kyrgyzstan (source: Global Water Partnership, 2014). For a color version of this figure, see www.iste.co.uk/mercier/climate.zip*

Indeed, electricity production requires water turbines in winter when demand is high and to store it in summer. This situation is incompatible with agriculture, which requires summer irrigation and has no use for winter water releases that sometimes flood agricultural land and cause damage to structures due to sudden ice breakup[5].

We find the same dispute between China and Kazakhstan in the case of the Irtysh and Ili river basin. The current economic development of Xinjiang is increasing the

5 The release of water on frozen canals causes ice breakup, which can damage hydraulic structures.

upstream withdrawals so that there is a decrease in water availability for Kazakhstan downstream.

While China's withdrawals from the Irtysh River until 2005 were in the range of 1.1 to 1.8 km^3/year, that is 18% of the flow, they have now reached 5 to 5.5 km^3/year since the commissioning of the transfer canals to the cities and industrial sites of Karamay and Urumqi. Such diversions affect the operation of hydropower stations and plants in the middle Irtysh valley, which constitutes the industrial axis of Kazakhstan. Similarly, the United Nations warns of the risk of the disappearance of Lake Balkhash if agricultural development projects and water transfer projects in Xinjiang continue in the Upper Ili, which supplies 73% of the lake.

This opposition between upstream and downstream countries is a classic pattern of large international basins: in Central Asia, as elsewhere, a compromise on water sharing is yet to happen because of the divergence of points of view between the position of the downstream States, which claim an "acquired right" to water according to the law of "prior appropriation", that is the first historical user is given priority over water, and upstream States, which are engaged later in resource development, claim rights to water in the name of a more equitable sharing of the resource at the watershed level. Kyrgyzstan and Tajikistan, for example, consider water as a national resource and a commodity to be monetized for downstream countries, which consider it freely "offered" for the use of all.

There is therefore no lack of causes of disagreement and tension over water, but, despite the Cassandras, no water war has broken out among Central Asian states. It is not in their interest to engage in a conflict, so riparian countries have found a *modus vivendi* through international institutions which, although they have not been able to establish closer cooperation in the field of water, have managed to avoid all the conflicts announced by some observers.

After years of diplomatic tensions, the regional situation has even eased since Uzbekistan decided in 2017 not to oppose the construction of dams in neighboring Kyrgyzstan and Tajikistan. The Uzbek government has even declared its intention to participate in the construction of the Kambarata-I dam, an old Kyrgyz governmental project that had shaped the discord between the two countries in the 2000s. The combined impact of population growth and global warming on water resources is increasingly worrying as states are being pushed towards the necessary cooperation.

9.2. The impact of climate change on water resources

9.2.1. *Recession of the cryosphere*

Since the second half of the 20th Century, there has been a general retreat of snow cover and glaciers almost everywhere on the Earth's surface as a result of climate change (IPCC 2019). The mountains of Central Asia are no exception to this global process, as depicted in Figure 9.8, which shows the mass balance of five glaciers in the Tien Shan and one in the Pamir mountain ranges (Abramov).

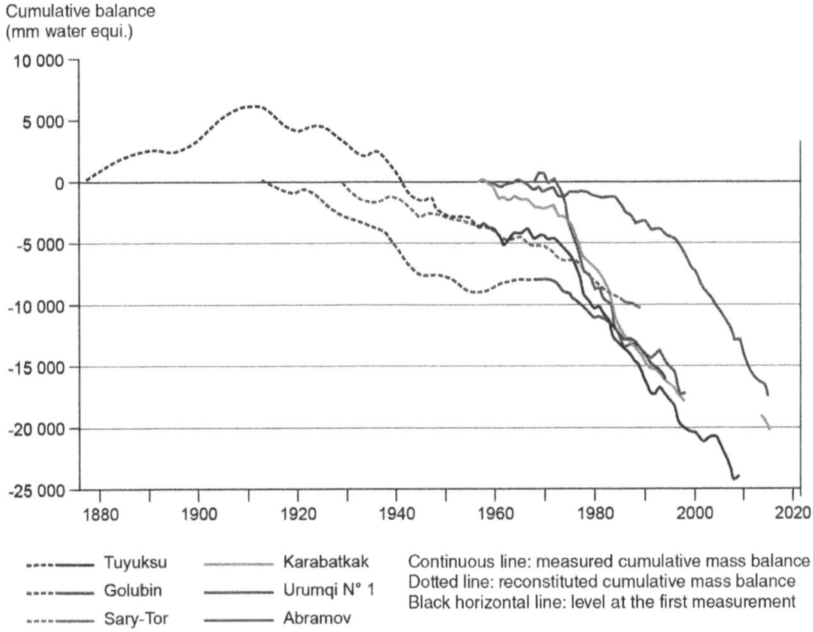

Figure 9.8. *Cumulative mean mass balances of six glaciers in the Tien Shan and Pamir mountain ranges (source: based on Sorg* et al. *2012; Hoelzle* et al. *2017). For a color version of this figure, see www.iste.co.uk/mercier/climate.zip*

Recent data are rather incomplete because the few glacier monitoring stations set up during the Soviet period were abandoned by the new Central Asian republics. However, the scientific community is mobilizing to restore measurements, including the installation of automatic stations and cameras (Hoelze *et al.* 2017). To the east of the Tien Shan range, on the Chinese side, the Urumqi glacier No. 1 is regularly monitored.

The recorded data show a general trend towards a reduction in surface area and a loss of glacier mass. For example, between 1957 and 2012, the Tuyuksu Glacier had an average annual mass balance of –13.44 mm/yr, a reduction in thickness of about 25 m (see Figure 9.9). The same is true for Urumqi Glacier No. 1, whose mean annual mass balance was –17.18 mm/yr between 1980 and 2012, equivalent to a loss of 15 m of ice thickness (Deng *et al.* 2019). However, the evolution is not linear over time because, after a certain stability or slight glacial retreat observed between the 1950s and the mid-1970s, melting accelerated from 1976–1977 (Aizen *et al.* 2006).

Figure 9.9. *The Tuyuksu glacier on the north slope of the Tien Shan (Zailiyskiy Alatau range). With a surface area of 5 km², the glacier is fed by a collector surrounded by peaks over 4,000 m high. Easily accessible, it has been the subject of regular observations since 1957. Since 1985, the glacier tongue has receded by more than 600 m (source: photograph by A. Cariou, 2017). For a color version of this figure, see www.iste.co.uk/mercier/climate.zip*

In Central Tien Shan (Kyrgyzstan), the area of glaciers in the Akshirak Massif decreased by 4% from 1943 to 1977 and by 26% from 1977 to 2003, corresponding to a loss of 10 km^3 of ice. Despite a variable glacial retreat between the more continental inner and eastern ranges and the more humid outer ranges, the overall area of the Tien Shan glaciers has shrunk from 14,152 km^2 in the 1970s to 12,949 km^2 in 2009, a reduction of 8.5%. Over the same period, the surface area of the Pamir glaciers increased from 12,449 to 11,834 km^2 (Aizen and Aizen 2013). This evolution is changing the glacial landscape, especially below 5,000 m where the recession is widespread. Many small glaciers have disappeared, especially those less than 2 km^2, while others have formed as a result of the fragmentation of larger glacial systems into several isolated branches. On the other hand, above 5,000 m, the situation is relatively stable for medium (between 2 and 10 km^2) and large (100 km^2) glaciers, with observed losses of less than 2% over the last 40 years. At the same time, there has also been a decrease in the snow cover since the second half of the 20th Century. For the Tien Shan area as a whole, average values calculated between 1940 and 1991 show that the maximum depth of the snow cover has decreased by about 0.1 m, while the average duration of the cover has been reduced by nine days as a result of the melting that occurs earlier in late spring (Sorg *et al.* 2012).

Consistent with climate change, this generalized retreat of the cryosphere is mainly linked to an increase in air temperature. Since the 1970s, mountain weather stations have clearly recorded a general trend of temperature increases of around 0.2°C per decade. Warming is particularly pronounced in spring and fall (April, September), extending the normal period of melting (June to August) of snow cover and ablation of glaciers.

On the other hand, the monitoring of annual precipitation does not show significant variation. No clear trend emerges from a review of the literature where the data are very heterogeneous: the increase in annual rainfall is observed in western Pamir and northern Tien Shan. In eastern Pamir and Tien Shan, winter precipitation has increased below 2,000 m and decreased above 2,000 m. It is currently difficult to draw any tangible conclusions on the evolution of rainfall at high altitudes due to the lack of weather stations above 3,000 m. The lack of meteorological stations above 3,000 m makes[6] it difficult to draw any tangible conclusions on the evolution of rainfall at high altitudes. Consequently, air temperature is a more important climatic factor than precipitation in explaining the negative mass balance of glaciers.

9.2.2. *The consequences of cryosphere retreat on hydrology*

The melting of mountain glaciers and the reduction in snow cover due to global warming directly affect water resources. Firstly, data from hydrological stations clearly show a constant increase in river flow over the last few decades due to the increased melting of water stored in the mountains. The increase in the average annual flow is between 5 and 20% depending on the catchment area (see Figure 9.10).

---

6 Of the eight stations installed at altitudes of more than 3,000 m by the Soviet Union in Pamir and Tien Shan, only three remained operational after the fall of the USSR.

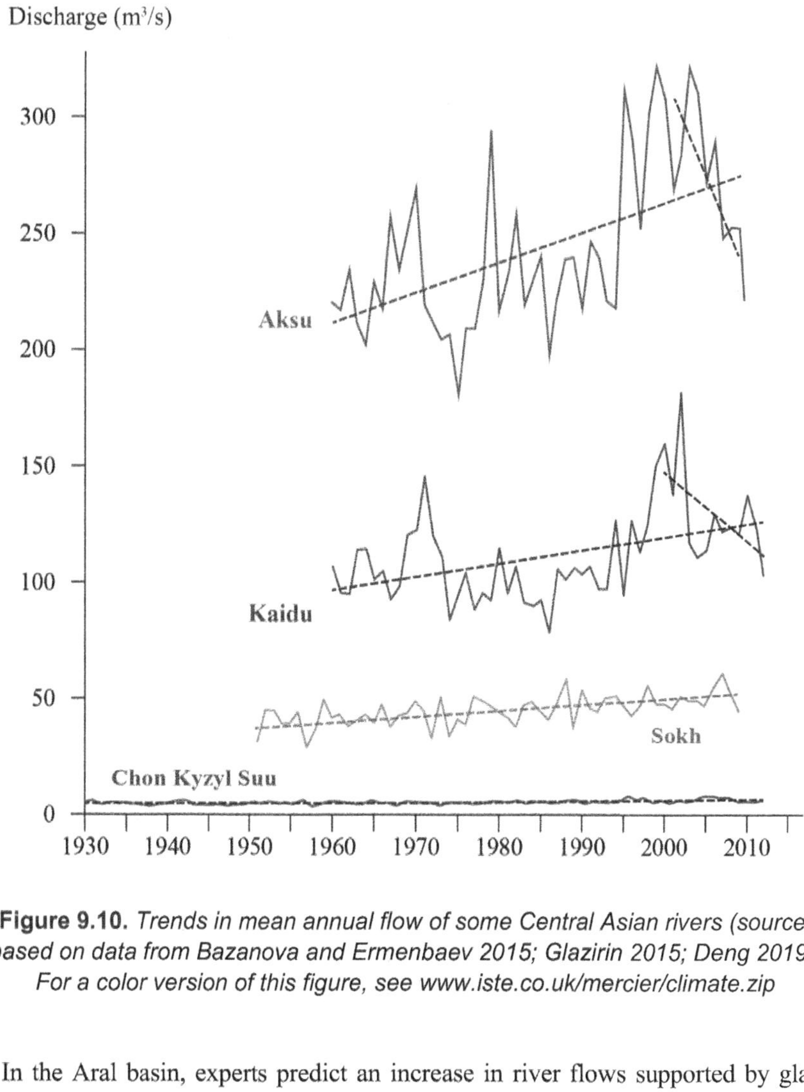

Figure 9.10. *Trends in mean annual flow of some Central Asian rivers (source: based on data from Bazanova and Ermenbaev 2015; Glazirin 2015; Deng 2019). For a color version of this figure, see www.iste.co.uk/mercier/climate.zip*

In the Aral basin, experts predict an increase in river flows supported by glacial melting until the middle of the 21st Century (Siegfried *et al.* 2012). In Xinjiang, the decrease in river flows seems to have already started, as in the Aksou, Kaidu and Urumqi catchments, where glaciated areas decreased by 29.7%, 64% and 57.7%, respectively, between 1960 and 2010 (Deng *et al.* 2019). This melting resulted in increased flows until the mid-1990s, which then decreased as a result of glacial retreat. However, the recession of the cryosphere is not expected to lead Central Asia to scarcity through a drying up of its waterways. Indeed, in the case of a climate scenario where precipitation would not change, as suggested by current data, the decrease in glacially induced flows would be compensated by the increase in runoff due to liquid precipitation linked to the rise in mountain temperatures.

In other words, the most significant impact of climate change is not quantitative but is mainly temporal with the evolution of hydrological regimes that are gradually shifting from glacial or nivo-glacial to nivo-pluvial regimes. In the future, high waters will be progressively less centered in the core summer as they gradually shift towards spring due to rainfall and early snowmelt. It is estimated that by 2050, the shift could be 30 to 60 days from the current maximum in late spring/early summer. On the other hand, a decrease in river flows in late summer and an increase in flows in autumn/early winter would be observed (Siegfried *et al.* 2012). As a result, the mountain would play less of a "water tower" role, as the increasing scarcity of solid precipitation would reduce its storage capacity. Ultimately, while the impact of climate change is not expected to disrupt the overall availability of water in the region, it will affect its temporal availability, a problem to which the population will have to adapt.

9.2.3. *Human societies facing the challenge of climate change*

The substantial change in the seasonality of river flows has far-reaching consequences for the economy and societies of Central Asia, especially for irrigated agriculture, which is still one of the pillars of the regional economy, mainly in Uzbekistan where more than a third of the working population is engaged in the irrigation of 4.3 million hectares. With decreasing flows in summer and increasing flows in spring and fall, the hydrological regime is no longer in phase with the full growing season. Reservoir dams on major rivers will not be able to store the full volume of the spring peak, so water will be "lost" in endorheic depressions. While this is positive for the restoration of wetland ecosystems degraded by decades of over-exploitation of rivers, it will result in an increased risk of spring flooding. On the other hand, it will mean reduced availability during the summer season, a peak period for water consumption mainly due to irrigated agriculture, which accounts for more than 80% of total annual withdrawals. Moreover, with the disappearance of the regulating role played by glaciers in regional hydrology, the region may face an increased risk of droughts. This is because glaciers accumulate excess annual precipitation in cool, wet years and deliver abundant meltwater in hot, dry years, resulting in above-average runoff. This regulating role is manifested by a reduction in the interannual variability of flows despite the interannual variability of precipitation, a phenomenon that is particularly marked in arid regions such as Central Asia. Consequently, glacial retreat is synonymous with greater vulnerability of populations to extreme events, particularly in terms of food security. The risk of loss of agricultural production due to insufficient irrigation water in summer and spring flooding is added to a particularly worrying situation of food insecurity in certain regions of Tajikistan and Uzbekistan, where population growth is already leading to a shortage of land and water (Ferghana, Khorezm). Finally, glacial retreat is also accompanied by new risks linked to the multiplication of moraine-dammed lakes and their catastrophic glacier lake outburst. In view of the particularly active seismicity in Central Asia, the emptying of proglacial lakes by rupture of the frontal moraine is a direct threat to populations and infrastructures[7]. Between 1952 and 2007, more than 70 cases of glacial lake overflows were recorded in Kyrgyzstan. The most catastrophic event occurred in 1998 on Lake Ikedavan, which was created by the melting of the Alaudyn glacier. The rupture of the moraine dam caused 50,000 m^3 of debris-flow to flood into the Shahimardan valley (in Tajikistan and Uzbekistan), destroying five villages and killing more than 100 people. According to the Ministry of Emergency Situations of the Kyrgyz Republic, about 287 glacial lakes are at the risk of bursting. A similar situation exists in Tajikistan where about 335 lakes in the Pamir range are considered potentially dangerous (Thurman 2011).

Faced with the challenges posed by this evolution of regional hydrology, the construction of reservoir dams seems the most appropriate response. However, construction projects are primarily concerned with hydroelectric production. Indeed, the upstream countries that have the most suitable mountain sites for such structures are mainly seeking to make up their domestic energy deficit and also aim to export their electricity surpluses to Pakistan, India or Uzbekistan. This is why Kyrgyzstan wishes to complete the hydroelectric cascade of the Naryn River (it comprises six operating stations with a total capacity of about 3,200 MW) by building the Kambarata I and Kambarata II reservoirs located upstream of the Toktogul dam. The same applies to Tajikistan, which wants to exploit the potential of the two main tributaries of the Amu Darya through the completion of the Roghun dam on the Vakhsh and the launching of the major Dashtijum project on the Pandj. Although the challenge of the new structures is mainly hydroelectric, the increase in storage capacity should nevertheless bring greater flexibility in the use of water, which could also benefit agriculture within the framework of a true integrated management of resources at the scale of the catchment areas. However, climate change is likely to reduce the efficiency and lifetime of dams. Indeed, in these mountainous areas, the melting of permafrost and the increase in stormy rains to the detriment of solid precipitation logically risks increasing erosion and slope instability, causing landslides and mudflows. These natural hazards are already very frequent in inland areas where the catchment areas are characterized by young, often bare soil and therefore very

7 See Chapter 2 and particularly Figure 2.14 for these paraglacial risks of moraine dam failure.

vulnerable to erosion (see Figure 9.11). As a result, all reservoirs will face a more rapid reduction in their useful volume due to the effect of alluvial deposits.

While dams are essential for balancing the irregularity of water resources, they alone cannot solve all the problems. An alternative would be to develop the use of groundwater, which has so far been much less exploited compared to the relative abundance of surface water. In the desert depressions where most irrigated agriculture is concentrated, groundwater reserves are far from negligible. In the foothills area, the pediments formed by thick detritic sediments provide good quality groundwater, while the exploitation of the large alluvial aquifers is more delicate because of their greater or lesser salinity.

Figure 9.11. *The Bartang River, a tributary of the upper Amu Darya River (Tajikistan). The incision of the rivers and the aridity of the Pamir valleys produce steep, bare slopes that are very sensitive to erosion and landslides at the slightest rainfall (source: photograph by A. Cariou, 2016). For a color version of this figure, see www.iste.co.uk/mercier/climate.zip*

The future of the region depends on a necessary water-saving policy, and there is considerable room for maneuver in this area. Central Asian countries have some of the world's most inefficient irrigation systems. Most irrigation and drainage systems are in a state of disrepair as they were mainly built during the Soviet era, especially between 1950 and 1980. About 80% of them are dug into the ground, with high losses through infiltration and evaporation, resulting in irrigation efficiency of around 50%[8]. In

8 This means that 50% of the water withdrawn from streams does not reach the field.

Xinjiang, as in the Central Asian Republics, irrigation is based almost exclusively on surface gravity systems or flood systems, which are very expensive in water. Sprinkler or drip techniques, which are much more water-efficient (75–90% efficiency), are being introduced only on small experimental areas. In addition, increased use of agricultural drainage wastewater is one of the simplest and most affordable solutions for increasing water availability. In the Amu Darya basin alone, 14 km^3 of water are discharged annually into lakes (7.39 km^3) or directly into the river (4.94 km^3). At least 2 km^3 can be easily recycled (SIC ICWC 2018).

However, the issue of water saving cannot be reduced to a technical question. Changes are also needed in irrigation management so that farmers have a much better level of control and use of water. This implies a change in farming practices and mentalities often characterized by a lack of awareness of water saving.

Despite the introduction of water pricing in Uzbekistan and Kazakhstan (but not in Kyrgyzstan, Turkmenistan and Xinjiang), it is difficult to change behavior that is the result of agricultural policies that have for decades distributed water without counting.

Finally, the latter must evolve by promoting agricultural production with a low "water footprint" and high added value. In Uzbekistan, cotton cultivation, a major water consumer, still occupies an important place despite the reduction in surface area since the country's independence in favor of cereal crops. The development of more intensive market gardening and food crops would make it possible to make better economic use of water while guaranteeing jobs and basic food products for the population. Compatible with the most efficient irrigation methods (drip irrigation), this precision agriculture also has the advantage of being able to respond to the challenge of the temporal evolution of water resources thanks to short cultivation systems that allow a shift in sowing dates.

Ultimately, making water management and use more efficient could offset some of the negative effects of climate change.

9.3. Conclusion

Although it is currently difficult to quantify the effects of climate change on the water resources in Central Asia, a reduction in absolute annual availability seems unlikely in the coming decades. However, it would be prudent for decision-makers and populations to prepare for reduced seasonality of river flows and increased risk of extreme events such as droughts and floods. The hydrological upheavals associated with the cryosphere recession should provide an opportunity to raise awareness of the urgency of implementing new water governance. For, more than climate change and physical water scarcity, the possible shortage is above all due to strong population growth fueled for decades by an unsuitable development model. In Central Asia, the myth of modernity and progress has been based on the transformation of millions of hectares of desert land into agricultural land on the basis of technical solutions that consider water resources to be unlimited. In the end, this Promethean program is evidence of a form of underdevelopment in view of the populations' over-dependence on irrigated agriculture and the largely wasted cryosphere water. This is why the quest for water security presupposes the implementation of new development programs based on a redeployment of the economy towards the tertiary and secondary sectors that are less water-intensive and have higher added value. The future of the region depends on the ability of its leaders to develop efficient water use strategies based on the need for integrated resource management at the watershed level.

9.4. References

Aizen, V. and Aizen, E. (2013). Modern and past climate and environmental change impact on cryosphere/water resources in Central Asia. *Glacial Flooding and Disaster Risk Management, Knowledge Exchange and Field Training in Huaraz*, July, 11–24.

Aizen, V., Kuzmichenok, V., Surazakov, A., Aizen, E. (2006). Glacier changes in the central and northern Tien Shan during the last 140 years based on surface and remote-sensing. *Annals of Glaciology*, 43, 202–213.

Cariou, A. (2015a). *L'Asie centrale. Territoires, sociétés et environnement*. Armand Colin, Paris.

Cariou, A. (2015b). L'eau et l'aménagement du territoire en Asie centrale. Une ressource fondamentale pour un développement à repenser. *Cahiers d'Asie Centrale*, 25, 19–58.

Deng, H., Chen, Y., Li, Y. (2019). Glacier and snow variations and their impacts on regional water resources in mountains. *Journal of Geographical Sciences*, 1, 84–100.

Falkenmark, M. (1989). The massive water scarcity now threatening Africa: Why isn't it being addressed? *Ambio*, 2, 112–118.

Francou, B. and Vincent, C. (2007). *Les Glaciers à l'épreuve du climat*. IRD, Montpellier.

Hoelze, M., Azisov, E., Barandun, M., Huss, M., Farinotti, D., Gafurov, A., Hagg, W., Kenzhebaev, R., Kronenberg, M., Machguth, H., Merkushkin, A., Moldobekov, B., Petrov, M., Saks, T., Salzmann, N., Schöne, T., Tarasov, Y., Usubaliev, R., Vorogushyn, S., Yakovlev, A., Zemp, M. (2017). Re-establishing glacier monitoring in Kyrgyzstan and Uzbekistan, Central Asia. *Geoscientific Instrumentation Methods and Data Systems*, 6, 397–418.

IPCC (2019). Special report on the ocean and cryosphere in a changing climate [Online]. Available at: https://www.ipcc.ch/srocc/.

Kaser, G., Grosshauser, M., Marzeion, B. (2010). Contribution potential of glaciers to water availability in different climate regimes. *Proceedings of the National Academy of Sciences of the USA*, 47, 20223–20227.

Kemmerikh, A. (1972). The contribution of glaciers to the river runoff of Central Asian rivers [Rol' lednikov v stokerekSredneyAzii]. *Data of Glaciological Studies*, 20, 82–94.

Lambrecht, A., Mayer, C., Aizen, V., Floricioiu, D., Surazakov, A. (2014). The evolution of Fedchenko glacier in the Pamir, Tajikistan, during the past eight decades. *Journal of Glaciology*, 220, 233–244.

Savoskul, O. and Smakhtin, V. (2013). Glacier systems and seasonal snow cover in six major asian river basins: Hydrological role under changing climate. Study report 150, International Water Management Institute, Colombo.

SIC ICWC, Scientific Information Center of the Interstate Coordination Water Commission of Central Asia (2018). Future of the Amu Darya Basin in the context of adaptation to climate change. In *16th International Conference "Europe-INBO 2018"*, Seville.

Siegfried, T., Bernauer, T., Guiennet, R., Sellars, S., Robertson, A., Mankin, J., Bauer-Gottwein, P., Yakovlev, A., (2012). Will climate change exacerbate water stress in Central Asia? *Climatic Change*, 3, 881–899.

Sorg, A., Bolch, T., Stoffel, M., Solomina, O., Beniston, M. (2012). Climate change impacts on glaciers and run in Tien Shan (Central Asia). *Nature Climate Change*, 2, 725–731.

Thurman, M. (2011). Natural disaster risks in Central Asia: A synthesis. Study, UNDP/BCPR, Regional Disaster Risk Reduction Advisor, Europe and CIS.

10

Spatial Impact of Climate Change on Winter Droughts in the Mediterranean and Consequences on Agriculture

Florian RAYMOND and Albin ULLMANN
University of Burgundy, Dijon, France

10.1. Climate variability and change in the Mediterranean basin

The Mediterranean basin is a singular and very complex geographical area. Its latitudinal position places it as a transition zone between the tropical climate and that of the middle latitudes. This area is also original because of the presence of the Mediterranean Sea, literally *the sea in the middle of the land*, small, rather warm, and semi-enclosed. Moreover, the geographical complexity of the basin, in terms of topography (with numerous mountainous massifs along the sea), coastal orientation and land use, leads to great climatic variability. The vulnerability of the Mediterranean territories to this climate variability is especially observed for regions where the lowest levels of economic development are combined with already difficult climatic conditions, especially in terms of water availability (surface and groundwater, drinking water and for irrigation). The Mediterranean basin is also a climate change hotspot (IPCC 2013). It is indeed a region of the globe where this change (taken in the strict sense of the term, i.e. when climate variability definitively goes beyond the range of natural climate variations) has already been proven since the second half of the 20th Century (Hawkins and Sutton 2012; IPCC 2013). It is also one of the regions of the world that will be the most exposed to future climate

Spatial Impacts of Climate Change,
coordinated by Denis MERCIER.

change, particularly in terms of reduced precipitation (Giorgi 2006; Polade *et al.* 2017; Lionello and Scarascia 2018).

The Mediterranean basin has no official boundaries. It can be seen as the hinge area between Southern Europe, Asia and North Africa. From a climatic point of view, it can be defined, in a somewhat reductive way, as the area under the influence of the "Mediterranean climate" of the Köppen classification (a classification of world climates dating from the beginning of the 20th Century, based on temperature and rainfall). Most of the Mediterranean basin is under the influence of this "Mediterranean climate", which is characterized by hot, dry summers and mild, wet winters. By including the Mediterranean Sea in this definition, the Mediterranean basin can be considered as all the sectors in the vicinity of and under the influence of the Mediterranean Sea and the Mediterranean climate. However, the "mildness" of winter is only very relative and actually hides particularly intense atmospheric manifestations. The Mediterranean climate is fantasized above all for its summer, but the crucial period for societies and ecosystems is the period from autumn to spring. This long, so-called "winter" period is decisive for the recharge of water resources, which is absolutely essential between two summer droughts. All these characteristics make the Mediterranean basin an area regularly subjected to so-called extreme climatic events. One of the most threatening extreme events for ecosystems and societies, apart from the well-known intense rainfall, is the drought in the so-called "rainy" season (from autumn to spring), characterized in particular by the excessively prolonged absence of rain when it should normally rain. Although drought can be approached in different ways, such as hydrological drought (lower groundwater, river and lake levels) or agricultural drought (low soil moisture slows plant growth, reduces yields and endangers livestock), the main cause of all types of drought is meteorological drought with a prolonged rainfall deficit (see Figure 10.1). Moreover, it is the latter that is the main source of concern in the context of climate change.

In the Mediterranean, long dry spells in rainy seasons are the main cause of meteorological droughts in this climate which is in such need of water. They even constitute the most important climatic hazard on the scale of the Mediterranean basin because they are directly linked to water resources, whether for agrosystems, energy, tourism, but also and simply for drinking water, especially due to the demographic growth in the southern and eastern countries of the basin. For example, cities such as Algiers and Tel Aviv have multiplied their population by five between 1950 and 2010 (PNUE/PAM 2016).

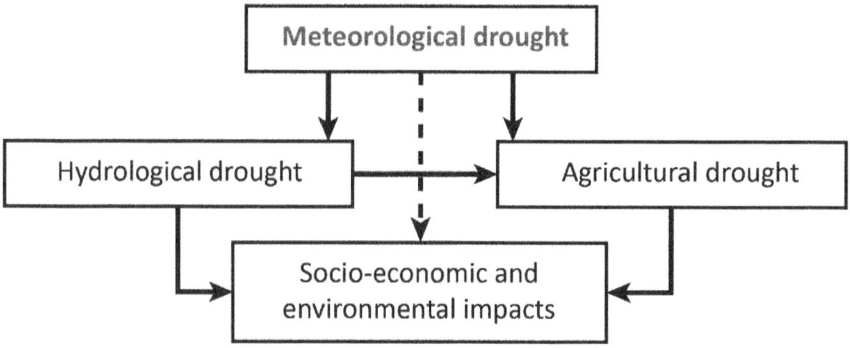

Figure 10.1. *Schematic representation of the different types of drought and their interactions, all of which result in socio-economic and environmental impacts (source: design and production F. Raymond). For a color version of this figure, see www.iste.co.uk/mercier/climate.zip*

In this geographical, climatic and societal context, questioning spatiotemporal scales and making use of spatialization makes sense in an attempt to understand where, when and how long dry episodes in rainy seasons have varied and will vary in the Mediterranean basin. In other words, it is the spatialization of climate change at the service of territorialization that is at issue here.

10.2. Droughts during rainy seasons

10.2.1. *Rainfall drought: the absence of rain in time and space*

There are various methods commonly used to understand meteorological drought: (1) studying deficits in cumulative precipitation (liquid: rain, and solid: snow and hail) on an annual, seasonal or monthly scale (Hoerling *et al.* 2012; Hertig *et al.* 2013); (2) using synthetic monthly or seasonal indices based on precipitation alone (the Standardized Precipitation Index: Zakhem and Kattaa 2016; Hertig and Tramblay 2017) or on precipitation and temperature (the Standardized Precipitation-Evapotranspiration Index: Vicente-Serrano *et al.* 2011; Drumond *et al.* 2017; the Palmer Drought Severity Index: Sousa *et al.* 2011; Dubrovsky *et al.* 2014); and (3) study the evolution of the number of dry days per year (Croitoru *et al.* 2012; Polade *et al.* 2014) and the maximum number of consecutive dry days (Kostopoulou and Jones 2005).

In addition to these various studies, the objective here is to define and conceptualize rainfall droughts in the Mediterranean as "climatic and geographical objects" in its own right, that is, very long sequences without recurrent and redundant rainfall on the scale of the Mediterranean basin and characterized by location, duration and spatial extension. This then makes it possible to study the geographical properties of rainfall droughts. This approach is of particular interest for this field of study, as it directly concerns the temporal distribution of rainfall potentially independent of cumulations, an important parameter to be taken into account regarding the impacts on societies and the environment.

Indeed, with equal accumulation over a given period, an accumulation of rain-free days over long sequences should have greater impacts than rain-free days more dispersed in time. For example, consider two so-called "dry" situations of 40 days, receiving a total of 24 mm of rainfall (see Figure 10.2). Studying the temporal distribution of rainfall makes it possible to differentiate between these two situations: situation number 1 receives 24 mm of rain over three days dispersed over the 40 days, considerably reducing the effects of the agricultural drought. Situation number 2, receiving 24 mm over the last three days, is characterized above all by an absence of rain for 37 consecutive days, marking an atmospheric and therefore agricultural drought that is much more pronounced than in situation number 1. However, if drought is approached via deficits in total rainfall, synthetic indices or the number of dry days, these two periods would show droughts with comparable characteristics.

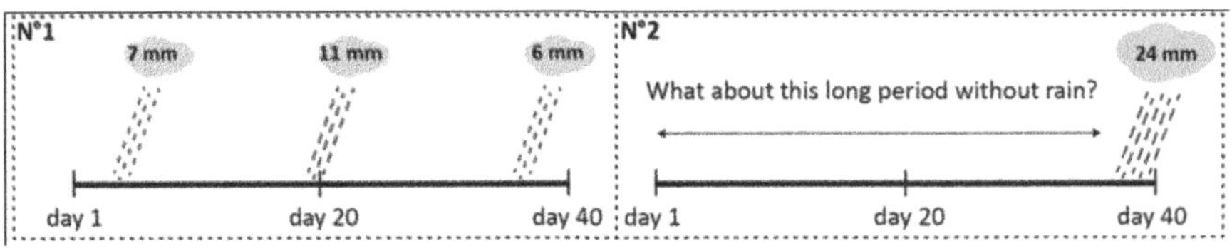

Figure 10.2. *Two so-called "dry" 40-day situations (N°1 and N°2), identical in terms of total rainfall received (24 mm) and number of dry days (37), but showing a significant difference in rainfall distribution (source: design and production F. Raymond). For a color version of this figure, see www.iste.co.uk/mercier/climate.zip*

10.2.2. *Detection of very long dry events in the Mediterranean Sea*

The methodology set up to detect long dry sequences during rainy seasons at the Mediterranean basin scale is precisely detailed in the article by Raymond *et al.* (2016). This method is based on daily rainfall totals from the E-OBS database (Haylock *et al.* 2008), which includes rainfall data from more than 2,300 meteorological stations across Europe, West Asia and North Africa. However, the lack of robust observation series over different Mediterranean sectors does not allow the E-OBS database to cover satisfactorily North Africa and the eastern Mediterranean basin (Brunet *et al.* 2014). As a result, the entire Mediterranean basin is unfortunately not considered in this study of very long dry spells (VLDS) in the rainy season. The detection of VLDS goes through three stages. The first stage involves detecting all the sequences without precipitation, from the shortest to the longest. The second step aims to select the longest non-precipitation sequences, which are the rarest in terms of frequency of occurrence, but which concern many days when they occur, and which have the greatest impact. The third step consists of selecting only events with a certain spatial extension in the Mediterranean basin. These different stages then make it possible to define the spatiotemporal properties (location, spatial extension and duration) of the VLDS affecting the Mediterranean basin.

10.2.3. *Spatial and temporal characteristics of the main event patterns of very long dry spells*

A total of 76 VLDS have been detected in the Mediterranean basin since the 1950s. The VLDS are climatic events that are not necessarily fixed in space. During their respective duration, their spatial extension may evolve, or they may move slightly. In addition, different yet remote regions can be affected by a VLDS synchronously. Indeed, as shown in the paper by Raymond *et al.* (2018a), these climatic events are mainly attributable to anticyclonic blockages on a regional scale capable of generating long periods without rain at the same time in several quite distinct sub-areas within the Mediterranean basin. However, typical patterns of such events affect the same territories in a recurrent and redundant manner.

Since the 1950s, there have been essentially three recurrent configurations of VLDS at the scale of the entire Mediterranean basin. The first sector regularly affected by VLDS mainly concerns the Balkans, with 11 events grouped in this configuration (see Figure 10.3a). In particular, Serbia, Montenegro, Northern Macedonia and Romania are the most exposed to these 11 VLDS in this region (almost 100% of the 11 VLDS affected these countries). The 11 VLDS in this "Balkan" configuration cover an average of more than 566,000 km^2, which is more than the surface area of Romania, Bulgaria, Serbia and Hungary combined. These VLDS persist for an average of 66 consecutive days, with the shortest lasting more than a month (38 consecutive days without rain) and the longest lasting more than four months without rain (130 consecutive days) in these sectors.

The second recurrent configuration of VLDS mainly affects the west of the Iberian Peninsula, with no less than 46 episodes affecting all or part of the Mediterranean basin. The 15 events that belong to this configuration mainly concern the Iberian Peninsula (see Figure 10.3b). It is especially the center of the peninsula that is exposed to the VLDS, especially the Spanish provinces of Toledo and Ciudad Real. These VLDS cover an average of more than 474,000 km^2, which is slightly less than the total surface area of Spain. When they occur, they persist for an average of 56 consecutive days, although they range from 14 days for the shortest to more than 80 days for the longest episode. Beyond this recurrent pattern, the Iberian Peninsula is the region of the Mediterranean basin most often affected by VLDS, with no less than 46 episodes affecting all or part of the territory since the 1950s.

The third recurrent configuration of VLDS at the Mediterranean basin scale mainly affects the Levant, with 25 VLDS since 1950 (see Figure 10.3a). It is mainly the southern Levant that is affected, with almost all of the 25 events involving Israel, Jordan and northwest Saudi Arabia. They cover an average of more than 385,000 km^2, which is more than the total area of Syria, Jordan, Lebanon, Israel and the West Bank combined. They persist for about 69 consecutive days, although they range from 40 days to nearly 110 days.

Of all the VLDS detected, 25 are not, however, associated with one of these three major recurrent configurations, being located in a more random manner and, above all, being less recurrent in the Mediterranean basin. The map in Figure 10.3b shows the spatial distribution of all the VLDS detected since the 1950s in the Mediterranean basin. In addition to highlighting the Mediterranean regions most regularly exposed to VLDS (with the Iberian Peninsula at the top of the list), it also highlights the territories most exposed to these extreme events on a local scale.) The territory most regularly subjected to VLDS is located south of the Levant (in brown in Figure 10.3b), with 25 events recorded in the same geographical area (the great majority of which are grouped together in the recurrent configuration known as the "Levant"). The Levant is subject to an arid climate and rainfall is clustered between the months of October and April only. Thus, the events detected in September and October in this sector mainly reflect a late start to the rainy season. They are therefore dry events that can be described as "seasonal", typical of a dry summer season that is more pronounced than usual (see Raymond *et al.* 2016). The second territory most regularly subjected to drought is the western Mediterranean basin, with the southern Iberian Peninsula and part of the Maghreb. This area has been affected by about 20 VLDS since the 1950s, with a large part of these events grouped together in the configuration known as the "Iberian Peninsula". Finally, the last territory particularly subject to VLDS since the 1950s is the southeastern Balkans and northwestern Turkey, with nearly 15 VLDS recorded (northeastern Greece, eastern Bulgaria and northwestern Turkey), the majority of which belong to the recurrent "Balkan" type configuration. Thus, the Mediterranean basin is regularly exposed to non-rainy episodes that can be described as "extremes" in terms of duration and spatial extension during the normally rainy season.

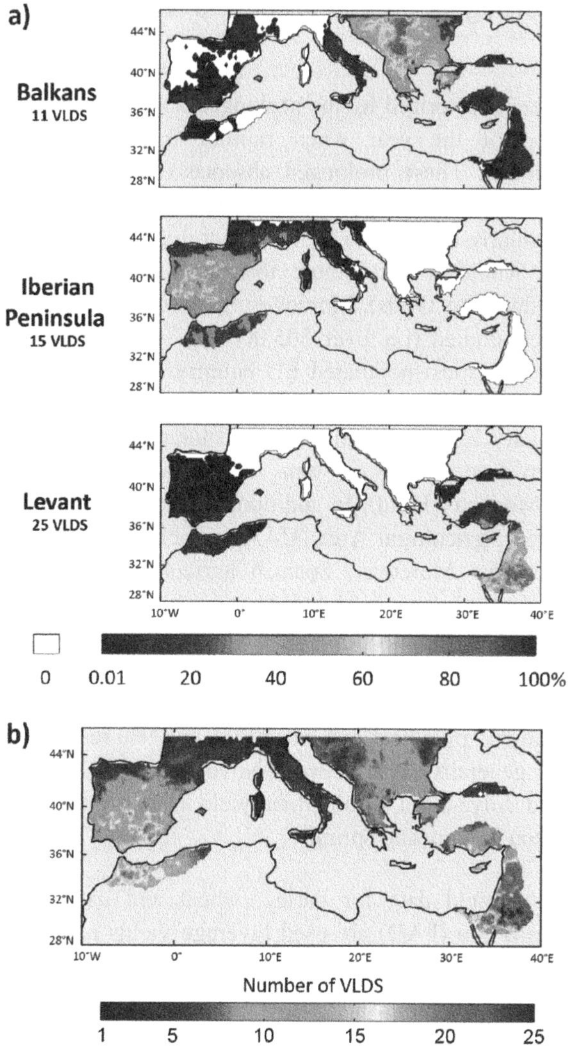

Figure 10.3. *(a) The three recurrent configurations of VLDS events observed in the Mediterranean basin since the 1950s: Balkans, Iberian Peninsula and the Levant. A percentage of 100% indicates that all the VLDS grouped in this configuration affect the sector in question. (b) Spatialization of the number of very long dry spell events observed over the whole of the Mediterranean territory studied since the 1950s (source: design and production F. Raymond). For a color version of this figure, see www.iste.co.uk/mercier/climate.zip*

10.3. Rainfall droughts in the Mediterranean: impacts on Spanish agrosystems

The VLDS are characterized by the prolonged absence of rain over a large area during the period of the year when rainfall is normally expected for the Mediterranean climate. These prolonged absences of rainfall generate numerous impacts on eco-agrosystems. Spain is chosen here to illustrate these impacts. Indeed, this country is regularly exposed to these meteorological hazards which frequently affect almost the whole country (more than 40 VLDS have been recorded on Spanish territory since the 1950s). Moreover, Spain alone covers nearly 13% of the Mediterranean area studied (i.e. over 505,000 km^2 out of the 3,900,000 studied). Spain is also the fifth most populated EU country (behind Germany, France, the United Kingdom and Italy, according to Eurostat), with about 46.9 million inhabitants on January 1, 2019. Spain was the 13th largest economy in the world and the 4th in the Euro zone in 2018 (in terms of gross domestic product, according to the International Monetary Fund). In addition, Spain has approximately 25 million hectares of Utilized Agricultural Area (UAA), which represents almost half of the country's surface area. Moreover, Spanish agriculture is largely dominated by non-irrigated crops, also known as "rainfed" crops, which are particularly dependent on the quality of the rainy season. In terms of production volume, cereals are the main crops in Spain, with a quarter of the UAA in 2019, and in particular barley (43% of the cereal agricultural area in 2018 according to the Spanish Ministry of Agriculture[1]), followed by wheat and oats (35% and 9% respectively). These three are winter cereals generally sown between November and December and harvested between June and July. As a result, their yield is dependent on rainfall and its distribution between autumn and spring.

Average annual yield data for barley, wheat and oats from the Food and Agriculture Organization (FAO) are used (average yields nationwide). Figure 10.4 shows the average yield anomalies (in percent) compared to average yields, firstly, for seasons not affected by one or more VLDS (without VLDS), and secondly, for seasons affected by at least one VLDS. For these three cereals, the yields associated with the seasons without VLDS are on average 6.5% to 8.5% higher than the average reference yields. The absence of extreme dry spells is therefore logically conducive to better yields. For seasons affected by at least one VLDS, yields are slightly lower. The impact on yields is particularly dependent on the duration and/or the number of VLDS that have occurred during the season. In fact, when more than 83 of the days between autumn and spring are affected by an VLDS, the yields show a sharp decrease in the average reference yields, in excess of 20% for barley, 19%

1 www.mapa.gob.es.

for wheat and 18% for oats. Finally, if we consider that the seasons affected by at least one VLDS have lower yields on average than the seasons without VLDS, it is the seasons with the longest and/or most frequent VLDS that experience the most severe reductions in winter cereal yields. Thus, when a seasonal rainfall deficit is due to one or more VLDS, the impact on agriculture will be much greater than if it is due to dry days scattered throughout the season (Raymond *et al.* 2018b).

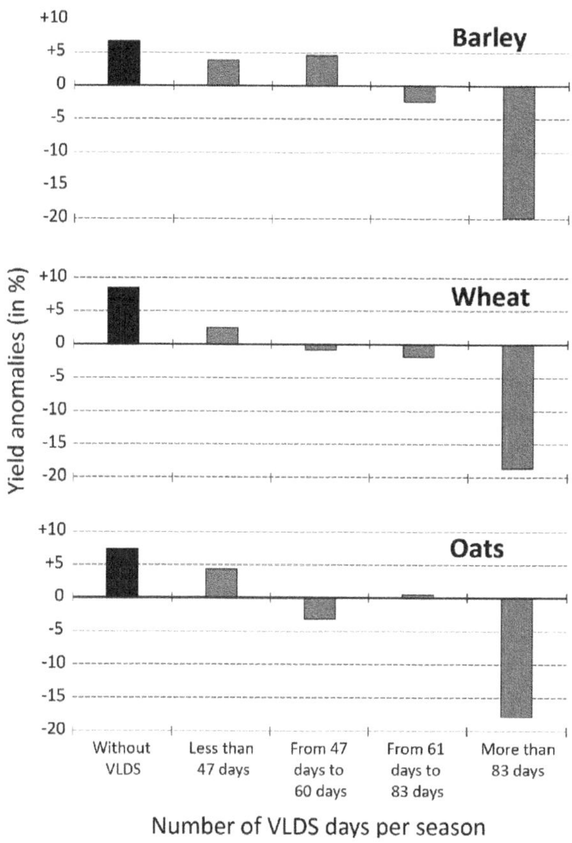

Figure 10.4. *Average yield anomalies in Spain for barley, wheat and oats (as a % of average yields calculated over the period 1961–2013), according to different classes relating to the number of days of VLDS per season*

COMMENT ON FIGURE 10.4.– *The first class (black histograms) groups the seasons that are not affected by any VLDS (without VLDS). The other four classes (gray histograms) concern seasons affected by at least 1 VLDS, and are based on quartiles calculated on the number of days of VLDS for all affected seasons (a day of VLDS is a day affected by a VLDS): (1) less than 47 days of VLDS per season; (2) between 47 and 60 days of VLDS; (3) between 61 and 83 days of VLDS; (4) more than 83 days of VLDS per season. (source: conception and realization F. Raymond).*

As an example, a significant meteorological drought hit Spain between December and February during the 2004–2005 rainy season (Garcia-Herrera *et al.* 2007). This historical drought is characterized more precisely by two events that were close in time with a total of 108 days without rain between autumn and spring. It is therefore mainly the succession of these two VLDS that had an impact on barley, wheat and oat yields, with production falling by 58%, 53% and 43%, respectively, compared to the usual average yields. Indeed, one of the peculiarities of the VLDS is that these episodes can have particularly strong impacts on agriculture throughout the rainy season in the Mediterranean climate. For example, during the autumn, the VLDS dry out the soils that are already particularly dry at the end of the summer, which is unfavorable to a good germination of the grains already planted. Once the plots have been sown and germination has taken place, a VLDS will temporarily slow down or even stop the germination of the seedlings, which will then find themselves in a situation of water stress. Later in the season, as spring approaches, VLDS will strongly affect grain formation and refolding, making it one of the main risks for winter grain yields.

10.4. Rainfall droughts in the Mediterranean: projections for the future

Will these people, with their strong impacts on societies and the environment, be more present in the future, in the current context of climate change? It is to this question that this section will attempt to propose an answer. For this purpose, climate models are used to try to make projections on the future variability of greenhouse gases. Climate models are numerical representations of the Earth's climate system. They are themselves made up of several models representing each of the different components of the climate system (atmosphere, ocean, sea ice, vegetation, rivers, etc.). These numerical tools, based on physical laws, model the functioning of each of the different environments of the climate system, their interactions, and the resulting exchanges of energy and water.

To study the future evolution of the Earth's climate, two parameters must be taken into account:

– the response of the Earth system to changes in natural radiative forcing (radiative forcing is the difference between the energy entering and leaving the atmosphere);

– future socio-economic developments that influence the chemical composition of the atmosphere through greenhouse gas (GHG) and aerosol emissions.

For this reason, the Intergovernmental Panel on Climate Change (IPCC) has called on the scientific community to develop scenarios of the evolution of anthropogenic radiative forcing of GHG emissions. These scenarios make it possible to simulate different climate evolutions, based on assumptions about future economic development. Thus, four radiative forcing trajectories, known as the "Representative Concentration Pathways" (RCPs; Moss *et al.* 2010) are retained in the fifth IPCC report (IPCC 2013). Among these four trajectories, two are used here to study the future variability of VLDS in the Mediterranean basin: 1) the RCP4.5 trajectory (global radiative forcing of +4.5 W/m^2 by 2100 compared to the preindustrial-era level) and 2) the RCP8.5 trajectory (+8.5 W/m^2). These two trajectories predict an average increase in the mean temperature at the Earth's surface of up to +2.6°C (RCP4.5) and up to +4.8°C (RCP8.5) by 2081–2100 by comparison with the historical period 1986–2005. These two trajectories are the most commonly found in the scientific literature. The RCP4.5 trajectory is described as optimistic, and represents a scenario of warming temperatures if GHG emissions increase until 2040 before stabilizing in the 2080s (Sanford *et al.* 2014). In contrast, RCP8.5 is a pessimistic trajectory, representing the warming temperature scenario if GHG emissions continue to grow steadily until 2100. At present, it is not possible to influence the projected temperature increase up to the 2050s (due to the high inertia of the climate system). On the other hand, it must be realized that all efforts made now will influence the evolution of temperatures in the second half of the 21st Century.

A set of eight regional climate models is used to study the future variability of VLDS in the Mediterranean basin for the two trajectories RCP4.5 and RCP8.5 (see Figure 10.5). Unlike global climate models, which cover the whole planet with coarse resolution, regional climate models focus on a given region and use finer resolutions, which enables them to better take into account the geographical characteristics of the region concerned (relief, coastlines, land use). Thus, regional models have the advantage of simulating the climate in a given region in a more realistic way compared to global models.

The Mediterranean basin is expected to see an increase in the number of VLDS days per season by the end of the 21st Century (up to 20 days for the RCP4.5 trajectory, up to 50 days for RCP8.5). This increase in the number of VLDS days per season translates into an increase in the number of events, but also in their duration and spatial extent (Raymond *et al.* 2019). However, it can be noted that for the

RCP4.5 trajectory, the models do not enable the proposing of evolutions qualified as robust, because the models do not show statistically significant evolutions.

Conversely, in the case of the RCP8.5 trajectory, the models show robust results as they are statistically significant over several regions of the Mediterranean basin (dashed black in Figure 10.5).

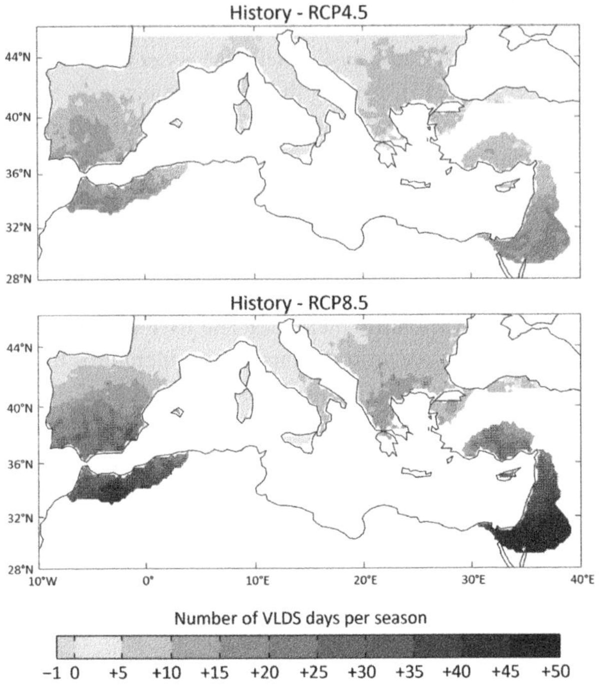

Figure 10.5. *Evolution of the average number of days of VLDS per rainy season for future trajectories (period 2066–2100) compared to the historical reference period (1971–2005). For a color version of this figure, see www.iste.co.uk/mercier/climate.zip*

COMMENT ON FIGURE 10.5.– *This evolution is shown for the RCP4.5 trajectory (historical – RCP4.5; top panel) and for the RCP8.5 trajectory (historical – RCP8.5; bottom panel). The future trajectories are simulated using 8 regional climate models (ALADIN52, CMCC_CCLM4, GUF_CCLM4, LMDZ4_NEMOMED8, WRF331F; RACMOE22, REMO2009, RCA4) and the result shown is an average of all eight models. The black dots indicate robust evolutions, for which at least six of the eight climate models simulate a statistically significant evolution (source: according to*

Student's T-test, threshold of 0.95) (source: design and realization F. Raymond, 2020).

Thus, two Mediterranean regions could be mainly affected by the future increase in the number of days of VLDS in the two trajectories studied: the east and the west of the basin (see Figure 10.5). In the eastern part of the basin (southern Turkey and the Levant), for which the VLDS are particular because they are "seasonal", the increase in the number of days of VLDS per season (up to 50 days per season, according to the RCP8.5 trajectory) will probably lead to a strong decrease in the beginning of the rainy season, but also to a decrease in the duration of this rainy period. In the western part of the basin, including the Iberian Peninsula and part of the Maghreb, the increase in the number of days of VLDS per season should reach up to 45 days according to the RCP8.5 trajectory. These results assume that more seasons will be affected by at least one VLDS in the future, and that future events will be even more intense. In view of the impacts of VLDS on Spanish agrosystems observed previously, these results do not aspire to optimism regarding the evolution of yields associated with rainfed agriculture.

10.5. Conclusion

In the Mediterranean basin, the so-called "rainy" period, from autumn to spring, is a crucial period for societies and ecosystems. This period between the dry summers of the Mediterranean climate should normally allow, at least partially, for water recharge. If drought prevails during this period, rain-fed agriculture will be the first activity directly affected by this climatic hazard. Due to the geographical complexity of the Mediterranean basin, climate variability is not homogeneous and univocal on the scale of the whole basin. More than elsewhere, and especially when it comes to understanding drought episodes, it is important to define and study recurrent, redundant and spatially coherent episodes within this basin: these are prolonged dry spells. These events are characterized here by their location, spatial extension and duration in terms of the number of consecutive days without rain. These typically geographical "climatic objects" also make it possible to answer the crucial question: who most often experiences the longest dry episodes? Indeed, the spatiotemporal characteristics of VLDS make it possible to locate and spatialize the affected territories and thus to tend towards impact and risk studies when these territories are vulnerable to the prolonged absence of precipitation: this is spatialization at the service of the territorialization of climate change.

Between autumn and spring, the VLDS are generally organized in three main spatial configurations: around the Levantine basin, on the Iberian Peninsula and in the Balkans. These episodes can last well over a month and affect areas of several thousand km^2. Climate models show that they would tend to be increasingly frequent during the 21st Century, but also increasingly intense (in terms of duration and spatial extent), especially for the western and eastern regions of the Mediterranean basin.

In terms of impacts, the VLDS significantly affect winter cereal yields (barley, wheat, oats) throughout Spain, with considerable yield reductions if the dry episode lasts a long time or if several episodes are recorded during the same rainy season. The characteristics of time (duration) and space (location and spatial extension) of the VLDS as defined in this chapter make it possible to get as close as possible to the spatiotemporal scales of agrosystem functioning. This thus offers real added value in terms of impact studies compared to work based on more "conventional" approaches to understanding droughts and their impacts.

Faced with the widespread drying up of the Mediterranean basin, predicted by most climate models, rain-fed agriculture could rapidly be in danger in many territories. This agricultural practice cannot, however, be replaced by irrigated agriculture, as the problem of water resource availability is already present in the Mediterranean basin and will become even more widespread with more frequent, longer and more extensive periods of water stress in the future. It is therefore important, as of now, to start thinking about the adaptation of agricultural practices in the face of current and future climatic changes, in order to eventually turn to species that may be less sensitive to long, dry periods and less water-intensive.

10.6. References

Brunet, M., Gilabert, A., Jones, P. (2014). A historical surface climate dataset from station observations in Mediterranean North Africa and Middle East areas. *Geoscienc. Data Journal*, 1, 121–128.

Croitoru, A.E., Chiotoroiu, B.C., Torica, V. (2012). Dry spells on the Romanian Black coast. In *Air and Water: Components of the Environment Conference*, Babeş-Bolyai University (ed.). Cluj-Napoca.

Drumond, A., Gimeno, L., Nieto, R., Trigo, R., Vicente-Serrano, S.M., (2017). Drought episodes in the climatological sinks of the Mediterranean moisture source: The role of moisture transport. *Global and Planetary Change*, 151, 4–14.

Dubrovsky, M., Hayes, M., Duce, P., Trnka, M., Svoboda, M., Zara, P. (2014). Multi-GCM projections of future drought and climate variability indicators for the Mediterranean region. *Regional Environmental Change*, 14, 1907–1919.

Garcia-Herrera, R., Paredes, D., Trigo, R.M., Franco-Trigo, I., Hernandez, I., Barriopedro, D., Mendez, M.A. (2007). The outstanding 2004/05 drought in the Iberian Peninsula: Associated atmospheric circulation. *Journal of Hydrometeorology*, 8, 483–498.

Giorgi, F. (2006). Climate change hot-spots. *Geophysical Research Letters*, 33, L0870.

Hawkins, E. and Sutton, R. (2012). Time of emergence of climate signals. *Geophysical Research Letters*, 39, L01702.

Haylock, M.R., Hofstra, N., Klein Tank, A.M.G., Klock, E.J., Jones, P.D., New, M.A. (2008). European daily high-resolution gridded data set of surface temperature and precipitation for 1950–2006. *Journal of Geophysical Research*, 113, D20119.

Hertig, H. and Tramblay, Y. (2017). Regional downscaling of Mediterranean droughts under past and future climatic conditions. *Global and Planetary Change*, 151, 36–48.

Hertig, E., Seubert, S., Paxian, A., Vogt, G., Paeth, H., Jacobeit, J. (2013). Changes of total versus extreme precipitation and dry periods until the end of the twenty-first century: Statistical assessments for the Mediterranean area. *Theoretical and Applied Climatology*, 111, 1–20.

Hoerling, M., Eischeid, J., Perlwitz, J., Quan, X., Zhang, T., Pigeon, P. (2012). On the increased frequency of Mediterranean drought. *Journal of Climate*, 25, 2146–2161.

IPCC, Intergovernmental Panel on Climate Change (2013). The physical science basis: Summary for policymakers (Contribution of WG I to the 5th Assessment Report of the IPCC). Summary, WMO/UNEP, New York.

Kostopoulou, E. and Jones, P. (2005). Assessment of climate extremes in the Eastern Mediterranean. *Meteorology and Atmospheric Physics*, 89, 69–85.

Lionello, P. and Scarascia, L. (2018). The relation between climate change in the Mediterranean region and global warming. *Regional Environmental Change*, 5, 1481–1493.

Moss, R.H., Edmonds, J.A., Hibbard, K.A., Manning, M.R., Rose, S.K., van Vuuren, D.P., Carter, T.R., Emori, S., Kainuma, M., Kran, T., Meehl, G.A., Mitchell, J.F.B., Nakicenovic, N., Riahi, K., Smith, S.J., Stouffer, R.J., Thomson, A.M., Weyant, J.P., Wilbanks, T. (2010). The next generation of scenarios for climate change research and assessment. *Nature*, 463, 747–756.

PNUE/PAM (2016). Stratégie méditerranéenne pour le développement durable 2016–2025. Report, Centre d'activité régionales, Plan Bleu, Valbonnes.

Polade, S.D., Pierce, D.W., Cayan, D.R., Gershunov, A., Dettinger, M.G. (2014). The key role of dry days in changing regional climate and precipitation regimes. *Scientific Reports*, 4, 4364.

Polade, S.D., Gershunov, A., Cayan, D.R., Dettinger, M.D., Pierce, D.W. (2017). Precipitation in a warming world: Assessing projected hydro-climate changes in California and other Mediterranean climate regions. *Scientific Reports*, 7, 10783.

Raymond, F., Ullmann, A., Camberlin, P., Drobinski, P., Château Smith, C. (2016). Extreme dry spell detection and climatology over the Mediterranean Basin during the wet season. *Geophysical Research Letters*, 43, 7196–7204.

Raymond, F., Ullmann, A., Camberlin, P., Oueslati, B., Drobinski, P. (2018a). Atmospheric conditions and weather regimes associated with extreme winter dry spells over the Mediterranean basin. *Climate Dynamics*, 50, 4437–4453.

Raymond, F., Ullmann, A., Camberlin, P. (2018b). Très longs épisodes secs hivernaux dans le bassin méditerranéen : variabilité spatiotemporelle et impact sur la production céréalière en Espagne. *Cybergeo: European Journal of Geography*, 858.

Raymond, F., Ullmann, A., Tramblay, Y., Drobinski, P., Camberlin, P. (2019). Evolution of Mediterranean extreme dry spells during the wet season under climate change. *Regional Environmental Change*, 19, 2339–2351.

Sanford, T., Frumhoff, P.C., Luers, A., Gulledge, J. (2014). The climate policy narrative for a dangerously warming world. *Nature Climate Change*, 4, 164–166.

Sousa, P.M., Trigo, R.M., Aizpurua, P., Nieto, R., Gimeno, L., Garcia-Herrera, R. (2011). Trends and extremes of drought indices throughout the 20th century in the Mediterranean. *Natural Hazards and Earth System Sciences*, 11, 33–51.

Vicente-Serrano, S.M., Lopez-Moreno, J.I., Lorenzo-Lacruz, J., El Kenawy, A., Azorin-Molina, C., Moran-Tejeda, E., Pasho, E., Zabalza, J., Begueria, S., Angulo-Martinez, M. (2011). The NAO impact on droughts in the Mediterranean region. In *Hydrological, Socioeconomic and Ecological Impacts of the North Atlantic Oscillation in the Mediterranean Region*, Vol. 46, Vicente-Serrano, S.M. (ed). Springer, Berlin.

Zakhem, B.A. and Kattaa, B. (2016). Investigation of hydrological drought using cumulative Standardized Precipitation Index (SPI 30) in the eastern Mediterranean region (Damascus, Syria). *Journal of Earth System Science*, 125, 969–984.

11

The Spatial Impacts of Climate Change on Viticulture Around the World

Hervé QUÉNOL[1] and Renan LE ROUX[2]

[1] *CNRS, Rennes, France*

[2] *CIRAD, Montpellier, France*

11.1. Introduction

Climate change threatens the long-term viability of agrosystems, especially in the field of viticulture (Porter and Xie 2014). The vine is extremely sensitive to climatic variations, with specific varieties producing wines of typicality and quality that are distributed over well-defined territories. The vine responds to global warming by modifying its annual growth cycle (e.g. earlier phenological stages and a shortening of the growing period) and significantly changing grape composition (e.g. increasing sugar levels, decreasing acidity in the berries and increasing alcohol content; see Duchêne and Schneider 2005; Van Leeuwen *et al.* 2009). There is no longer any doubt that climatic conditions marked by extreme events will be much more frequent in the future and will inevitably have an impact on global viticulture. Already, some wine-growing regions are subject to climatic conditions that are unfavorable for quality viticulture, while new wine-growing regions are emerging (Jones *et al.* 2005; Webb *et al.* 2008). This inevitably raises the question of how vineyards adapt to climate change, hence the need to carry out future climate simulations. However, even if the limits of the major climatic regions are dependent on the evolution of the global climate, the specificities of wines are often defined by the influence of local climates. In order to simulate the spatial impacts of climate change on viticulture in the world, it

Spatial Impacts of Climate Change,
coordinated by Denis MERCIER.

is therefore necessary to adopt a multi-scalar approach, that is from the global to the local scale. This approach requires the development of climate change models that take into account the interweaving of spatial scales and thus provide simulations of the spatial impacts of climate change on viticulture with a resolution fine enough for adaptation planning at the scale of the vineyard, that is, the scale at which the winegrower will be able to modify his practices.

11.2. Recent climatic trends in the world's wine-growing regions

Although the vine is a hardy plant, its ability to grow and produce a viable crop is defined by particular climatic conditions, particularly with regard to the availability of water and heat. The high latitude limit for viticulture is defined by conditions where the average annual temperature is not high enough to allow the grapes to ripen properly, whereas at low latitudes the temperatures are too high and often marked by periods of severe drought. The theoretical limit for vine cultivation lies in a latitudinal band between approximately 30° N and 50° N for the northern hemisphere and 20° S and 50° S for the southern hemisphere. This zoning corresponds to the 12–22°C isotherms of the average temperature of the vine growing season, that is, from April to October in the northern hemisphere and from October to April for the southern hemisphere (Jones 2006).

Contemporary climate change is causing this latitudinal band of possible vine cultivation to shift northwards in the northern hemisphere and southwards in the southern hemisphere. In Europe, vineyards are currently developing in England and Scandinavia. In North America, new wine-growing regions are emerging further north in Canada, while in South America, the southern lands of Patagonia are experiencing a large increase in vine planting. Although the climatic conditions in these new wine-growing regions are not yet optimal, future temperatures will certainly be favorable for the production of quality wines. Conversely, even if the low-latitude wine-growing regions do not disappear, the climatic conditions will be marked by excessively high temperatures and significant periods of drought. This is the case for the Mediterranean wine-growing regions of Southern Europe, for North Africa and for South Australia (Quénol 2014).

This latitudinal shift is not the only geographic transformation, as extension to higher altitudes also occurs (Delay *et al.* 2015). Indeed, in regions where climatic conditions become unfavorable, the rise in altitude for the cultivation of vines becomes an alternative. This is the case of the vineyards of Cape Province in South Africa, which have extended to higher altitudes in order to find more favorable thermal conditions. Like latitudinal displacement, the increase in temperature in recent decades has made it possible to develop high-altitude viticulture, for example, in the regions of

Cafayate and Jujuy (northwestern Argentina), where thermal conditions in high-altitude vineyards (between 1,500 m and 2,800 m) now enable the production of quality wines thanks to a better ripening of the grapes (Barroso *et al.* 2019). Altitudinal displacement does not only occur in mountainous regions. Until a few decades ago, the upper parts of the slopes did not have the required thermal conditions for good grape ripening. Now, higher temperatures due to climate change have led to better conditions for grape ripening and the quality of the wines has been significantly improved. This is particularly the case in the high slopes of Burgundy, where the vineyards are located between 350 and 400 m in altitude.

This change in the limits of vine cultivation characterizes recent climatic trends in the world's wine-growing regions. Generally speaking, a 1°C increase in the global average temperature is considered to be equivalent to a potential shift in viticulture of about 200 km to higher latitudes or 150 m in altitude. However, defining the limits of vine cultivation solely with the average temperatures during the growing season is not sufficient. For example, all tropical vineyards (Brazil, India, China, Vietnam, etc.) are outside this zoning and even if the climatic conditions do not seem favorable for vine cultivation, here these new vineyards are developing in a perennial way (Solomon 2017). It is therefore necessary to use other climate indicators related to vine cultivation and wine characteristics in order to carry out a precise agroclimatic zoning of viticulture in the world and then to simulate the spatial impacts of climate change on future viticulture for 2050–2100.

11.3. Climate zoning in viticulture

Numerous studies have shown the strong influence of climate (especially air temperature) on the functioning of the vine and the quality of the grapes. This research has led to the development of vine-specific bioclimatic indices that allow the division of wine-growing climates into different climatic regions (Tonietto 1999). Most of these indices are often based on the accumulation of temperatures (sum of degrees/day) during the growing season of the vine. These indices assume, firstly, that the vine will only develop once its base temperature has been exceeded. When the average daily temperature is higher than the base temperature, the difference accumulates day after day throughout the growth season. This assumes that the vine needs a specific total of accumulated degree-days to reach maturity before the harvest. These different indices are therefore used to define the aptitudes of a wine-growing region in terms of grape production, grape variety adaptation and to predict the course of the phenological cycle (Neethling 2016). The bioclimatic indices most commonly used in viticulture (Winkler and Huglin indices, see Box 11.1) make it possible to spatially characterize the types of viticultural climates in the world.

The two main indices used in viticulture are the Winkler Index and Huglin Index. The first refers to the concept of growing degree-days, which is calculated as the sum of daily average temperatures above 10°C for the period April 1 to October 31 in the northern hemisphere (October 1 to April 30 in the southern hemisphere). The base temperature of 10°C corresponds to the minimum temperature necessary for the physiological activity of the vine.

Five wine-growing climatic regions stand out in this classification, from the coolest to the warmest (see Table 11.1). The advantage of using the Winkler index is that the accumulated heat is very well correlated with the phenology of the vine.

Region	Value (°C)	Examples
V	> 2,205	Jerez (E), Hunter (Aus), Palermo (I), Fresno (USA)
IV	1,927–2,205	Venice (I), Mendoza (Arg), Stellenbosch (RSA)
III	1,650–1,926	Montpellier (F), Milan (I), Porto (P), Napa (USA),
II	1,389–1,667	Rioja (E), Côtes-du-Rhône (F), Barolo (I), Santiago (Chile)
I	< 1,371	Geisenheim (D), Champagne (F), Dijon (F), Bordeaux (F)

Table 11.1. *Winkler "degree-day" thermal index classes (source: based on Winkler* et al. *1974)*

The Huglin index differs because it is the sum of average and maximum temperatures above 10°C from April 1 to September 30 in the northern hemisphere (September 1 to April 30 in the southern hemisphere). This index gives more weight to daytime temperatures, where most of the vine development takes place, and is strongly correlated with the composition of the berries at harvest.

By integrating a latitude coefficient, the Huglin index (HI) takes into account the duration of sunshine, a very important factor in the growth of the vine (see Table 11.2).

Type	Value (°C)	Examples
Very hot	> 3,000	São Francisco Valley
Hot	2,400 ≤ HI ≤ 3,000	Malaga (Spain), Marsala (Italy)
Hot temperate	2,100 ≤ HI ≤ 2,400	Napa (USA), Montpellier (France)
Temperate	1 800 ≤ HI ≤ 2 100	Pau, Bordeaux (France)
Cool	1,500 ≤ HI ≤ 1,800	Colmar, Angers (France)
Very cool	HI ≤ 1,500	Quebec City (Canada), London (England)

Table 11.2. *Huglin heliothermal index classes (source: based on Huglin 1978; Tonietto 1999)*

Box 11.1. *Main bioclimatic indices used in viticulture*

By calculating bioclimatic indices using data from national weather station networks, spatial analyses of the world's different wine climates are carried out. Figure 1.1 shows the average Huglin index calculated over the period 1950–2000 and highlights the spatial variability of wine climate types (from "very cool" to "very hot") across the world's wine-growing regions. The calculation of bioclimatic indices over several growing seasons (at least over several 30-year climate normals) provides an indication of the spatial evolution of wine-growing climates in the context of climate change. Wine-growing regions with a Huglin index close to 3,000 D/J (in red, Figure 11.1) are clearly subject to extreme thermal conditions. The increase in temperature is characterized in particular by an increase in bioclimatic indices which may imply a change in the categories in the classifications of types of wine-growing climates. The mapping of the Huglin and Winkler indices between the periods 1980–2009 and 1950–1979 shows an increase in index values over the whole of Europe and North Africa (Santos *et al.* 2012) and confirms the evolution of optimal vine growing conditions in Europe. Compared to the period 1950–1979, the period 1980–2009 was characterized by an increase in the number of accumulated degree-days in almost all of Europe, with a greater increase in the western Mediterranean (see Figure 11.2).

Observed climate change but also modified phenological stages (earlier with a shortening of the vine's vegetative cycle) and changes in the characteristics of wines (increase in alcohol content, modification of aromatic qualities) have raised a number of questions about the impact of climate change on viticulture, hence the need to carry out simulations of future climate.

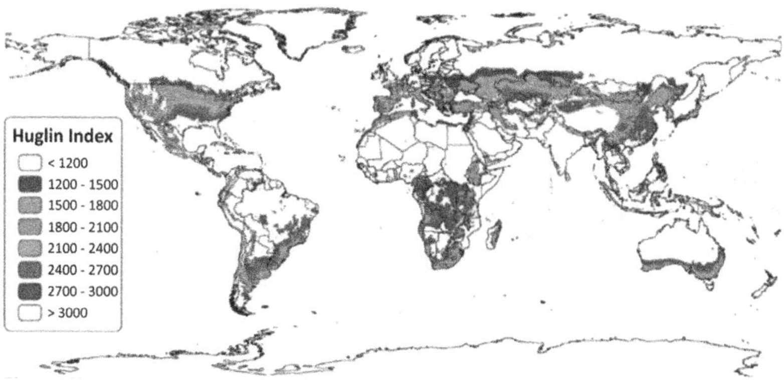

Figure 11.1. *Mapping of the average Huglin index for the world's wine-producing regions over the period 1950–2000 (Jones* et al. *2009) (source: Worldclim[1]). For a color version of this figure, see www.iste.co.uk/mercier/climate.zip*

1 https://www.worldclim.org/ .

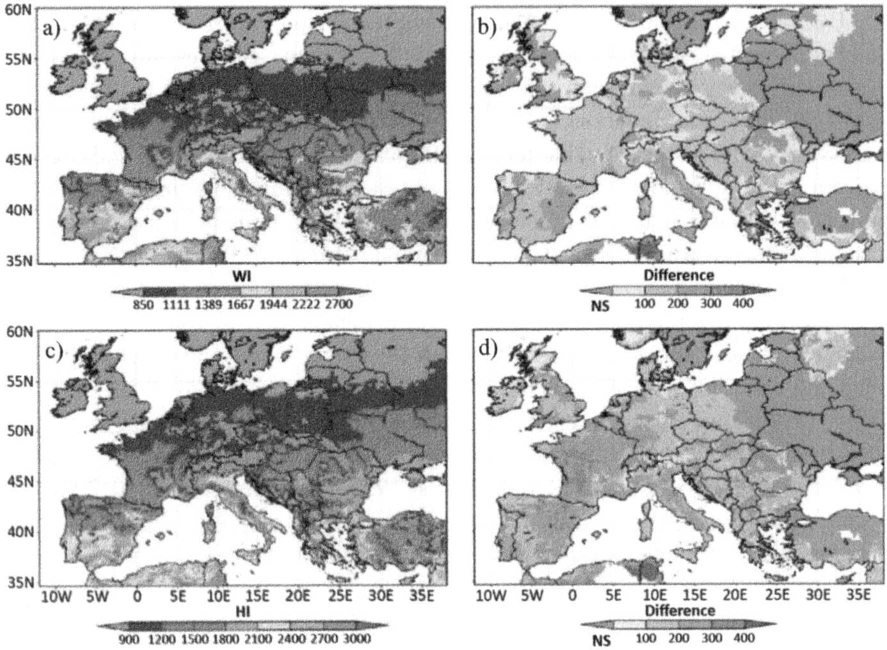

Figure 11.2. *(a) and (c) show the Winkler and Huglin indices in Europe over the period 1950–2009. (b) and (d) show the difference, for each index, between 1950–1979 and 1980–2009 (source: Santos* et al. *2012). For a color version of this figure, see www.iste.co.uk/mercier/climate.zip*

11.4. Impact of climate change: anticipating changes in the spatial distribution of vines

Recent climate changes and impacts on viticulture show that the adaptation of vineyards to climate change is crucial and must be based on simulations of the future climate. The calculation of bioclimatic indices from the outputs of future climate models according to the different IPCC scenarios aims to establish the limits of the vine's aptitude, in particular to simulate the geographical evolution of wine-growing regions in the world. However, many sources of uncertainty exist, particularly at the level of the climate model (type of modeling, spatial resolution of model outputs, etc.) as well as on the representativeness of bioclimatic indices which do not take into account the physiology of the vine and its capacity to adapt to climate change. These different sources of uncertainty must be identified and assessed within the framework of a climate zoning of wine-growing regions for the 2050–2100 horizon and the planning of adaptation strategies.

11.4.1. *Towards climate change modeling in wine-growing regions*

Most of the work on the future impacts of climate change on viticulture has been carried out using climate data from the model outputs of the coupled model intercomparison projects (CMIP) used in previous IPCC reports (mainly CMIP 3 and 5) (IPCC 2014). The quality and spatial resolution of future projections depend on the type of model used, since advances in computer technology and knowledge in atmospheric sciences have led to a significant improvement in the quality of model outputs, particularly between CMIP3 and CMIP5 (see Box 11.2). The first studies were carried out in the early 2000s with the outputs of global climate models with a spatial resolution of 50–100 km. For reasons of capacity and computation time, it was difficult to carry out climate simulations over an entire territory and, above all, to obtain a spatial resolution suited to wine-growing regions. Several studies consisted of calculating climatic indicators (bioclimatic indices, climatic extremes, etc.) on a few grid points in order to characterize future climatic conditions related to vine growing (Garcia de Cortazar 2006; Brisson and Levrault 2010). For example, Briche *et al.* (2014) studied the evolution of thermal conditions for vine growing (calculation of Huglin and Winkler indices) and the evolution of thermal extremes (spring frost and heat waves) in Champagne between 2001 and 2100 according to three scenarios (A1B, B1 and B2) from three grid points of the Arpège-climat model. Even if this initial work did not make it possible to spatially represent the impacts of climate change on the scale of wine-growing regions, they have provided the first analyses of possible developments in wine growing for 2050 and 2100.

Significant advances in computer computing capacity and access to regionalized climate data from the climate models of the 5th IPCC report (2014) have made it possible to map the spatial impacts of climate change on global viticulture. The regional models offer high spatial resolution (between 10 and 20 km) and take into account local parameters such as topography and land use. For example, the CORDEX (Coordinated Regional Climate Downscaling Experiment) project provides regionalized projections of climate change over the entire planet[2]. The most recent work is based on calculations of bioclimatic indices from different regionalized grid points (Hannah *et al.* 2013; Moriondo *et al.* 2013; Fraga *et al.* 2014; Cabre *et al.* 2016). This work showed significant potential changes in the future distribution of vineyards. The warming trend often results in an increase in bioclimatic indices, which leads to a change in the classification of wine climate types from one category to another.

On a global scale, this predicts significant changes in the distribution of vineyards by 2050–2100, with the disappearance of some wine-growing areas such as southern

2 http://www.cordex.org/.

Australia and some Mediterranean countries, as well as the emergence of new wine-growing areas such as in Northern Europe or South America (Hannah *et al.* 2013). The different studies carried out in Europe show a different evolution according to the geographical position of wine-growing regions (Santos *et al.* 2012; Moriondo *et al.* 2013; Tóth and Végvári 2016; Irimia *et al.* 2018). The Mediterranean wine-growing regions are already subject to problems of drought and very high maximum temperatures.

Global climate models are complex computer programs that aim to represent processes in the atmosphere through the application of fundamental equations of atmospheric physics. The planet is represented by a three-dimensional grid in which equations are applied to simulate these natural processes. These global models are used to calculate meteorological parameters (such as temperature, humidity, wind and precipitation) for each cell of the grid on different time scales. By integrating these equations over time, dynamic model systems attempt to predict the state of the atmosphere at a given time in the future based on its current state. The global climate models from CMIP 5 represent the entire planet with a spatial resolution of 50 km over the period 2000–2300. This spatial resolution makes it possible to simulate the influence of emission scenarios on the general circulation of the atmosphere and to estimate their impact on climate variables (such as temperature and precipitation) on a global scale. These models are the basis for any simulation of climate change and are the starting point for climate modeling at finer spatial resolutions. Global climate models have a spatial resolution of several tens of kilometers and do not allow the representation of finer-scale atmospheric phenomena such as those influenced by surface features. Downscaling methods are therefore used to incorporate the effects of complex surface variability and increase the spatial resolution of the models. *Regional climate models* are dynamic, downscaled global climate models that aim to regionalize global model outputs by using nested grids of models of increasing resolution. The downscaling provides a spatial resolution close to 10 km. However, the large computational capacity required to produce these models makes it difficult to obtain a finer spatial resolution over a long period of time. Downscaling using statistical methods requires less computational capacity than dynamic methods, which can provide better spatial resolution. *Statistical downscaling* involves the application of a number of statistical techniques to identify the relationship between a selected climate variable (such as air temperature) and land surface characteristics (such as elevation, aspect, slope or land cover). The advantage of statistical downscaling is that it requires far fewer computational resources, but it has the disadvantage that it no longer takes into account the dynamics between the atmosphere and surface conditions (see Figure 11.4).

Box 11.2. *Global and regional climate models*

In the short term and whatever the scenario, wine production in the southernmost countries will be negatively affected. Northern Europe seems to offer new opportunities with a general improvement in climatic conditions for vine cultivation. This could make it possible to extend the current production area. Between these two sectors, there is a transition zone where the climatic conditions for vine cultivation vary and are sometimes even contradictory depending on the output of the models used. These divergences according to the models mainly concern the simulation of future rainfall, where the results are very different, especially in this part of Europe (Ouzea *et al.* 2014). By 2050, the scenarios do not show strong differences in spatial impacts.

On the other hand, by 2100, the impacts will be significant, leading to a major transformation of European vineyards and a sharp reduction in production area in the Mediterranean regions. In France, the Huglin index (see Table 11.2) has been mapped using data from the regional Aladdin model (spatial resolution of 8 km). The maps were produced for the periods 1986–2005, 2031–2050 and 2081–2100 according to RCP scenarios 2.6, 4.5 and 8.5. Over the period (1986–2005), the northernmost wine-growing regions (Loire Valley, Champagne, Alsace, Burgundy) correspond to the "cool climate" class. The Bordeaux vineyards are in the "temperate climate" class and the Mediterranean vineyards are mainly in the "warm temperate climate" class.

For the period 2031–2050 and according to the three scenarios (RCP 2.6, 4.5 and 8.5), the simulations showed an increase in the Huglin index corresponding to a northward shift of about one climate class. This is the case for the Bordeaux region, which would theoretically move to the "warm temperate climate" class, for Burgundy, the Loire Valley and Alsace to the "temperate climate" class, and for Languedoc to the "warm climate" class. These findings would not lead to major changes for French viticulture except for the Mediterranean vineyards. On the other hand, by 2081–2100, this potential migration (from areas with a specific wine-growing style) will have a much greater impact, especially according to scenario 8.5. Most of the French wine-growing regions would be in the "warm climate" type with a shift to "very warm climate" for the Languedoc (see Figure 11.3) (Quénol *et al.* 2017). Although these maps were produced using modeled data, the results illustrate the range of possible trends (depending on the scenario) in the evolution of French viticulture and can help stakeholders make decisions on future strategies.

These different works based on the climatic adaptability of the vine according to climate change scenarios show that we can expect major upheavals on a global scale with a redistribution of wine-growing regions (Quénol 2017). Nevertheless, the calculation of bioclimatic indices based on climate change model outputs only provides a theoretical answer on the possible impacts of climate change on viticulture. There are many sources of uncertainty in climate change modeling, and it must be admitted that it is impossible to know exactly what the climate will be in 2050 or 2100, firstly because the models we use do not yet perfectly represent all the complex processes operating in the atmosphere, but also because the link between bioclimatic indices and vine behavior is biased. These indices only provide indications on the theoretical limits of vine cultivation and quality wine production, but these indications do not take into account some important factors, such as the physiological capacity of the vine to adapt to the future climate (Van Leeuwen *et al.* 2013).

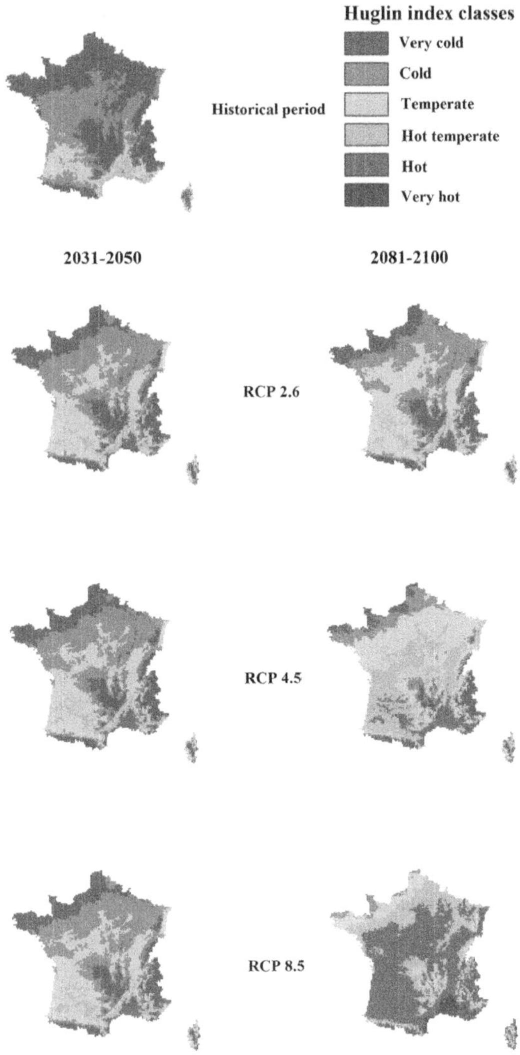

Figure 11.3. *Huglin index modeled on France over the periods 1986–2005, 2031–2050 and 2081–2100 under RCP scenarios 2.6, 4.5 and 8.5 with a spatial resolution of 8 km (sources: DRIAS). For a color version of this figure, see www.iste.co.uk/mercier/climate.zip*

Regionalized climate simulations make it possible to estimate the impacts of climate change in large wine-growing regions, but the spatial resolution of the models is not sufficiently precise to take into account local effects linked to roughness and the nature of the surface (topography, land use, etc.). However, atmospheric parameters at the boundary layer level are dependent on surface conditions (roughness and nature of the surface) and these can lead to high spatial climate variability over relatively small areas (of around a few kilometers to a few meters). It is these local climatic variations that determine the specificities of a wine and it is at the plot scale that the winegrower can implement his strategy of adaptation to climate change (Neethling *et al.* 2017, 2019). It is therefore necessary to integrate the spatial variability of climate at local scales into the outputs of regionalized climate change models in order to assess the spatial impacts at the vineyard scale (Quénol 2017).

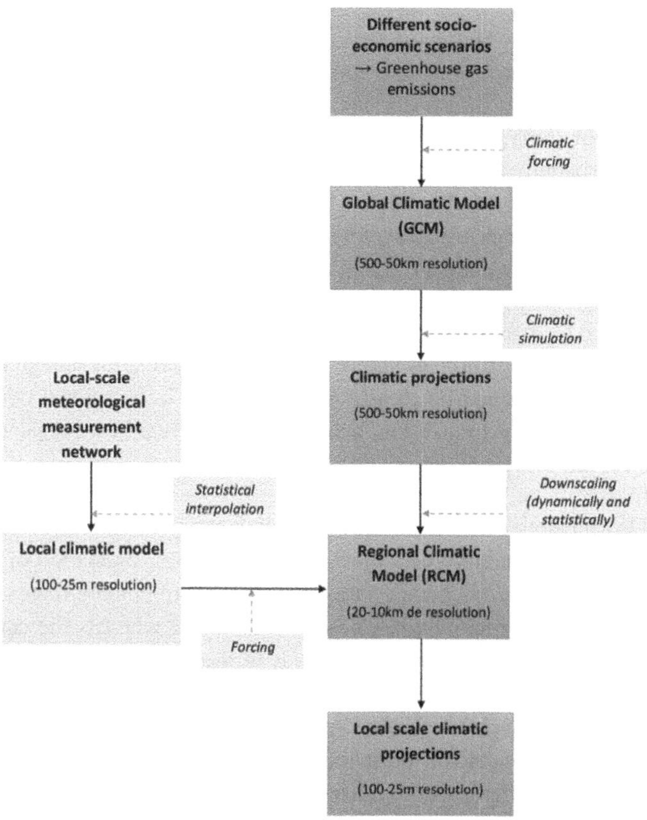

Figure 11.4. *Different stages of downscaling (from global to local) climate change projections (source: adapted from Daniels* et al. *2012). For a color version of this figure, see www.iste.co.uk/mercier/climate.zip*

11.4.2. *The need to take into account local factors*

Local factors cause climate variations that may be greater than the climate variability at larger scales, which are taken into account by global and regional models (Quénol 2017).

However, despite the significant progress made in terms of the spatial resolution of climate projections, they are not yet sufficiently accurate to consider the influence of local parameters. To overcome these limitations, advanced statistical methods (multiple linear regression, non-linear regression, Support Vector Regression, etc.) are used to spatially interpolate climate data obtained at local scales (Madelin 2004; Joly *et al.* 2009; Bonnardot *et al.* 2012; Bonnefoy 2013; Le Roux *et al.* 2017a, 2017b; Bois *et al.* 2018; Le Roux *et al.* 2018).

These methods are based on establishing the relationship between surface characteristics (landscape morphology and land use) and meteorological variables.

In this type of study, the relationship between climatic elements and topographic features is then spatially assessed at the study site. The application of this approach to spatializing the climate at the local scale using statistical methods requires the installation of meteorological measurement networks in the vineyards based on local factors likely to modify the climate locally (Quénol 2014).

This approach to spatializing climate on a local scale, based on field measurements, does not allow the estimation of future climate. However, the integration of local climate variability (validated by measuring actual data) into regional climate change models makes it possible to take into account the influence of local factors in model outputs (see Figure 11.4).

Figure 11.5 illustrates how local-scale modeling can be used to provide much more detail on the spatial variability of a bioclimate index. The calculation of the Winkler index from the regionalized climatic outputs of the future climate represents only six grids over the whole Pomerol/Saint-Emilion appellation, whereas the contribution of the spatial variability of the local climate leads to the influence of local factors to be integrated (indispensable for the characterization of these terroir wines) in future simulations. This clearly shows that the spatial variability of the local climate must be taken into account in models of the spatial impacts of climate change with a view to developing adaptation strategies for viticulture.

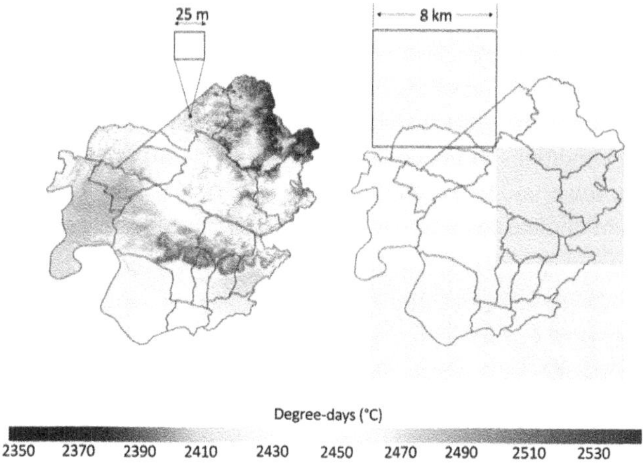

Figure 11.5. *Spatialization of the Winkler index in the Pomerol/Saint-Émilion region with a resolution of 8 km (rudder mesh, right) and at a resolution of 25 m (left) according to the RCP 8.5 scenario for the period 2081–2100 (source: adapted from Le Roux* et al. *2017). For a color version of this figure, see www.iste.co.uk/mercier/ climate.zip*

This approach of integrating spatial climate variability at the local scale into regional climate models has been followed for several European vineyards – Pomerol/Saint-Émilion (France), Val de Loire (France), Sussex (UK), Rheingau (Germany), Cotnari (Romania) and Rioja (Spain) – in the framework of the ADapatation of VIticulture to CLIMate change project: high resolution observations of adaptation scenarios for viticulture (LIFE-ADVICLIM; see Quénol and Bonnardot 2014; Quénol *et al.* 2014). In this case, the Euro-CORDEX regionalized outputs (12 km resolution) were forced by the local climate model (25 m resolution) for the periods 2031–2050 and 2081–2100 according to the RCP4.5 and RCP8.5 scenarios. The results showed that the spatial climate variability within the studied wine region is similar to the temperature increase (sum of degrees/days) between the current and future periods (2050 and 2100). For example, in the Rheingau wine-growing region (Germany), the increase in the modeled Huglin index, based on the regionalized outputs, would be 200 D/D and 800 D/D for the periods 2031–2050 and 2081–2100 respectively, according to the RCP8.5 scenario, compared to the reference period 1986–2005. The spatial variability within the Rheingau vineyard due to the steep slopes and the proximity to the Rhine is above 300 D/D (see Figure 11.6).

The very high spatial variability of the climate due to local effects therefore makes it possible to identify more clearly the sectors of the vineyard more or less favorable to quality viticulture in the context of climate change. On the Pomerol/Saint-Émilion site, regional-scale modeling of bioclimatic indices according to RCP8.5 scenarios in 2100 showed negative developments corresponding to unfavorable overmaturation conditions for the production of quality wines with current grape varieties. However, on a much finer scale, the great spatial variability of the Saint-Émilion area would still offer favorable conditions in the cooler parts of the region, such as the northern part, but also on the lower slopes of the vineyard. The contribution of knowledge of the spatial impacts of climate change on a local scale is clearly seen here.

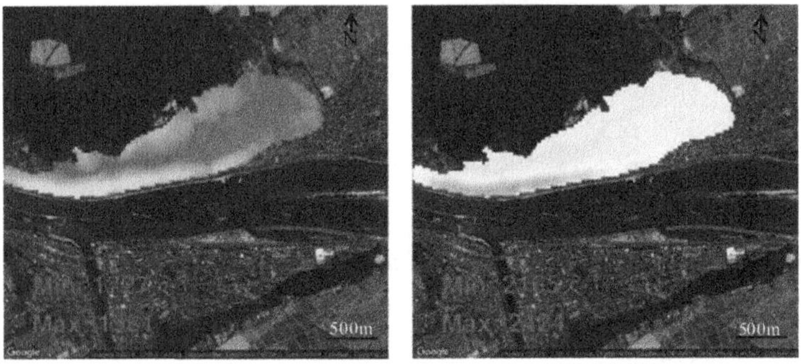

(2031-2050) (2081-2100)

1700 1800 1900 2000 2100 2200 2300 2400 2500 (°C/D)

Figure 11.6. *Modeling of the spatial variability of the Huglin index at the local scale in Rheingau vineyards (Germany) according to the RCP 8.5 scenario. For a color version of this figure, see www.iste.co.uk/mercier/climate.zip*

11.5. Conclusion

The climate has always had a strong impact on the functioning of the vine and the characteristics of wines. On a global scale, it is the climate that will define the potentialities of vine cultivation, whereas the characteristics and specificities of the wines will rather depend on the regional and local climates. The climatic evolution of recent decades has already had a significant impact on viticulture (early phenological stages, increase in the alcoholic degree, etc.) and has already modified the geographical distribution of wine-growing areas. This inevitably raises the question of the adaptation of vineyards to climate change, hence the need to carry out future climate simulations. The calculation of bioclimatic indices based on regionalized model outputs has made it possible to estimate the spatial impacts of climate change on global viticulture in the near (2050) and distant (2100) future according to different IPCC scenarios. These climate impact studies focus on potential changes in the world's major wine-growing regions, and the results suggest relatively sudden consequences, such as the disappearance of certain wine-growing regions (for the most pessimistic scenarios up to 2100). Combined with regional climate scenarios, the analysis of the spatial variability of the local climate has made it possible to refine the spatial resolution of the models and to propose adaptation methods that are reasoned at the farm level (wine-growing practices, rootstocks, grape varieties, etc.) rather than at the scale of the major wine-growing regions (Barbeau *et al.* 2015).

Integrating the spatial variability of the local climate into regional models has the advantage of providing information at the scale of the vineyard plot, that is, the scale at which the winegrower works. As the local models are built from real field data, they are validated, which reduces the biases linked to modeling. However, these models also have the disadvantage of being static. Ideally, in order to obtain local-scale simulations of future climate, downscaling from the global to the local level should be performed using the same method. At present, computational capabilities do not yet allow this type of approach to be applied to large wine-growing regions, but major advances in computer computation capacity will likely make this feasible in the near future.

11.6. References

Barbeau, G., Neethling, E., Ollat, N., Quénol, H., Touzard, J.M. (2015). Adaptation au changement climatique en agronomie viticole. *Revue AE&S*, 5(1).

Barroso, R., Ortiz, H., Malaniuk, M., Quénol, H., Murgo, M., Coria, C., Videla, R., Prieto, S., Manzano, H., Quini, C., Aruani, C. (2019). Vinos de altura del noroeste Argentino – Características físico-químicas y sensoriales. *42nd World Congress of Vine and Wine, BIO Web of Conferences*, Geneva, 15.

Bonnardot, V., Carey, V., Madelin, M., Cautenet, S., Quénol, H. (2012). Using atmospheric and statistical models to understand local climate and assess spatial temperature variability at fine scale over the Stellenbosch wine district, South Africa. *Oeno One*, 46(1), 1–13.

Bonnefoy, C. (2013). Observation et modélisation spatiale de la température dans les terroirs viticoles du Val de Loire dans le contexte du changement climatique. PhD thesis, Université Rennes 2, Rennes.

Bois, B., Joly, D., Quénol, H., Pieri, P., Gaudillière, J.P., Guyon, D., Saur, E., van Leeuwen, C. (2018). Temperature-based zoning of the Bordeaux wine region. *Oeno One*, 52(4), 1–16.

Briche, E., Beltrando, G., Somot, S., Quénol, H. (2014). Critical analysis of simulated daily temperature data from the ARPEGE-climate model: Application to climate change in the Champagne wine-producing region. *Climatic Change*, 123, 241–254.

Brisson, N. and Levrault, F. (2010). *Climate Change, Agriculture and Forests in France: Simulations of the Impacts on the Main Species. The Green Book of the CLIMATOR project (2007–2010)*. ADEME, Angers.

Cabre, M.F., Nunez, M., Quénol, H. (2016). Regional climate change scenarios applied to viticultural zoning in Mendoza, Argentina. *International Journal of Biometeorology*, 60, 1325–1340.

Daniels, A.E., Morrison, J.F., Joyce, L.A., Crookston, N.L., Chen, S.C., McNully, S.G. (2012). Climate projections FAQ. Technical report, USDA, 1–32.

Delay, E., Piou, C., Quénol, H. (2015). The mountain environment, a driver for adaptation to climate change. *Land Use Policy*, 48, 51–52.

Duchêne, E. and Schneider, C. (2005). Grapevine and climatic changes: A glance at the situation in Alsace. *Agronomy for Sustainable Development*, 25(1), 93–99.

Fraga, H., Malheiro, A.C., Moutinho-Pereira, J., Jones, G.V., Alves, F., Pinto, J.G., Santos, J.A. (2014). Very high resolution bioclimatic zoning of Portuguese wine regions: Present and future scenarios. *Regional Environmental Change*, 14(1), 295–306.

Garcia de Cortazar, A.I. (2006). Adaptation du modèle STICS à la vigne (*Vitis vinifera L.*). Utilisation dans le cadre d'une étude du changement climatique à l'échelle de la France. Thesis, École supérieure nationale d'agronomie, Montpellier.

Hannah, L., Roehrdanz, P.R., Ikegami, M., Shepard, A.V., Shaw, M.R., Tabor, G., Zhi, L., Marquet, P.A. and Hijmans, R.J. (2013). Climate change, wine, and conservation. *Proceedings of the National Academy of Sciences of the USA*, 110, 6907–6912.

Huglin, P. (1978). Nouveau mode d'évaluation des possibilités héliothermiques d'un milieu viticole. *Comptes rendus des séances de l'académie d'agriculture de France*, 64, 1117–1126.

IPCC (2014). Climate Change 2014: Synthesis report. Contribution of working groups I, II and III to the Fifth Assessment Report of the Intergovernmental Panel on Climate Change [Core Writing Team, Pachauri, R.K. and Meyer, L.A. (eds)]. Report, IPCC, Geneva.

Irimia, L.M., Patriche, C.V., Quénol, H., Sfîcă, L., Foss, C. (2018). Shifts in climate suitability for wine production as a result of climate change in a temperate climate wine region of Romania. *Theoretical and Applied Climatology*, 131(3–4), 1069–1081.

Joly, D., Brossard, T., Cardot, H., Cavailhès, J., Hilal, M., Wavresky, P. (2009). Interpolation par régressions locales : application aux précipitations en France. *L'Espace Géographique*, 38(2), 157–170.

Jones, G.V. (2006). Climate and terroir: Impacts of climate variability and change on wine. *Geoscience Canada*, 9.

Jones, G.V., Duchêne, E., Tomasi, D., Yuste, J., Braslavska, O., Schultz, H., Martinez, C., Boso, S., Langellier, F., Perruchot, C., Guimberteau, G. (2005). Changes in European winegrape phenology and relationships with climate. In *XIV International GESCO Viticulture Congress*. Geisenheim, August, 23–27, 54–61.

Jones, G.V., Moriondo, M., Bois, B., Hall, A., Duff, A. (2009). Analysis of the spatial climate structure in viticulture regions worldwide. *Le Bulletin de L'OIV*, 82, 507–518.

Le Roux, R., de Rességuier, L., Corpetti, T., Jégou, N., Madelin, M., Van Leeuwen, C., Quénol, H. (2017a). Comparison of two fine scale spatial models for mapping temperatures inside winegrowing areas. *Agricultural and Forest Meteorology*, 247, 159–169.

Le Roux, R., de Rességuier, L., Katurji, M., Zawar-Reza, P., Sturman, A., van Leeuwen, C., Quénol, H. (2017b). Analyse multi scalaire de la variabilité spatiale et temporelle des températures à l'échelle des appellations viticoles de Saint-Émilion, Pomerol et leurs satellites. *Climatologie*, 14, 1–17.

Le Roux, R., Katurji, M., Zawar-Reza, P., Quénol, H., Sturman, A. (2018). Comparison of statistical and dynamical downscaling the WRF model. *Environmental Modeling and Software*, 100, 67–73.

Madelin, M. (2004). L'aléa gélif printanier dans le vignoble marnais en Champagne. Modélisation spatiale à une échelle fine des écoulements de l'air et des températures minimales. PhD thesis, Université Paris-Diderot – Paris VII, Paris.

Moriondo, M., Jones, G.V., Bois, B., Dibari, C., Ferrise, R., Trombi, G., Bindi, M. (2013). Projected shifts of wine regions in response to climate change. *Climatic Change*, 119(3/4), 825–839.

Neethling, E. (2016). Adaptation de la viticulture au changement climatique : vers des stratégies à haute résolution. PhD thesis, Université Rennes 2, Rennes.

Neethling, E., Petitjean, T., Quénol, H., Barbeau, G. (2017). Assessing local climate vulnerability and winegrowers' adaptive processes in the context of climate change. *Mitig. Adapt. Strateg. Glob. Change*, 22(5), 777–803.

Neethling, E., Barbeau, G., Coulon-Leroy, C., Quénol, H. (2019). Spatial complexity and temporal dynamics in viticulture: A review of climate-driven scales. *Agricultural and Forest Meteorology*, 276, 107618.

Ouzeau, G., Déqué, M., Jouini, M., Planton, S., Vautard, R., Jouzel, J. (2014). Le climat de la France au XXIe siècle. Report, Direction générale de l'énergie et du climat [Online]. Available at: www.Developpement–durable.gouv.fr.

Porter, J.R. and Xie, L. (2014). Food security and food production systems. In *Climate Change 2014: Impacts, Adaptation, and Vulnerability*, Field, C.B., Barros, V.R., Dokken, D.J., Mach, K.J., Mastrandrea, M.D., Eren Bilir, T., Chatterjee, M., Ebi, K.L., Estrada, Y.O., Genova, R.C., Girma, B., Kissel, E.S., Levy, A.N., MacCracken, S., Mastrandrea, P.R., White, L.L. (eds). IPCC, Geneva, 485–533.

Quénol, H. (2014). *Changement climatique et terroirs viticoles.* Lavoisier, Paris.

Quénol, H. (2017). Viticulture – experimentation or adaptation ? In *Adaptating to Climate Change*, Thiebault, S., Laville, B., Euzen, A. (eds). ediSens, Paris.

Quénol, H. and Bonnardot, V. (2014). A multi-scale climatic analysis of viticultural terroirs in the context of climate change: The "TERADCLIM" project. *International Journal of Vine and Wine Sciences*, 23–32.

Quénol, H., Grosset, M., Barbeau, G., Van Leeuwen, K., Hofmann, M., Foss, C., Irimia, L., Rochard, J., Boulanger, J.P., Tissot, C., Miranda, C. (2014). Adaptation of viticulture to climate change: High resolution observations of adaptation scenario for viticulture. The Adviclim European project. *Bulletin de l'OIV*, 87, 395–406.

Quénol, H., de Cortazar Atauri, I., Bois, B., Sturman, A., Bonnardot, V., Le Roux, R. (2017). Which climatic modeling to assess climate change impacts on vineyards? *Oeno One Journal*, 51(2), 91–97.

Salomon, J.N. (2017). Nouveaux vignobles et évolution des anciens face à la mondialisation. *Les Cahiers d'Outre-Mer*, 231–232.

Santos, J.A., Malheiro, A.C., Pinto, J.G., Jones, G.V. (2012). Macroclimate and viticultural zoning in Europe: Observed trends and atmospheric forcing. *Climate Research*, 51(1), 89–103.

Tonietto, J. (1999). Les macroclimats viticoles mondiaux et l'influence du mésoclimat sur la typicité de la syrah et du muscat de Hambourg dans le sud de la France : méthodologie de caractérisation. PhD thesis, École nationale supérieure agronomique, Montpellier.

Tóth, J.P. and Végvári, Z. (2016). Future of winegrape growing regions in Europe. *Australian Journal of Grape and Wine Research*, 22(1), 64–72.

Van Leeuwen, C., Bois, B., Cellie, N., Tregoat, O., Roby, J.P. (2009). Les modifications de l'expression du terroir induites par les changements climatiques nécessitent une adaptation du matériel végétal et des techniques viticoles. *Revue française d'œnologie*, 235, 10–14.

Van Leeuwen, C., Schultz, H., Garcia de Cortazar-Atauri, I., Duchêne, E., Ollat, N., Pieri, P., Bois, B., Goutouly, J.P., Quénol, H., Touzard, J.M., Malheiro, A.C., Bavaresco, L., Delrot, S. (2013). Why climate change will not dramatically decrease viticultural suitability in main wine-producing areas by 2050. *Proceedings of the National Academy of Sciences of the USA*, 110(33), E3051–E3052.

Webb, L., Whetton, P.H., Barlow, E.W.R. (2008). Climate change and winegrape quality in Australia. *Climate Research*, 36, 99–111.

Winkler, A., Cook, J., Kliewer, W., Lider, L. (1974). *General Viticulture*. University of California Press, Berkeley.

12

Climate Change in the Amazon: A Multi-scalar Approach

Vincent DUBREUIL[1], Damien ARVOR[2], Beatriz FUNATSU[3], Vincent NÉDÉLEC[1] and Neli DE MELLO- THÉRY[4]

[1] *University of Rennes 2, France*

[2] *CNRS, Rennes, France*

[3] *CNRS, Nantes, France*

[4] *University of São Paulo, Brazil*

12.1. Introduction

The Amazonian space can be defined in different ways. As its key feature is the Amazon River, the most convenient is to define it by its watershed, totaling nearly 6.2 million km^2 spread over nine countries, more than 60% of which is in Brazil. The Amazon is also by far the largest hydrological system on the planet in terms of both its length (over 7,000 kilometers) and average flow (nearly 200,000 m^3/s, or 15% of the freshwater that flows into the oceans). A second way of defining this area is to consider the Amazonian biome: it is not only forested but it is also the largest tropical forest in the world with nearly 6.5 million km^2. The two definitions diverge in the south, where the watershed is largely occupied by savannahs (*cerrado*), in the north, where the Amazonian biome spills over into the Orinoco basin and the Guianas, and in the west into the Andean peaks. A third definition is based on administrative divisions set up by the different countries that share this space: in Brazil, for example, the legal Amazon, delimited in 1953 to determine the regions eligible for development aid, corresponds to the states of the northern region

(Acre, Amapá, Amazonas, Pará, Rondônia and Roraima), the states of Mato Grosso, Tocantins and Maranhão west of the 44th meridian, that is more than 50% of the country's surface area. Whatever the definition chosen, the area is therefore gigantic (larger than Western Europe), largely dominated by Brazil and located mostly in the southern hemisphere.

Since its discovery by the Spanish and then by the Portuguese, the Amazon has been a coveted and emblematic territory (Théry 1997; Droulers 2014). It is still so today in a context of climate change and strong anthropic pressure. This pressure is exerted firstly in terms of deforestation and the exploitation of forest resources (wood) but also its subsoil (minerals). The scale of these transformations (5,000 to 28,000 km^2 deforested each year in the Brazilian Amazon alone) thus raises a series of major questions and issues: in terms of biodiversity erosion, as it is estimated that Amazonian ecosystems host 10 to 15% of the Earth's biodiversity (Marengo *et al.* 2018); in terms of carbon storage or emission as Amazonian forests store between 150 and 200 billion tons of carbon (Saatchi *et al.* 2011) and in the mid-2000s, deforestation was responsible for nearly 75% of Brazil's CO_2 emissions. Finally, in terms of climate, the question of the decrease in the process of water recycling by the forest, as well as the impact of global warming and changes in surface conditions, raises the question of the role of the Amazon in the global water and carbon cycles (Aragão *et al.* 2014). However, the question is not only of a global scale as climate change is also observed at the local scale, mainly due to the artificialization of surfaces (urbanization and deforestation). In this chapter, we will therefore first present the Amazonian climate system before addressing the observed changes and then the questions on future changes and political stakes.

12.2. The Amazonian climate system

12.2.1. *Heat, humidity and regional diversity*

The French expression "*touffeur équatoriale*" (Droulers 2014), referring to hot, "stuffy" conditions, describes the Amazonian climate well at first glance. The Amazon is indeed the largest warm and humid continental domain on Earth. The average rainfall in the watershed is almost 2,200 mm per year, but it often reaches 2.5 meters, with records exceeding 3 meters in Brazil and more than 6 meters on the eastern slopes of the Andes (see Figure 12.1). Everywhere, except at high altitudes, annual thermal averages are above 25°C and few months average less than 18°C (the traditional warm-zone limit). Seasonal and interannual variability, however, adds some nuance to these generalizations.

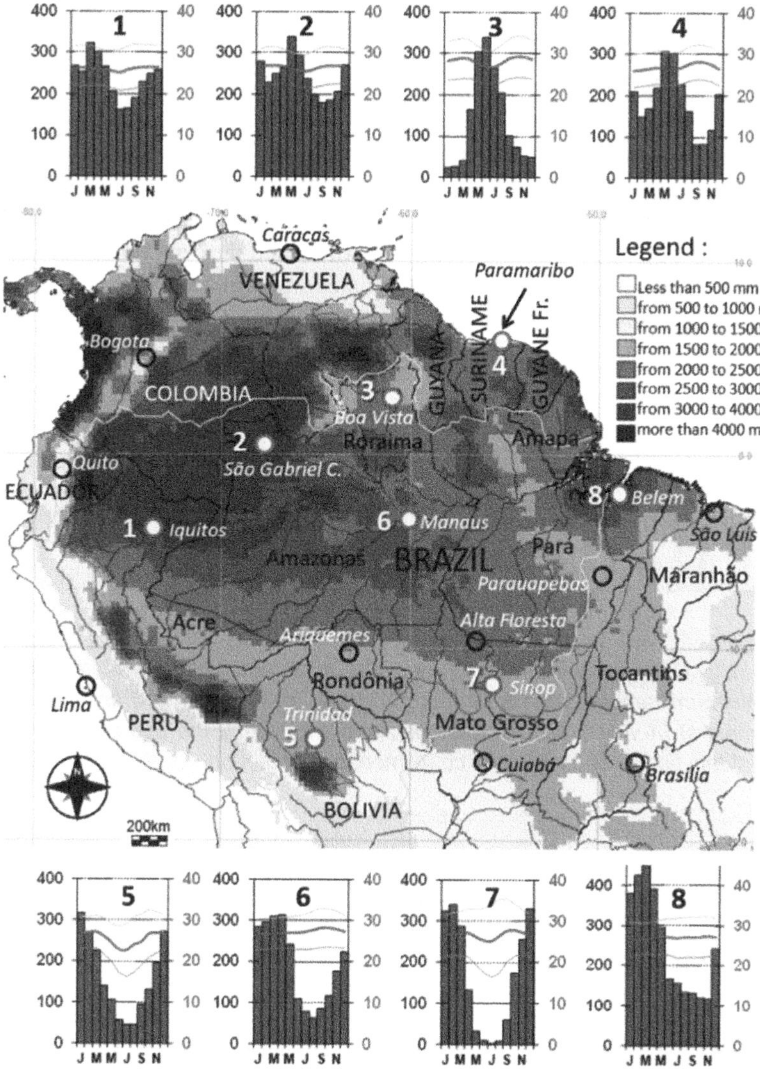

Figure 12.1. *Location and synthesis map of the Amazonian climate. For a color version of this figure, see www.iste.co.uk/mercier/climate.zip*

COMMENT ON FIGURE 12.1. – *Main map: mean annual precipitation (TRMM data 1998–2015); red line = boundary of the Amazon biome; light blue line = boundary of the Amazon watershed. From 1 to 8, mean climate diagrams (period 1971–2010); precipitation (histogram in millimeters, in blue) and mean (red), minimum (light blue) and maximum (orange) temperatures (source: designed and created by the authors).*

Temperatures thus show a certain variability even if the average annual temperature amplitudes are low (3 to 8°C). The diurnal rhythm, above all, is quite marked, especially in the dry season: at 33–36°C (rarely more) during the day, lower temperatures follow during the night, most often between 18 and 24°C (see Figure 12.2). The rainy season is more monotonous due to cloud cover, which further limits diurnal thermal contrasts. In the southern part of the basin, cool air descents (in latitude) are not unknown: these "so called" *friagem* episodes take place due to the absence of a topographic barrier between the Andes and the central Brazilian plateau and bring temperatures down below 10°C, sometimes north of 10° S in low regions, which is another singularity in this context of tropical regions.

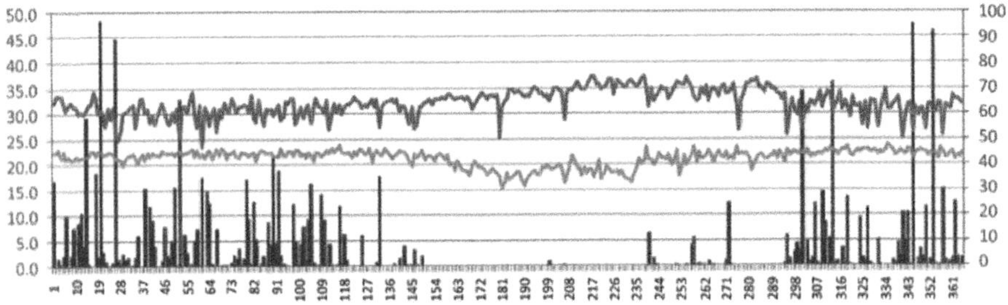

Figure 12.2. *Daily temperature and precipitation in Alta Floresta (Mato Grosso) in 2014: minimum temperature (in degrees) in blue, maximum in red and precipitation (in millimeters) in violet (source: designed and created by the authors). For a color version of this figure, see www.iste.co.uk/mercier/climate.zip*

Precipitation is mostly of the convective type, brought by cumuliform clouds, whose diurnal cycle usually shows a maximum in the afternoon. If the cells are isolated and small in size, the showers are brief but intense; most often they are organized in larger structures in the form of squall lines or mesoscale convective systems (from a few tens to a few hundred kilometers in extent). Convection develops at the end of the morning with the heating of the lower atmospheric layers which creates an unstable thermal stratification (Funatsu *et al.* 2012). The higher the humidity, the more vigorous is the upward motion and the stronger the convergence of the lower layers. In the east (Belém, French Guiana), sea breeze fronts reinforce this mechanism and help explain the higher rainfall. Precipitation variability increases towards the south and east and all regions are likely to be hit by long periods of abnormally low rainfall, such as during the droughts that affected the basin in 2005 or 2010.

In Amazonia, the major climatic rhythms are therefore dictated by rainfall. Most often, the dry (or less rainy) season is opposed to the rainy season which lasts from December to March in the south and from May to October in the north. The region has humid tropical climates with more or less contrasting seasons: the Af, Am and Aw types of the Köppen classification (Dubreuil *et al.* 2019). The dry season, which is short or even absent in the central western Amazon, develops towards the north (Roraima), south (Mato Grosso) and east (Maranhão) at the same time as the average annual rainfall total decreases. We then leave the equatorial environment of the rainforest to enter the world of the savannahs. At the scale of the entire Amazon basin, the maps show the weak representation of bimodal rainfall patterns: the equatorial type regime with 2 maxima (March–June and October–January) is only found in certain sectors of Guyana (very slight retreat in February-March in Paramaribo) and Colombia (in the intra-Andean valleys). Everywhere else, it is the tropical regime with a single summer rainy season that prevails. For the majority of Brazilian authors (Nimer 1989), it is above all the presence and duration of the dry season that distinguishes the so-called "equatorial" from "tropical" climates. In almost all of the Amazon, there is therefore a marked minimum rainfall (one dry month) or even a true dry season, the presence of which results from global circulation mechanisms.

12.2.2. *Radiation balance and general circulation*

The Amazon depends on a particular system to the Americas for the set of rainfall mechanisms, particularly the Intertropical Convergence Zone (ITCZ), which differs from Africa and does not have the same north-south asymmetries and a clear bioclimatic zonation. The annual net energy balance shows a surplus for the Amazon as a whole of the same order as that of the Congolese basin (between 40 and 60 W m^{-2}) and the region thus appears as a source region on a global scale. The low storage capacity on the continent also leads to a rapid return of energy to the atmosphere in the form of intense convective activity, followed by energy redistribution in the form of meridian (Hadley) and zonal (Walker) cells to restore the balance between latitudes and between the eastern and western sides of the oceans.

The southern summer meridional circulation is marked by the dominance of the northeasterly trade winds, which penetrate far into the interior of the continent, keeping an eastern component, often assimilated to a monsoon flow. The zonal circulation is marked by a strong continental convection over the whole of the Amazon and a convergence of low layers flux from the Atlantic, while the divergence in altitude (Bolivian anticyclone) and the subsidence at low layers concern the Atlantic and the northeast of Brazil. Towards the southeast, convective systems develop along the South Atlantic Convergence Zone (SACZ). This forms mainly during spring and summer in the southern hemisphere and is oriented NW-SE along the same axis as the subtropical jet. It is a zone of convergence in the humid and unstable air mass that connects the Amazon with the frontal systems from high latitudes.

Generally speaking, the Amazonian topographic system of a vast basin open to the Atlantic with the modest relief of the Guyana (1,280 m in Suriname, 2,810 m in the Roraima) and the Brazilian plateau (rarely more than 600 m), separated from the Pacific by high and continuous cordilleras, has decisive consequences: the low-level Atlantic flow penetrates into the interior of the continent, unlike the Pacific flow, and rainfall on the eastern slope of the Andes has an Atlantic origin despite the distance separating them. Furthermore, the eastward concave shape of the western cordillera favors the convergence of the monsoon flow at low levels.

Several studies have shown the sensitivity of precipitation in this region to oceanic forcing and in particular to the El Niño Southern Oscillation (ENSO), or the Atlantic meridional Sea Surface Temperature gradient. In the positive phase of ENSO, from December to February, a decrease in precipitation is generally observed over the eastern Amazon (from Amapá to Roraima) and northeast Brazil (Espinoza *et al.* 2009). On the contrary, rainfall is in excess over the whole domain when the Pacific is colder (La Niña) and the South Atlantic is warmer, reinforcing the intensity of the wet flow of the lower layers (Michot *et al.* 2018). In detail, correlations between sea surface temperatures (SST) and precipitation remain modest, generally around 0.6. Indeed, the effects of the two ocean basins may or may not be cumulative: for example, the presence of warm anomalies over the Atlantic mitigates the effects of a positive ENSO and part of the Amazon may then experience excess rainfall. Seasonally, the Atlantic thermal gradient is fairly well associated with rainfall between March and June while the links with the Pacific increase in the following months. In simplified terms, correlations with SSTs are better with the northern Amazon than with the southern part where the relationships are not very significant. It is in the latter region, on the other hand, that the climate-forest links are most acute.

12.2.3. *The forest-climate interaction issue*

In order to quantify the role of forests on climate and the magnitude of climate change due to deforestation, various experimental devices have been set up in Amazonia, often within the framework of international collaborations such as the Large-Scale Biosphere Atmosphere Experiment in Amazonia (LBA) program, a joint program between INPE (*Instituto Nacional de Pesquisas Espaciais*, the Brazilian National Institute for Space Research) and NASA (the US National Aeronautics and Space Administration). Several sites have been instrumented, sometimes with very heavy means such as the gigantic flow tower of more than 300 meters, installed near Manaus. These devices have enabled progress to be made in understanding forest-atmosphere interactions (Nobre *et al.* 1991), sometimes challenging certain preconceived ideas.

Thus, the Amazon is often presented as the "green lung" of the planet. The image is very inaccurate (because a lung emits CO_2 and consumes oxygen and not the other way around!) and in reality, even false: indeed, the Amazonian forest consumes as much oxygen as it produces! It is true that the growth of vegetation leads to the fixation of carbon in the soil, woody biomass and a release of oxygen, but in a forest in equilibrium, the death and then decomposition of this vegetation will consume almost as much oxygen. Of course, deforestation stops the process of growth and release of oxygen through the leaves of the trees and fires inject ash and CO_2 into the atmosphere. But if we consider the whole natural forest cycle from an oxygen perspective, the Amazon is not the "green lung" of the planet as it has long been believed. But the action of vegetation does not stop there and is more complex.

The most important fact is the essential role played by tropical forests on the climate system via the water cycle, an ancient idea as Buffon, already in the 18th Century, said that "forests make it rain". It is estimated that in the Amazon, for example, 80 to 90% of the available radiative energy is used for evapotranspiration. The remaining 10–20% is used to warm the ambient air. In this region, the latent heat transfer flux is therefore much higher than the sensible heat flux. Moreover, in the Amazon, the evapotranspiration flux (1,000 to 1,200 mm) corresponds on average to 50% of the rainfall: a good part of the moisture in the atmosphere therefore comes from the water vapor provided by the forest. On a local and regional scale, the Amazonian forest controls rainfall and temperature through evapotranspiration, according to a process called "moisture recycling" that varies, depending on the authors, from 35 to 80% (Marengo *et al.* 2018).

More recently, Makarieva *et al.* (2013) proposed the "biotic pump" theory, suggesting that local evaporation and condensation can have a major influence on atmospheric dynamics: the principle is that large forested landmasses such as the Amazon sustain high rainfall, but also that water vapor transferred to the atmosphere by evaporation from forests represents a potential energy reserve available to accelerate turbulent flows and winds, thus exercising control over the circulation itself. The contribution of the Amazon rainforest thus goes far beyond the water cycle alone (local and regional) but has a wider impact on the entire climate system.

Thus, the moisture generated by the Amazonian forest is also redistributed more widely, particularly towards southeast Brazil, via a low layer flow located on the eastern flank of the Andes and often compared to a flying river. Although climate change may affect this flow of moisture from the Amazon, particularly to the Plata Basin, deforestation resulting from intensive land-use activities poses a more immediate threat to these regions (Marengo *et al.* 2018).

12.3. A changing system: deforestation, warming and drying?

12.3.1. *Pioneering dynamics: rise and (provisory?) decline*

Before studying the climate dynamics observed over the last 40 years, it should first be recalled that the Amazon has undergone profound transformations in terms of land-use or landscape. Deforestation in the Brazilian Amazon represents a cumulative area of nearly 750,000 km^2 since 1980 (INPE-PRODES 2019), that is nearly one-fifth of its total surface area.

To this gross deforestation (clear-cutting of the forest) should be added probably at least an equivalent area of forest degraded by fire or logging. This process is particularly advanced in its eastern and southern parts, especially in the states of Mato Grosso, Pará and Rondônia (Dubreuil 2002). Throughout this region of the deforestation arc, forests and savannahs have been replaced by agricultural commodities (soya, maize, cotton) but above all by extensive grazing for meat production. These lands were opened up as part of public and private colonization projects based on roads and urban infrastructures set up by military governments, particularly from 1970 onwards as part of the national integration program. The impacts on the environment and traditional societies have been considerable and have long been relegated to the background.

Since 1988, INPE has been conducting the program for monitoring deforestation by satellite (PRODES). Using mainly LANDSAT data, it provides the Brazilian government with reliable figures on the progress of deforestation, aggregated by municipalities, states, etc. The graph in Figure 12.3 shows the strong progression of deforestation until 2004 (nearly 28,000 km^2 cleared), then the decrease until 2012 (4,500 km^2) and an increase in recent years (7,500 km^2 in 2018). A second real-time satellite detection program (DETER) was set up in 2006 to provide IBAMA (*Instituto Brasileiro do Meio Ambiente e dos Recursos Naturais Renováveis)* with the capacity to catch violators of environmental legislation in the act. Other satellite observation programs provide interesting data, particularly on fires, used to burn forest or renovate pasture plots. Few regions in the world are monitored in this way and provide regularly updated data.

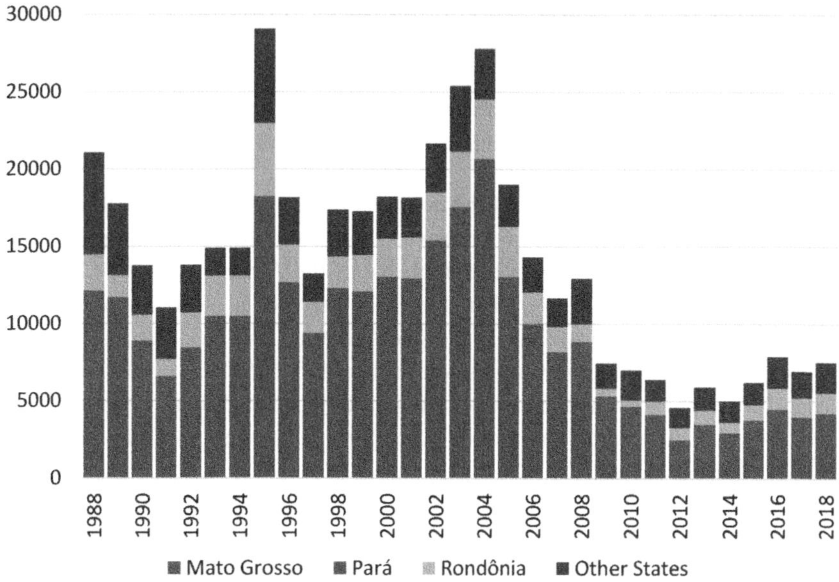

Figure 12.3. *Annual deforestation in the Brazilian Amazon in km*^{2} *(source: INPE 2019). For a color version of this figure, see www.iste.co.uk/mercier/climate.zip*

The factors that explain the annual variations in deforestation are multiple and combine economic logic (world demand for *commodities,* in particular), government incentives (first for the development and then for the preservation of the Amazon with the PPCDAM – *Plano de Ação para Prevenção e Controle do Desmatamento na Amazônia Legal* – in 2004) and international pressure (moratorium on soy and beef in 2006 and 2008). Far from being inevitable, deforestation in the Amazon is thus largely the result of public policies and the action (or inaction) of successive Brazilian governments, economic logics, international pressure and the role of NGOs in promoting sustainable development projects (Le Tourneau *et al.* 2013; Nepstad *et al.* 2014). In recent years, these evolutions have also become part of the debates on the reform of the forestry code and the progressive implementation of the Rural Environmental Registry (in Portuguese CAR for *Cadastro Ambiental Rural*). More generally, these transformations contribute to the artificialization of land whose impacts on the climate can be observed at different spatial scales.

12.3.2. *Increase in temperature and decrease in rainfall*

The question is whether there is already an observable climate trend in recent decades and whether this trend can be linked to anthropogenic actions such as increasing greenhouse gas concentrations and deforestation.

Concerning temperatures, observations show an increase in the average temperature in the Amazon of 0.6°C between 1973 and 2013 with a more marked warming during the dry season in the southeast of the Amazon (Alves *et al.* 2017). The signal is somewhat weaker than for other continental surfaces of the globe (Dubreuil 2018) but, as elsewhere in the world, the last two decades have been the warmest, with 2016 being the warmest year followed by 1998, two years marked by a strong ENSO phenomenon. Surveys carried out in 13 Amazonian communities by Dubreuil *et al.* (2017) show that the most frequent responses (58% of respondents) refer to a warming climate.

Precipitation provides a more complex pattern. In contrast to temperatures, studies show no statistically significant, global and unanimous rainfall trend over the last thirty or forty years in the region (Espinoza *et al.* 2009; Marengo *et al.* 2014; Silva Junior *et al.* 2015; Alves *et al.* 2017; Funatsu *et al.* 2019). However, rainfall has tended to increase in northwestern Amazonia since 1990, while southern Amazonia shows a seasonal trend of decreasing rainfall (Debortoli *et al.* 2015; Dubreuil *et al.* 2017).

The results and their interpretations may differ from one study to another depending on the data used and the period considered: the major problem being the low density of the observation network and the gaps in the series (Delahaye *et al.* 2015; Michot *et al.* 2019). Studies using satellite data have made it possible to refine the trends but over periods that remain too short to draw definitive conclusions (Fu *et al.* 2013; Arvor *et al.* 2017a). The study of the frequencies of annual climate types using the Köppen classification also made it possible to show the decrease in the frequency of wet year types (Af and Am) in favor of extended dry season types (Aw) between 1964 and 2015 (Dubreuil *et al.* 2019). In all cases, the length of the dry or wet season represents a key question.

12.3.3. *The dynamics of the start and end dates of the rainy season*

Since seasonal climatic rhythms are dictated by the alternation between the rainy and dry seasons, it is legitimate to look at whether climate change is manifested not only in terms of precipitation amount but also when and how the rainy season begins or ends. Various authors have pointed to a lengthening of the dry season in the region, mainly in the southern Amazon. The reasons for this are still under discussion, as this trend may be related to the large-scale influence of Atlantic SST gradients, or may be the result of reduced water recycling over cleared areas (Fu *et al.* 2013; Khanna *et al.* 2017; Sena *et al.* 2018).

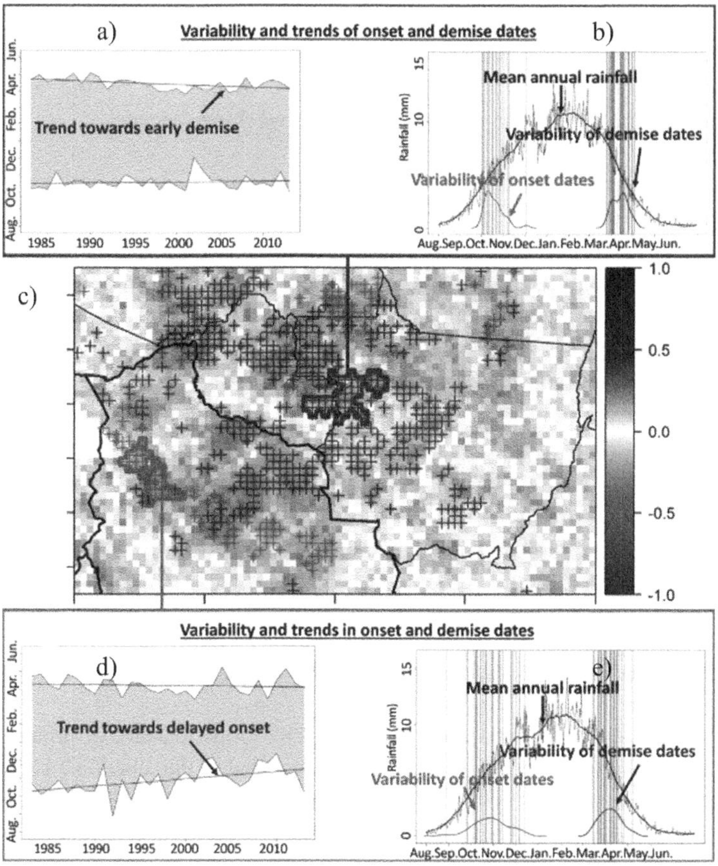

Figure 12.4. *Evolution of rainy season onset and demise dates in the southern Amazon from 1983 to 2014. For a color version of this figure, see www.iste.co.uk/mercier/climate.zip*

COMMENT ON FIGURE 12.4.– *Central map (c) shows trends in the length of the rainy season (t-value), with significant trends of late onset (green crosses) and early termination (blue crosses). The upper and lower parts show the 31-year variability and trends for the start and end dates in Bolivia (d) and (e) and at the border between the states of Rondônia and Mato Grosso (a) and (b) (source: Arvor* et al. *2017a).*

Thus, using data from ground stations, (Debortoli *et al.* 2015) showed for the period from 1971 to 2010 that 88% of the rainfall stations in the southern Amazon (from Rondônia to Tocantins) had experienced an increase in the length of the dry season, the majority of them with an earlier end of the rainy season. The use of remote sensing data over three decades (1983–2014) allowed Arvor *et al.* (2017a) to confirm significant trends towards a shortening of the rainy season in the southern Amazon, mainly related to earlier demise dates in Mato Grosso and Rondônia and later onset dates in Bolivia (see Figure 12.4). According to the authors, the lengthening of the dry season can reach almost a week per decade.

These changes are significant because the timing of the onset and demise of the rainy season is one of the major factors in controlling ecosystems (phenology, hydrology) and also imposes strong constraints on agricultural practices, as most of the production is rainfed. In regions of intensive agriculture, such as Mato Grosso, there is thus a strong correlation between the length of the rainy season and the proportion of adoption by producers of the double cropping system (i.e. a soybean harvest followed by a maize or cotton harvest during the same rainy season (Arvor *et al.* 2014). Current trends thus highlight a major paradox concerning the evolution of agriculture in Mato Grosso: indeed, the generalization of double cropping systems in this State, initially made possible by favorable rainfall conditions, would indirectly (as a factor of deforestation) cause the decline of agriculture by contributing to the establishment of a shorter rainy season, which could precisely prevent the adoption of double cropping systems.

12.3.4. *Local effects of land-use changes*

At the local scale, the replacement of forest by crops, pasture or urbanized areas first has an impact on the radiation balance. By increasing the albedo (0.11 to 0.14 for forests compared to 0.18 to 0.22 for pastures; albedo values in cities are highly variable and dependent on the type of surface) and decreasing roughness, deforestation modifies the flux transfers between the surface and the atmosphere. The replacement of tree vegetation by crops also results in a decrease in latent heat flux and an increase in sensible heat flux, especially during the dry season: during this period, in fact, the tree roots allows them to draw water from the soil and continue to ensure a latent heat flux comparable to the wet season (Dubreuil *et al.* 2012).

At the local scale, the increase in the proportion of sensible heat flow at the expense of latent heat flow thus results in an increase in temperatures in cleared areas. In cities, the absence of vegetation reduces the latent heat flow and most of the energy is available to heat the atmosphere: this phenomenon is amplified at the beginning of the night, when buildings release the heat stored during the day.

For example, measurements carried out in the Alta Floresta region in northern Mato Grosso show significant differences between stations located in different land-use contexts. The annual average varies from 24.8°C under forest cover to 27.2°C in the city center; in the pastures the values are intermediate, generally around 26.5°C. The annual rhythm of the minimum and maximum values (see Figure 12.5) helps to understand these contrasts. The minima in the pastures (21.2°C) are similar to the forest (21.2 and 21.0°C), except for the month of August when it is slightly warmer in the cleared sectors (18.2°C for the pastures and 17.5°C in the forest); for the minima, there is no difference between the temperature under forest cover and that measured in the forest clearing.

The maximum temperatures show more contrasts but the curves for the four stations studied here remain parallel to each other, that is, the presence of the dry or wet season does not change the order of the hottest and coldest sectors. The pastures are always warmer (31.7°C) than the forest, where the difference becomes very significant between the station under forest cover (28.7°C) and the forest clearing (31.2°C). This difference is even greater during dry season, with the warmest month (August) having a difference of nearly 4°C between the forest clearing (35.1°C, almost as much as in pasture) and the forest (31.6°C). Overall, the temperature differences in the forest remain less pronounced (7.6°C between the maximum and minimum temperatures) than in the cleared areas (10 to 10.5°C) regardless of the season, but this is mainly related to the reduction in maximum temperatures under forest cover. The effect on minima is negligible and deforestation is therefore not accompanied by a significant drop in night temperatures: the difference is therefore essentially marked during the day and during dry season, which is also reflected in the relative humidity curves.

Urbanization, which is often forgotten when we talk about climate change in the Amazon, is far from being a negligible phenomenon. In the Amazon, as elsewhere in Brazil, nearly 85% of the population lives in cities. Apart from the historic river cities founded during the Portuguese domination, most of the cities were founded in the 1970s as roads and pioneer fronts were opened. Some of them are already large cities: Ariquemes in Rondônia, Sinop in Mato Grosso or Parauapebas in Pará are cities founded in the 1970s whose population already exceeds 100,000 inhabitants. The meteorological records in Sinop, for example, clearly show the urban heat island effect with average temperature differences of over 4°C at night and during the dry season between the center and the rural periphery (see Figure 12.6). At the local level, urbanization is thus also proving to be a powerful agent for transforming the Amazonian climate. This process will probably continue and be reinforced in the future with continued urban expansion and global changes.

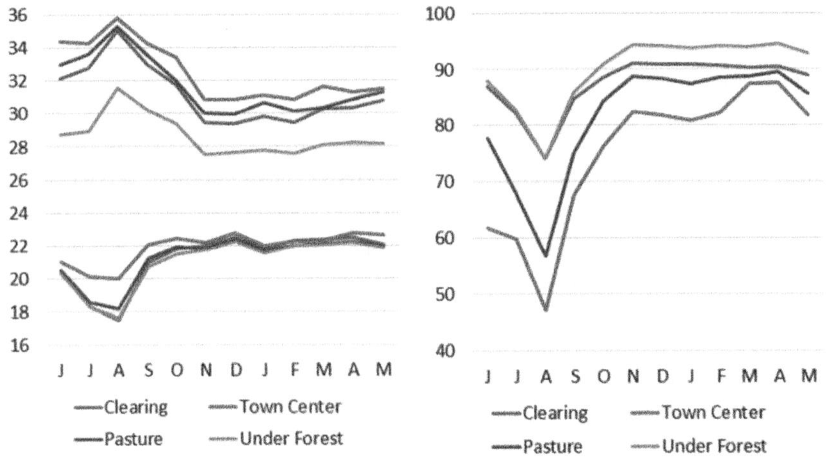

Figure 12.5. *Temperatures (left, in°C) and relative humidity (right, in %) from June 2013 to May 2014 in Alta Floresta (Mato Grosso) for different land use types (source: based on Dubreuil* et al. *2015). For a color version of this figure, see www.iste.co.uk/mercier/climate.zip*

Time	J	F	M	A	M	J	J	A	S	O	N	D	AN
00:00	1.1	0.5	1.1	1.4	2.8	2.3	1.9	3.2	2.0	1.7	0.7	0.9	**1.6**
01:00	1.0	0.5	1.1	1.2	2.7	2.3	2.2	3.2	2.0	1.7	0.6	0.9	**1.6**
02:00	0.9	0.4	0.9	1.2	2.5	2.2	2.6	3.2	2.0	1.6	0.6	0.8	**1.6**
03:00	0.8	0.4	0.9	1.1	2.4	2.2	2.8	3.3	2.1	1.6	0.5	0.9	**1.6**
04:00	0.8	0.4	0.8	1.0	2.2	2.1	2.8	3.4	2.2	1.5	0.4	0.8	**1.5**
05:00	0.8	0.4	0.8	1.0	2.1	2.1	2.8	3.4	2.2	1.4	0.4	0.8	**1.5**
06:00	0.8	0.4	0.7	0.9	2.0	1.9	2.7	3.4	2.1	1.2	0.3	0.6	**1.4**
07:00	1.1	0.5	1.0	1.2	2.6	2.1	2.6	2.8	1.7	0.7	0.4	0.8	**1.5**
08:00	1.1	0.4	1.3	1.0	2.0	3.0	2.4	0.8	0.0	0.1	0.0	0.1	**1.0**
09:00	0.3	-0.3	0.6	0.5	0.9	1.8	1.1	-0.2	-0.7	-0.3	-0.4	-0.4	**0.2**
10:00	-0.4	-0.6	0.2	0.2	0.4	0.6	0.2	-0.7	-1.3	-0.6	-0.5	-0.6	**-0.3**
11:00	-0.4	-0.7	0.1	0.2	0.0	-0.1	-0.3	-0.7	-1.1	-0.6	-0.4	-0.6	**-0.4**
12:00	-0.7	-1.1	-0.1	0.4	0.5	-0.1	-0.3	-0.7	-1.1	-0.3	-0.9	-0.4	**-0.4**
13:00	-0.3	-0.8	0.0	0.1	0.3	-0.1	-0.5	-0.5	-1.1	0.0	-0.8	-0.6	**-0.4**
14:00	-0.8	-1.0	-0.6	0.2	0.5	0.0	-0.2	-0.3	-0.8	0.9	-0.7	-0.4	**-0.3**
15:00	-0.6	-0.6	0.0	0.5	1.0	0.0	0.0	0.7	-0.5	1.5	0.1	-0.1	**0.2**
16:00	-0.3	-0.1	0.8	0.6	1.2	0.5	0.6	1.2	0.5	2.1	0.4	0.4	**0.7**
17:00	0.0	0.1	1.3	1.0	1.4	1.2	1.4	2.9	1.1	2.2	0.9	0.7	**1.2**
18:00	0.6	0.2	1.3	1.5	2.5	2.1	2.6	4.7	1.9	2.4	1.1	1.0	**1.8**
19:00	1.3	0.7	1.8	2.2	3.5	3.2	3.5	4.9	2.2	2.3	1.4	1.4	**2.3**
20:00	1.7	0.6	2.0	2.3	3.6	3.2	3.1	4.2	2.1	2.2	1.4	1.5	**2.3**
21:00	1.7	0.6	1.8	2.1	3.3	2.9	2.3	3.6	2.1	2.3	1.2	1.3	**2.1**
22:00	1.7	0.6	1.6	1.9	3.0	2.7	1.9	3.2	1.9	2.1	1.0	1.2	**1.9**
23:00	1.3	0.4	1.3	1.6	2.7	2.4	1.8	3.0	2.0	2.0	0.8	1.1	**1.7**
00:00	1.1	0.4	1.1	1.4	2.7	2.4	1.9	3.1	1.9	1.7	0.7	1.0	**1.6**
	0.6	**0.1**	**0.9**	**1.1**	**1.9**	**1.7**	**1.7**	**2.2**	**1.0**	**1.3**	**0.4**	**0.5**	**1.1**

Figure 12.6. *Urban heat island intensity in Sinop in 2018: average temperature difference (in degrees) per hour (row) and per month (column) between the center and the periphery (source: designed and created by the authors). For a color version of this figure, see www.iste.co.uk/mercier/climate.zip*

12.4. Uncertainties of future changes, perceptions and adaptations

12.4.1. *Savanization and tipping points*

To study future climate change, in the Amazon as elsewhere, the use of global and regional physical models is essential: their results may show significant differences because feedback processes remain poorly known and the quantification of certain parameters (root depth, soil water content) uncertain. Some climate models suggest that the impact of deforestation on climate in the long run could depend on a threshold of deforested area and the spatial organization of remaining forest patches. Thus, Sampaio *et al.* (2007) suggested that a 40% threshold value of regional deforestation could constitute a tipping point followed by a sudden change of climate in the region towards a warmer and drier climate, thus generating a savanization. This phenomenon would be of particular concern for the southern part of the Amazon, which is already the least rainy and most exposed to deforestation. Recent extreme climatic events in the region, such as droughts (2005, 2010, 2016) and floods (2009, 2012, 2014), changes in rainy and dry seasons, increased risk of fires and their impacts on climate, health and biodiversity would thus be the first signs of what could happen in the Amazon with further climate change (Alves *et al.* 2017; Boers *et al.* 2017; Marengo *et al.* 2018). The combination of severe droughts and floods would put additional stress on the Amazonian forests, the consequences of which are still poorly understood, including in terms of biodiversity (Barlow *et al.* 2016).

However, we must remain cautious about future developments because the relationship between Amazonia and climate is not necessarily linear and is sometimes counter intuitive. For example, (Debortoli *et al.* 2016) showed that there were no significant correlations between the trends observed in rainfall stations in the southern Amazon and the proportion of land cleared in buffer zones of varying radii around these same stations. The correlations, although low in most cases, do indicate, however, that the larger the buffer zone, the higher the degree of correlation between land use and rainfall.

Numerical simulations have also shown that local deforestation could lead to a vegetation-related breeze effect, resulting in a local increase in rainfall in deforested areas (Lawrence and Vandecar 2015). Thus, fragmented deforestation could encourage an initial increase in rainfall, which would locally protect drier ecosystems. However, after a certain level of deforestation, typically beyond higher patches of 30–50 km (Debortoli *et al.* 2016), this relationship can be reversed, leading to drying out and reduced rainfall by altering the process of water recycling through the forest (Makarieva and Gorshkov 2010).

Finally, several studies show that the effects of deforestation are greater in times of drought, as fires started to clear land can become uncontrollable and burn larger areas, especially forests that have already been logged. Forest fires, drought and logging increase the vulnerability of forests, while deforestation and smoke can inhibit rainfall and in turn increase fire risk, triggering a catastrophic feedback process (Marengo *et al.* 2018).

12.4.2. *An overall impact which is certain, but which remains to be specified*

The combined effect of large-scale climate change and local deforestation thus remains difficult to assess (Costa and Foley 2000; Cox *et al.* 2004). The fragmentation of the Amazon forests and the nature of the vegetation cover after deforestation are important factors in the intensity of future climate change. Numerical simulations have sought to quantify the impact on the climate of various deforestation scenarios (generally 40% to 100% of the Amazonian forest) and various more or less optimistic future greenhouse gas concentrations scenarios (see Figure 12.7). Global models show average temperature increases of around 4°C and up to 6°C for the end of the 21st Century for the worst-case scenario (RCP8.5). The decrease in precipitation would be most significant in the southern and eastern Amazon, up to 15% in 2020–2030 and up to 40–55% by 2100 (Sampaio *et al.* 2018).

Total deforestation scenarios would lead to a 10–20% decrease in annual rainfall across the Amazon as a whole and less moisture recycling, reducing rainfall by up to 40% in some areas (Swanna *et al.* 2015). Climate models show that with continued high deforestation rates (i.e. prior to 2004), a reduction of 8.1–10.4% in average annual rainfall in the Amazon basin could be observed by 2050. Spracklen and Garcia-Carreras (2015) show that tropical deforestation leads to reduced evapotranspiration, increasing surface temperatures by 1 to 3°C and modifying the circulation in the boundary layer, increasing precipitation in some areas and reducing it elsewhere.

Thus, a 40% deforestation scenario for the Amazon would result in a 12% reduction in rainfall in the rainy season and a 21% reduction in rainfall in the dry season for the entire Amazon basin. These heterogeneous changes in time and space will therefore have variable impacts on the Amazonian populations.

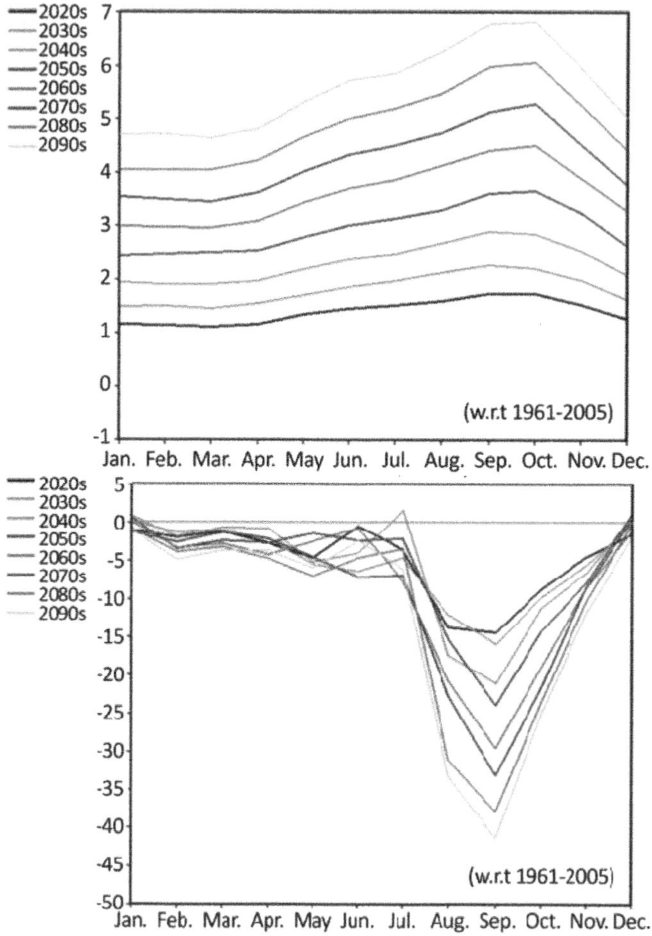

Figure 12.7. *Multi-model averages of the projected change in air temperature (top,°C) and precipitation (bottom, %) in the southern Amazon for the RCP8.5 scenario per decade over the period 2020–2100 relative to the baseline period 1961–2005 (source: Sampaio* et al. *2018). For a color version of this figure, see www.iste.co.uk/mercier/climate.zip*

12.4.3. *Perceptions and adaptations by local populations*

The Amazon is also widely studied because of its original settlement, where Amerindian communities, traditional populations living off forest and river resources, and a large migrant population attracted by the resources and recently opened spaces of the pioneer fronts coexist (Dubreuil 2002). These populations are therefore in a prime position to observe, and possibly experience, climate change. They also observe the weather to plan certain activities. Their perception of possible changes is therefore an interesting indicator in terms of climate, but also a prerequisite for the implementation of adaptation strategies. However, it is not easy to compare rainfall measurements (considered as "objective") with social perception data on rainfall because many biases (generational, cultural, social, etc.) can be put forward (Dubreuil *et al.* 2017). However, several studies conducted in the Amazon have shown the interest of such approaches.

Studying 13 communities in the Brazilian Amazon, Funatsu *et al.* (2019) show that 72% of the sampled population perceived some form of climate change, and that there is a robust signal of increasing perception with age. In terms of perception of precipitation, the average of responses across all sites gives a higher percentage (20%) in favor of shifting the start or end of the rainy season. In general, the responses also generally converge towards a decrease in precipitation (15.6%) associated with an increase in intensity (16.9%) and unpredictability (17.4%). However, an equally large proportion of the population (15.7%), on the contrary, does not observe any change in precipitation (see Figure 12.8). These disparate results indicate that it is difficult for populations to assess changes in rainfall. In all cases, the delay in the beginning or end of the wet season is therefore the most important change perceived by the population, a point also well established by the observations.

The surveys also show that perceptions of climate change and measured rainfall variations differ from region to region. Only in the southern Amazon are measured and perceived changes in rainfall patterns consistent with decreased rainfall. Thus, while perceptions are varied and the consistency with measured data is not always good, the deforestation arc in the southern Brazilian Amazon clearly shows a concomitance of decreasing rainfall and the higher perception of rainfall change by communities (Dubreuil *et al.* 2017).

In relation to the perception of these changes, few sites reported changes in recent practices or types of production related to environmental changes. The main changes mentioned concern Mato Grosso and correspond rather to a better adaptation to environmental conditions. Thus, several producers have started to irrigate their plots to secure production and make it effective throughout the year. In some regions, the conversion to milk production has led farmers to try to cushion the drop in production (and thus the loss of income) during the dry season by building water reservoirs or providing additional fodder. Finally, in intensively cultivated areas (Sorriso), in addition to the generalization of direct sowing, the double cropping systems are being extended to make the land productive throughout the rainy season;

for certain crops that may lack water at the beginning of the dry season (beans), the practice of irrigation is also becoming more widespread.

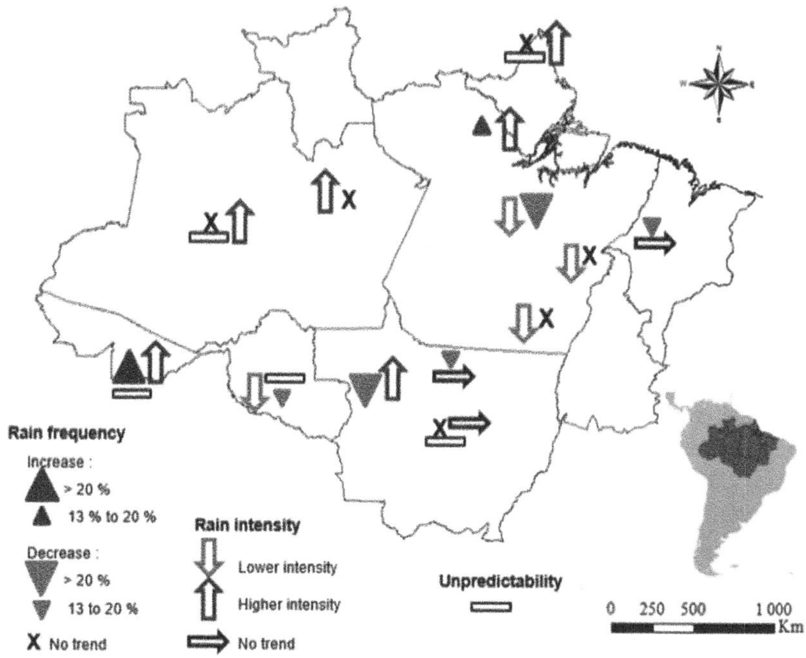

Figure 12.8. *Synthesis of perceptions of climate change in 13 Amazonian communities (source: based on Funatsu* et al. *2019). For a color version of this figure, see www.iste.co.uk/mercier/climate.zip*

At the local institutional level, the interviews conducted over the past few years point out that, although municipal authorities reaffirm their mission and value the objectives of economic, social and sustainable development, and public institutions and actors repeat in their speeches the main conditions of international agreements and protocols (COP 21), little has changed in their institutional behavior.

12.5. Conclusion: a stake in the global negotiations

The Amazon region is considered one of the world's most exposed regions to the risks associated with climate variability and change. Changes are observed and perceived at different scales and affect deforested areas and, even more so, urbanized regions. The share of different forcings, whether global (increase in greenhouse gases) or local (change in land-cover), is still debated in the scientific community: it is clear, however, that the risk to the Amazon is not only due to future climate change, but also to possible interactions with other processes, such as deforestation, forest fragmentation and degradation and fires. In such a context, many scientists believe that the region is approaching a tipping point that threatens the entire ecosystem and more broadly the entire continent.

Awareness of this risk in the mid-2000s led to the implementation of strong policies to protect the forest while negotiating its safeguard at the international level and actively participating in discussions on the fight against climate change. In Brazil, however, the National Climate Change Policy (PNMC in Portuguese) was only approved in 2009 and, while the decrease in deforestation from 2005 onwards has led to a significant drop in CO_2 emissions, those of methane and nitrous oxide have remained significant (De Mello-Théry and Dubreuil 2017). In 2010, Brazil also adopted the ABC plan (*Agricultura de Baixa Emissão de Carbono*), a multisectoral mechanism for mitigation and adaptation to climate change, of which several programs directly concern the Amazon: recovery of degraded pastures, integrated crop-livestock-forest or agroforestry systems, dissemination of no tillage practices, among others (Arvor *et al.* 2017b).

At COP 21 in 2015, Brazil committed to the sustainable development goals and resolutions adopted at the United Nations Conference on Sustainable Development (Rio + 20) in 2012. Brazil's Nationally Determined Contribution (NDC) under the Paris Agreement aims to reduce greenhouse gas emissions by 37% from 2005 levels by 2025 and by 43% from 2005 levels by 2030. This commitment was based on Brazil's implementation of the REDD+ (Reducing Emissions from Deforestation and Forest Degradation) scheme from 2016, whose objectives for the Brazilian Amazon are to move towards zero illegal deforestation by 2030 and to offset greenhouse gas emissions by restoring and reforesting 12 million hectares of forests by 2030. At the same conference in Paris, the government of the State of Mato Grosso (Brazil's leading state for soybean and cotton production!) presented the PCI (*Produzir, Conservar e Incluir*) strategy, whose objective is to raise funds for the State of Mato Grosso to increase and improve the efficiency of agricultural and forestry production, preserve remaining forests, reduce environmental liabilities, fight against deforestation and develop a low-carbon economy allowing the sequestration of 6 GT of CO_2.

The Brazilian institutional crisis and subsequent changes in government put these commitments on hold and Brazil, without formally withdrawing from the Paris

Agreement, gave up hosting COP 29 as it had previously committed to do. At the same time, deforestation is now officially approved by the Brazilian government. In July and August 2019, the INPE announced a massive resumption of deforestation (almost four times higher in July 2019 than in the same month of 2018) and forest fires (an 83% increase since the beginning of 2019 compared to the same period in 2018) in the Brazilian Amazon, leading to the dismissal of the INPE director (questioned by the President himself on the veracity of the figures provided by his institute!) and triggering a global controversy on the action of the Brazilian government. The scale of the debate shows the special place that the Amazon continues to play not only in terms of climate but also diplomatically.

12.6. References

Alves, L.M., Marengo, J.A., Fu, R., Bombardi, R.J. (2017). Sensitivity of Amazon regional climate to deforestation. *American Journal of Climate Change*, 6, 75–98.

Aragão, L.E.O.C., Poulter, B., Barlow, J.B., Anderson, L.O., Malhi, Y., Saatchi, S. (2014). Environmental change and the carbon balance of Amazonian forests. *Biological Review*, 89, 913–931.

Arvor, D., Dubreuil, V., Ronchail, J., Meirelles, M.S., Funatsu, B. (2014). Spatial patterns of rainfall regimes related to levels of double cropping agriculture systems in Mato Grosso (Brazil). *International Journal of Climatology*, 34, 2622–2633.

Arvor, D., Funatsu, B., Michot, V., Dubreuil, V. (2017a). Monitoring rainfall patterns in the Southern Amazon with PERSIANN-CDR data: Long-term characteristics and trends. *Remote Sensing*, 9, 889–909.

Arvor, D., Tritsch, I., Barcellos, C., Jegou, N., Dubreuil, V. (2017b). Land use sustainability on the Southeastern Amazon agricultural frontier: Recent progress and the challenges ahead. *Applied Geography*, 80, 86–97.

Barlow, J., Lennox, G.D., Ferreira, J., Berenguer, E., Lees, A.C., MacNally, R. (2016). Anthropogenic disturbance in tropical forests can double biodiversity loss from deforestation. *Nature*, 535, 144–147.

Boers, N., Marwan, N., Barbosa, H.M.J., Kurths, J. (2017). A deforestation-induced tipping point for the South American monsoon system. *Nature Scientific Reports*, 7, 41489.

Costa, M.H. and Foley, J.A. (2000). Combined effects of deforestation and doubled atmospheric CO_2 concentrations on the climate of Amazonia. *Journal of Climate*, 13, 18–34.

Cox, P.M., Betts, R.A., Collins, M., Harris, P.P., Huntingford, C., Jones, C.D. (2004). Amazonian forest dieback under climate-carbon cycle projections for the 21st century. *Theoretical and Applied Climatology*, 78, 137–156.

Debortoli, N., Dubreuil, V., Funatsu, B., Delahaye, F., Henke, C., Rodrigues Filho, S., Saito, C., Fetter, R. (2015). Rainfall patterns in the southern Amazon: A chronological perspective (1970–2010). *Climatic Change*, 132(2), 251–269.

Debortoli, N., Dubreuil, V., Hirota, M., Rodrigues Filho, S., Lindoso, D., Nabucet, J. (2016). Detecting deforestation impacts in southern Amazonia rainfall using rain gauges. *International Journal of Climatology*, 37(6), 2889–2900.

Delahaye, F., Kirstetter, P.E., Dubreuil, V., Machado, L.A., Vila, D. (2015). A consistent gauge database for daily rainfall analysis over the Legal Brazilian Amazon. *Journal of Hydrology*, 527, 292–304.

Droulers, M. (2004). *L'Amazonie, vers un développement durable*. Armand Colin, Paris.

Dubreuil, V. (ed.) (2002). *Environnement et télédétection au Brésil*. Presses universitaires de Rennes, Rennes.

Dubreuil, V. (2018). Le changement climatique. In *Géographie des environnements*, Arnould, P., Simon, L. (eds). Belin, Paris.

Dubreuil, V., Debortoli, N., Funatsu, B., Nédélec, V., Durieux, L. (2012). Impact of land-cover change in the southern Amazonia climate: A case study for the region of Alta Floresta, Mato Grosso, Brazil. *Environmental Monitoring and Assessment*, 184(2), 877–892.

Dubreuil, V., Segouin, M., Nédélec, V., Racape, A., Funatsu, B. (2015). Étude du gradient thermique foret-pâturage en Amazonie brésilienne : exemple de la saison 2013–2014 dans la région d'Alta Floresta. In *Actes du XXVIIIème Colloque de l'Association Internationale de Climatologie*. AIC, Liège, 263–268.

Dubreuil, V., Funatsu, B., Michot, V., Nasuti, S., Debortoli, N., de Mello-Théry, N.A., Le Tourneau, F.M. (2017). Local rainfall trends and their perception by the Amazonian communities. *Climatic Change*, 143(3), 461–472.

Dubreuil, V., Fante, K.P., Planchon, O., Sant'anna Neto, J.L. (2019). Climate change evidence in Brazil from Köppen's climate annual types frequency. *International Journal of Climatology*, 39, 1446–1456.

Espinoza, J.C., Ronchail, J., Guyot, J.L., Gérard, C., Naziano, F., Lavado, W. (2009). Spatio-temporal rainfall variability in the Amazon basin countries (Brazil, Peru, Bolivia, Colombia, and Ecuador). *International Journal of Climatology*, 29, 1574–1594.

Fu, R., Yin, L., Li, W., Arias, P.A., Dickinson, R.E., Huang, L. (2013). Increased dry-season length over southern Amazonia in recent decades and its implication for future climate projection. *Proc. Natl. Acad. Sci. U.S.A.*, 110, 18110–18115.

Funatsu, B.M., Dubreuil, V., Claud, C., Arvor, D., Gan, M.A. (2012). Convective activity in Mato Grosso State (Brazil) from microwave satellite observations: Comparisons between AMSU and TRMM datasets. *Journal of Geophysical Research*, 117, 1–16.

Funatsu, B., Dubreuil, V., Racape, A., Debortoli, N., Nasuti, S., Le Tourneau, F.M. (2019). Perceptions of climate and climate change by Amazonian communities. *Global Environmental Change*, 57, 101923.

INPE-PRODES (2019). Taxas de desmatamento anuais [Online]. Available at: http://www.obt.inpe.br/OBT/assuntos/programas/amazonia/prodes.

Khanna, J., Medvigy, D., Fueglistaler, S., Walko, R. (2017). Regional dry-season climate changes due to three decades of Amazonian deforestation. *Nature Climate Change*, 7, 200–204.

Lawrence, D. and Vandecar, K. (2015). Effects of tropical deforestation on climate and agriculture. *Nature Climate Change*, 5, 27–36.

Le Tourneau, F.M., Marchand, G., Greissing, A., Nasuti, S., Droulers, M., Bursztyn, M., Lena, P., Dubreuil, V. (3013). The DURAMAZ indicator system: A cross-disciplinary comparative tool for assessing ecological and social changes in the Amazon. *Phil. Trans. R Soc. B*, 475, 465–474.

Makarieva, A.M. and Gorshkov, V.G. (2010). The biotic pump: Condensation, atmospheric dynamics and climate. *International Journal of Water*, 5, 365–385.

Marengo, J.A., Souza Jr. C.A., Thonicke, K., Burton, C., Halladay, K., Betts, R.A., Alves, L.M., Soares, W.R. (2018). Changes in climate and land use over the Amazon region: Current and future variability and trends. *Frontiers in Earth Science*, 6, 228.

de Mello-Théry, N.A. and Dubreuil, V. (2017). Políticas de adaptação às mudanças climáticas nas práticas agrícolas. In *Amazônia, olhares sobre o territorio e a região*, Da Costa, J.M. (ed.). Universidade Federal do Amapa, Macapá.

Michot, V., Vila, D., Arvor, D., Corpetti, T., Ronchail, J., Funatsu, B.M., Dubreuil, V. (2018). Performance of TRMM TMPA 3B42 V7 in replicating daily rainfall and regional rainfall regimes in the Amazon basin (1998–2013). *Remote Sensing*, 10(12), 1879.

Michot, V., Arvor, D., Ronchail, J., Corpetti, T., Jégou, N., Lucio, P.S., Dubreuil, V. (2019). Validation and reconstruction of rain gauge-based daily time series for the entire Amazon basin. *Theoretical and Applied Climatology*, 1–17.

Nepstad, D.C., McGrath, D., Stickler, C., Alencar, A., Azevedo, A., Swette, B., Bezerra, T. (2014). Slowing Amazon deforestation through public policy and interventions in beef and soy supply chains. *Science*, 344, 1118–1123.

Nimer, E. (1989). *Climatologia do Brasil.* IBGE, Rio de Janeiro.

Nobre, C.A., Sellers, P.J., Shukla, J. (1991). Amazonian deforestation and regional climate change. *Journal of Climate*, 4, 957–988.

Saatchi, S.S., Harris, N.L., Brown, S., Lefsky, M., Mitchard, E.T., Salas, W. (2011). Benchmark map of forest carbon stocks in tropical regions across three continents. *Proc. Natl. Acad. Sci. U.S.A.*, 108, 9899–9904.

Sampaio, G., Borma, L.S., Cardoso, M., Muniz Alves, L., von Randow, C., Andrés Rodriguez, D. (2018). Assessing the possible impacts of a 4°C or higher warming in Amazonia. In *Climate Change Risks in Brazil*, Nobre, C.A., Marengo, J.A., Soares, W.R. (eds). Springer, Berlin.

Sena, E.T., Silva Dias, M.A.F., Carvalho, L.M.V., Silva Dias, P.L. (2018). Reduced wet season length detected by satellite retrievals of cloudiness over Brazilian Amazonia: A new methodology. *Journal of Climate*, 31(24), 9941–9964.

Silva Junior, C.H.L., Almeida, C.T., Santos, J.R.N., Anderson, L.O., Aragão, L.E.O.C., Silva, F.B. (2015). Spatiotemporal rainfall trends in the Brazilian Legal Amazon between the years 1998 and 2015. *Water*, 10, 1220.

Spracklen, D.V. and Garcia-Carreras, L. (2015). The impact of Amazonian deforestation on Amazon basin rainfall. *Geophys. Res. Lett.*, 42, 9546–9552.

Swanna, A.L.S., Longob, M., Knoxc, R.G., Leed, E., Moorcroftd, P.R. (2015). Future deforestation in the Amazon and consequences for South American climate. *Agricultural and Forest Meteorology*, 214–215, 12–24.

Théry, H. (ed.). (1997). *Environnement et développement en amazonie brésilienne*, Belin, Paris.

13

The Impacts of Climate Change on the Distribution of Biomes

Delphine GRAMOND
Sorbonne University, Paris, France

The current climate is experiencing detected changes in several variables (temperature, precipitation) and benchmark values (averages, extremes, etc.), potentially modifying the environmental filters on which flora and fauna depend in their distribution and biomass. According to the Intergovernmental Panel on Climate Change (IPCC), not only rising temperatures, but also the occurrence of prolonged and more frequent extreme events are considered to affect the biosphere at different levels (species distribution, phenology, ecosystem structure, inter- and intra-specific interactions, etc.) (Hoegh-Guldberg *et al.* 2019). This chapter proposes to assess to what extent (conceptually and operationally) the biome, the structural unit of the biosphere at the global scale, can be considered as an integrative indicator of the direct, indirect and even induced impacts of current climate change on biotic communities. A synthesis of the ecological manifestations identified in response to climate change (early spring, very likely global increase in the number of warmer days and nights, etc.) will be presented and then illustrated by feedback from work carried out on the Arctic biome (tundra/taiga ecotone). Questioning the biome as a "sentinel" of climate is a relevant approach that should not exclude taking into account the complexity of environmental trajectories at the different scales considered.

Spatial Impacts of Climate Change,
coordinated by Denis MERCIER.

13.1. Biomes, a representation of life on a global scale

A biome is, on a global scale, a unit of the biosphere, that circumterrestrial layer – which can be up to 20 km thick – where organic matter is potentially present. On a vertical plane, living organisms have been identified up to a depth of more than 8,000 m and a passive presence of seeds or spores can be detected up to the upper limit of the troposphere (about 10 km), the lower layer of the atmosphere where most of the climate processes interacting with the biosphere take place. Horizontally, the biosphere is subdivided into biomes, that is, macro-units characterized by plant assemblages and animal populations linked by common abiotic factors. According to Walter and Box, the most structuring factor in the subdivision and distribution of biomes is climate (Walter and Box 1976). If biomes are a "reliable" reflection of the global climate system, to what extent can they reveal the impacts of climate change on the biosphere? The relevance of the biome as an indicator is a point to be discussed at first glance. In this chapter, only issues related to terrestrial biomes will be addressed and will serve as a reference pattern for discussion, as current knowledge on marine biomes is still very fragmentary.

Successive conceptions of biomes first proposed a pure descriptive model with the aim of identifying biogeographical units, then evolved towards explanatory models, which are now used as predictive tools. Thus, the evolution of the biome concept has been marked by three key stages: the first works formulated a solid theoretical corpus mobilizing the principle of zonality as a major control variable at the global scale (see section 13.1.1). Then, this knowledge, in parallel with the development of calculation tools, made it possible to formalize models based on mechanical links between ecosystem functioning and spatial distribution (see section 13.1.2). Currently, work is focusing on the search for common ground between macroecological and macrobiogeographical phenomena, without neglecting internal dynamic trajectories and thus enriching the deterministic perspective of the original concept (see section 13.1.3). Before identifying how biomes react to the impacts of climate change, and thus defining relevant measures and indicators, it is therefore necessary to clearly delimit the meaning of the concept of "biome", to discuss its scales of apprehension and to specify the characteristics of this unit, which strictly speaking does not refer to species composition but to other observation reference frames (structure, functionalities, etc.). On a global scale, biological variations in flora and fauna are complex with repercussions on the resources available to support biodiversity, biogeochemical cycles and also human needs. The potential risks of climate change impacts on the biosphere are multiple and the issues are particularly crucial at a time when the erosion of biodiversity is considered a threat to humanity according to the report of the Intergovernmental Science-Policy Platform on Biodiversity and Ecosystem Services (IPBES) 2019 (IPBES 2019).

13.1.1. *The biome, an indicator of climatic context: what are the realities?*

On continents, biomes can be likened to vast, relatively homogeneous, small-scale geographical units, essentially identifiable by a plant base with similar characteristics in terms of physiognomy (forest, savannah, grassland, etc.). Vegetation is indeed the significant biological element that first materializes and then spatially particularizes terrestrial biomes, for two reasons: flora is a relatively static component of terrestrial biomes (compared to fauna) and autotrophic plants make up 99% of terrestrial biomass (compared to 0.9% for heterotrophic plants and 0.1% for animals). In the oceans, aquatic "biomes" are not defined in terms of flora, which is mainly composed of unicellular algae with no specific structure. Instead, aquatic systems are identified according to physical criteria such as depth, salinity, oxygen concentration, etc., which are not directly related to climatic factors alone. We will therefore only focus here on terrestrial biomes – the "geobiosphere" – which it is accepted that it is effectively distributed according to climatic control variables, forcing at the global and zonal scales, more stochastic at the largest scales.

The spatial links between climate and vegetation are among the first ecological observations described by Theophrastus as early as the 15th Century BC. Ecology and biogeography work still identify climate as the environmental factor explaining the forces and forms of vegetation on the continents, since the three ecological parameters of light, heat and water are the key resources necessary for plant growth and maintenance. Threshold effects are manifested by changes in the presence and/or physiognomy of plant formations, for example, when monthly temperatures never exceed 0°C, plants are unable to establish themselves because permanent frost is a critical factor in sustaining plant cell structure. Certain thermal thresholds are thus considered to be botanical limits, such as the Köppen line (isotherm + 10°C in the warmest month, a thermal value that is imperative for tree development), which traditionally serves as a biogeographical boundary between boreal forests and tundra in the northern hemisphere. Thus, through its physiognomic description, the vegetation framework makes it possible to define major types of biomes (forested/non-forested) qualified according to climatic criteria (equatorial, tropical, temperate, polar biomes, etc.), which are reflected in plant landscapes. In this context-dependent logic, the biomes are mostly delimited according to a principle of climatic zonation; the emergence of this model deserves to be re-examined in order to grasp the foundations of the conceptual knowledge and reference frameworks and assess their scope in the current context of thinking on the impacts of climate change on the biosphere and its components. Moreover, this global outlook should not overshadow a more territorial approach that considers that the flora is not just a passive component of the geographical context. The retroactive effects of vegetation on climate mechanisms must also be discussed in order to adjust the conclusions about the impacts of climate change on the geography of biomes (Foley *et al.* 2003).

13.1.2. *From the roots of a globalizing concept to the emergence of an operational scale*

Biomes are a representation of the heterogeneous structure of the biosphere on a global scale. The term "biome" was first suggested in 1916 by the American botanist F.E. Clements – father of the concept of "climax" – at the opening of the first meeting of the American Ecological Society in 1916. He then predefined the term as synonymous with "biotic community" (Clements 1917). The notion of "biome" – and not yet the concept – then carries the heritage of prodromes that are the concepts of association[1], (plant) formation,[2] biocenosis[3] and especially "life zone" proposed by the American zoologist C.H. Merriam in 1894 and which prefigures the principle of the biome on a global scale. This work will anchor the hypothesis that biotic units are part of a specific geographical context. In fact, C.H. Merriam notably described correspondences between changes observed in biotic communities as a function of the increase in latitude at a constant altitude and those observed during an increase in altitude at a constant latitude. He described the spatially explicit relationship between the distribution of biotic communities and a geographical context (Merriam 1894). At the same time, the Franco-German geobotanist Andreas F.W. Schimper proposed an explanatory global model of vegetation as a function of climatic parameters such as temperature and water (Schimper 1898). Schimper thus positioned the concept of formation in a decisive manner by demonstrating its usefulness on a global scale: the physiognomy of a plant formation materializing the identity of a biome ("open"/ "dense" forest, meadow, etc.).

---

1 Term relating to phytosociology. The association qualifies a plant community recognized by a specific assemblage of characteristic species (von Humboldt and Bonpland 1805).

2 Introduced by the German botanist A.H.R. Grisebach in 1838. In his reference work of 1872, he classified plants according to their climatic region.

3 Term invented and introduced into the scientific literature by the German biologist K.A. Möbius in 1877, which then shows the need to study species in a specific spatial context, that is in interactions with other species, taking into account coexistence factors.

The foundations of the biome concept were then established, with climate being clearly identified as a determining factor in the structuring of biomes on a global scale. Based on the work of A.F.W. Schimper, for whom the macroclimate plays a primary role in the distribution of "zonal formations", but also on the work of the soil scientist V.V. Dukushaev, who published a world classification of soils (Dukushaev 1899) based on the theory of zonality, or on world climate maps (Thornthwaite 1933; Köppen 1900) of a statistical type based on precipitation, temperature or evapotranspiration. The Russian-German botanist Heinrich Walter published a first synthesis in 1954 and presented a zonal/azonal translation of the biomes, which was not purely bioclimatic, but which also took into account factors such as soils and altitude (Walter 1954). His research, published in German, remained relatively confidential until the 1970s, when a great deal of work was done on the classification of biomes on a global scale.

We will mention here two classifications that are still used as references, one based on an analytical approach, the other on a more systemic approach. These two conceptions are symbolic of the two complementary scientific approaches that are then expressed:

– an American approach that defines plant formations through a geographical and climatic prism ("humid equatorial forest");

– and a European approach based on the physiognomy of vegetation as a grid for reading formations ("dense forest", "flat steppe", etc.), the key to identifying biomes.

For example, the American ecologist R.H. Whittaker (1970) proposed a simple climate diagram that analytically allows us to draw theoretical biome boundaries based on average temperatures and cumulative annual precipitation (Whittaker 1970). The Russian-German botanist H. Walter's diagrams – which transcended linguistic boundaries when translated into English in 1976 – are based on a distinction of climatic regions according to their annual seasonality and reflect the hypothesis of a factor conditioning the forms of plant growth (Walter and Box 1976). Thus, the temperature and precipitation values used to define the climatic zones correspond, in his model, to the humidity and cold-related stress conditions, which are major determinants of plant physiognomy. Its more holistic approach conditions the understanding of a biome by reading the interrelationships that make it up (climate, soil, physiognomy and zonation). The mapping he proposes brings out the concept of "zonobiomes" which, going beyond a purely descriptive approach, emphasizes causal relationships between dominant environmental factors and the presence of characteristic biomes (see Figure 13.2).

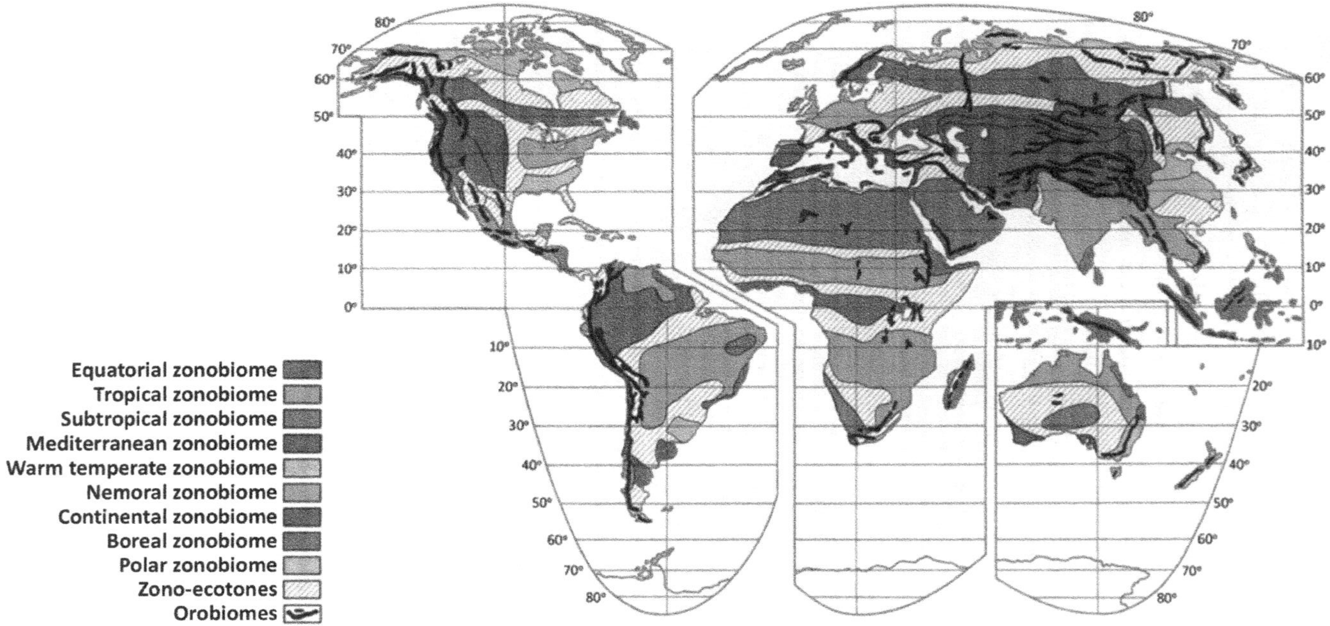

Figure 13.1. *Global map of terrestrial zonobiomes according to Walter (1976) classification (source: Heinrich and Hergt 1993). For a color version of this figure, see www.iste.co.uk/mercier/climate.zip*

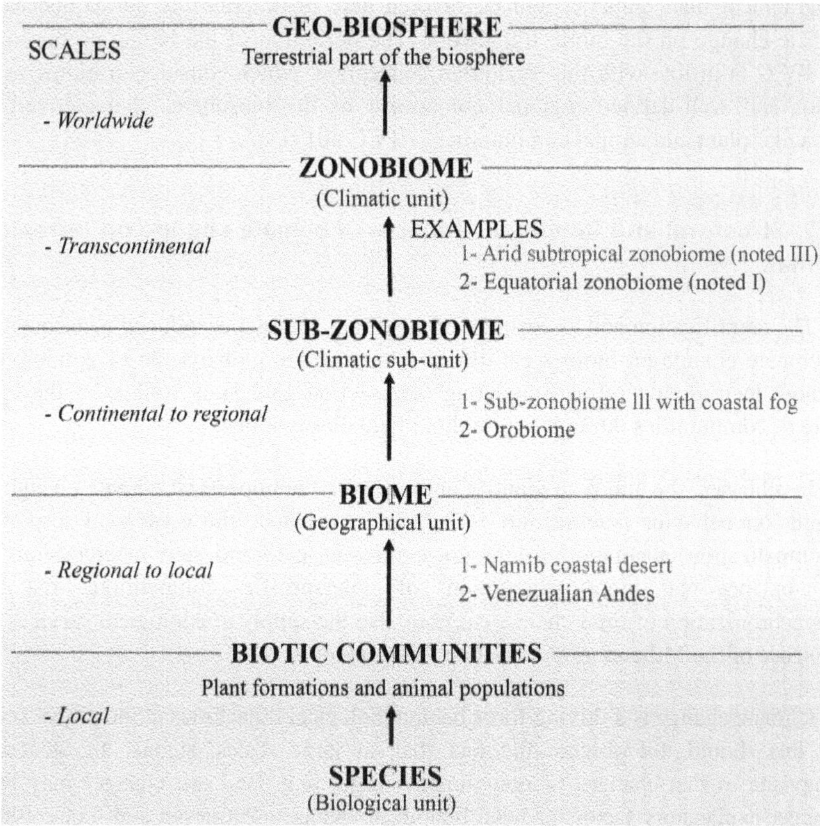

Figure 13.2. *From zonobiomes to biomes: a hierarchy of concepts (source: Mucina 2019, modified from Walter and Box 1976. For a color version of this figure, see www.iste.co.uk/mercier/climate.zip*

Walter's biome design introduces a hierarchy based on a scale of understanding from the global scale zonobiome to the sub-continental to regional scale biome (see Figure 13.2). Thus, starting from the synthetic vision, which sees the biome as "content in a container", it is necessary to engage in a systems thinking that sees the biome rather as a "flow in a matrix" (based on the theory of transient systems in dynamic equilibrium), that is, the product of dynamic processes, whose disturbances, feedbacks, etc., are mainly identifiable and identifiable at zonal, or better regional scales. It is in this sense that the Walterian conception of the biome, more geographical than climatic, will be favored here to describe the consequences of climate change on the biotic framework. The definition of the biome proposed by the IPCC is in line with this Walterian conception, which considers a biome as "a major and well-defined regional component of the biosphere, characterized by particular plant and animal communities" (IPCC 2019).

13.2. Structural and functional impacts of climate change on terrestrial biomes

The identification and assessment of the direct, indirect or induced consequences of climate change on biomes can be monitored at the global scale of zonobiomes through increasingly refined modeling (see section 13.2.3) as well as at the local scale of communities through longitudinal field observations.

In all cases, the aim is to identify and measure the impacts of climate change on vegetation behavior (see section 13.2.2) and to deduce the ecological responses (acclimatization, adaptation, migration, extinction, etc.) and their repercussions on the balance of biota (disruption of interspecific competition, risk of desynchronization of food chains, etc.) but also the supply of ecological services (in the sense of the Millennium Ecosystem Assessment[4]).

Climate change is a driving force behind biological trajectories at the global scale, but this should not obscure the fact that, at large scales, signals are generally composite in that changes of non-climatic origin (e.g. land use change) may be a stronger explanatory factor for local biological changes (Parmesan and Yohe 2003). On a small scale, the major biotic changes observed and generalized to zonobiomes then reflect a global climate indicator, lasting responses to the global increase in the number of hot days and nights, the increase in frequency and duration of hot periods or the increase in areas of high rainfall intensity and/or quantity (IPCC 2019).

13.2.1. *From bioclimatic bathing to modification of ecological processes*

For vegetation, the factors of light, water and temperature are crucial to the development, organization and maintenance of species within biotic communities.

---

4 The Millennium Ecosystem Assessment (MEA) of 2005 was the world's first global program assessing the interactions between economic, social and environmental issues. The concept of ecosystem services emerged from this program.

The climatic parameters “water” and “temperature” are factors that regulate the rudimentary processes of feeding, growth, reproduction, etc., and each species has tolerance limits, resulting in an environmental selection that partly explains the presence of a species in a given territory (case of neutral models without taking into account interspecific competition, for example).

The intensity of life processes is conditioned in particular by ecological spectra, with extreme temperatures (<0°C or >60°C) inhibiting the presence of vegetation on continents, for example at very high latitudes (permanent winters in polar climates) or subtropical deserts (marked by permanent aridity). Shelford’s Law of Tolerance can be applied at the species or biome level and is the basis for species distribution models based on climate parameters alone (Holdrige and Whittaker models, see section 13.1.2).

The sensitivity of a species, and by extension a biome, to current climate change will depend in part on its adaptability to new environmental conditions. In the face of current (and future) climate change, species have four types of potential responses: acclimatization, adaptation, change in geographic distribution (i.e. migration) or extinction through loss of favorable habitat (Feeley *et al.* 2012).

Before illustrating the biological responses at different scales to current climate change, the identification of the main causes and potential consequences on the biomes will allow the characterization of reactive behaviors that serve as a basis for recent predictive modeling.

13.2.2. *Identifying changes: from global diagnosis to biological responses*

The main direct causes of the observed changes in vegetation are an increase in global temperature, changes in the spatial and temporal distribution patterns of precipitation, and an increase in CO_2 concentration. Assessing the respective weights of these triggers of potential changes on the biomes is problematic insofar as the impacts of each factor cannot be isolated and the assimilation capacities of vegetation also have different time steps according to the species and scales considered (the degrees of reactivity are still poorly documented).

It is, for example, difficult to separate the influence of the increase in CO_2 – which is almost uniform on a global scale – from a rise in spring temperatures, which is often more marked on a regional scale. Both stimuli can lead to earlier flowering, but it is difficult to estimate the respective influence of each in triggering the ecological process (Parmesan and Yohe 2003).

To these direct causes we must add indirect causes such as changes in biotic interactions, changes in fire regimes, the increase in the active layer above the permafrost, changes in the availability of water, nutrients, etc. (Garamvölgyu and Hufnagel 2013), which make it even more complex to identify the causal links between climate change and biological responses.

In order to synthesize the current and potential impacts of climate change on biomes, it is necessary to integrate the fact that these impacts can occur at the species, community and biome (and by extension the zonobiome) levels and that the response mechanisms mainly imply a modification to their ecological spectrum along three non-exclusive axes:

– time (adjustment of the lifecycle to new climatic conditions, including phenology and diurnal rhythms);

– space (disperse to areas where the habitat is more favorable or change position on a habitat scale);

– or themselves (physiological and/or morphological changes) (Bellard *et al.* 2016).

The changes identified are often interdependent, including hierarchically, as changes at the biome level (e.g. latitudinal shift in a species' range) potentially affect species at the population level, that is, hybridization capacity as a result of species migration (Scheffers *et al.* 2016; see Figure 13.3).

Responses are not labelled as "positive" (acclimatization, adaptation) or "negative" (stress, extinction) because this qualification may vary according to the scale of analysis; for example, adaptation (genetic) may be characterized as positive at the community (or biome) level but negative at the species level if this adaptation is accompanied by a reduction in genetic variation and due to the species' ability to respond to other stressors.

Similarly, a species migration may generate differentiated interactions within the new habitat, some of which may lead to timely (coexistence) but also lethal competition (decisive competition favorable to the "migrating" species).

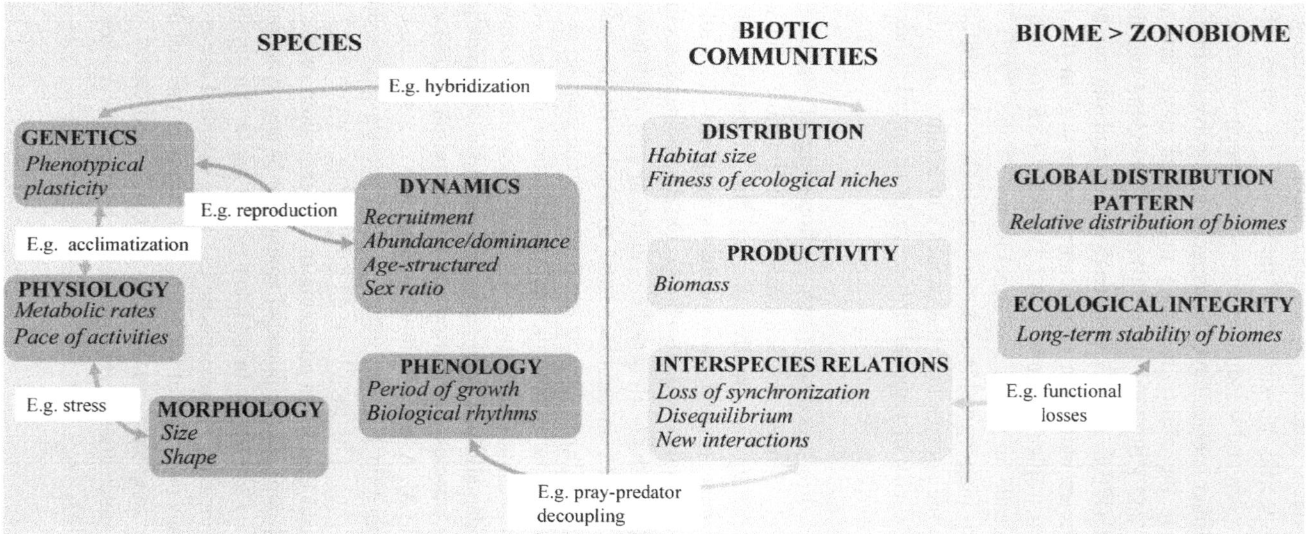

Figure 13.3. *Types of major impacts identified on terrestrial biomes and hierarchical scale involved (source: design and production D. Gramond). For a color version of this figure, see www.iste.co.uk/mercier/climate.zip*

The biological effects induced by climate change are notably linked to the mixing of species-specific physiological thresholds in terms of tolerance to temperature (annual mean and seasonal extremes) and precipitation (annual total and seasonal rhythms). Either species have the capacity to survive despite environmental changes or they have the ability to move to habitats more favorable for their survival. Latitudinal and altitudinal shifts in boreal, temperate and tropical biomes have been observed (IPCC 2019). Meta-analyses carried out on both plant and terrestrial animal species indicate a significant upward shift in the biomes of the boreal, temperate and tropical regions of 6.1 km per decade towards the poles and about 6.1 m per decade in altitude (Parmesan and Yohe 2003). This shift in geographic ranges has biogeographic consequences both to the north, but also to the southern limits, as some species lack the capacity to acclimatize, adapt or migrate, resulting in a potential risk of regional imbalance of an interdependent intensity of ecosystem resilience.

For most species, life processes (such as flowering of plants, hibernation of mammals) are closely linked to seasons and interannual climate variations. However, current change is rapidly altering the rhythms and intensity of fluctuations. On the genetic level, two mechanisms of plasticity allow species to survive when environmental conditions change. On the one hand, acclimatization, of a physiological nature, which allows the species to adjust to new variations in temperature, humidity, etc.; on the other hand, to another degree, adaptation of an ecological nature, which allows the species, from its gene pool, to evolve towards a new species if the temperature, rainfall, etc., modifications persist. Acclimatization is therefore rapid in relation to adaptation. Acclimatization leads to fewer changes in communities than adaptation. These responses can lead to cascading effects on population dynamics, species distribution and beyond on the nature and stability of biomes (see Figure 13.3). A well-documented example of a cascade effect is that of potential desynchronization between interdependent species (plants and pollinators, prey-predators, etc.), caused by the observed early spring conditions (2 to 3 days less per decade), which can lead to an alteration or reorganization of the functioning of the biotic community, through the potential emergence of new synchronies (Scheffers *et al.* 2016; Lavorel *et al.* 2017). In addition to this earlier spring, global warming (accompanied by an increase in atmospheric CO_2) has extended the growth period of many plant populations over half of the Earth's surface over the past 30 years, which may lead to short- or long-term instabilities (changes in food chains, biogeochemical cycles, interspecific interactions, etc.) in some biomes (Reyes-Fox *et al.* 2014; Buitenwerf *et al.* 2015).

Studies on the history of biodiversity have shown that habitat "tracking" – that is, migration to favorable conditions – is the key response of species to ensure their survival (Lavorel *et al.* 2017), which is why spatialization of range changes at different scales is a diagnostic tool. It makes it possible to assess the rates and mechanisms of potential structural and functional reorganization of biota. As vegetation often reacts slowly to changes in environmental conditions, the time lag between a climate change and a change in vegetation, while it can be observed at the scale of a community over several decades, is difficult to observe at the biome scale, hence the widespread use of modeling despite the uncertainties it induces.

13.3. Spatializing change: biome modeling

To spatially characterize biological responses to the effects of climate change, biomes represent practical reference areas for capturing ecological processes on a global scale. The first mobilized, fragmentary models made it possible to establish spatial relationships between biome types and climate types (Yates *et al.* 2000). One example is the Holdrige model; the first simple global model to assess the sensitivity of vegetation to climate change, which only considers bioclimatic parameters such as mean annual temperature, mean annual precipitation and evapotranspiration potential (Holdridge 1947). Box's model is based on Plant Functional Types (PFTs) and is considered representative of plant communities for the specificity of their physiological and morphological traits (Box 1981). Its correlational model hypothesizes behavioral scenarios linked to climatic constraints characterized by indicators that take seasonality into account. Thus, its model predicts the occurrence of (structurally defined) forms of growth by means of their hypothetical climatic envelope. Box's first model, while being a precursor, was quickly criticized for its overly empirical approach, making it difficult to couple it to currently available global climate models (Box 2019). In this chapter, we present a recent synthesis proposed at the global scale (Gonzales *et al.* 2010) and a focus on the Arctic region (Grimm *et al.* 2013), both based on the same model (see section 13.1.1).

13.3.1. *Observed and projected global impacts*

The deployment of analysis tools and calculation methods has made it possible to envisage increasingly complex modeling, taking into account different variables to refine predictions. Modeling no longer has a purely descriptive objective (that of reporting changes), but an explanatory, even functional one (seeking causality to the forms and forces of the changes identified). Numerous uncertainties still reside in the simulations because the vegetation models used differ, as do the climate models, the emission scenarios, the historical thickness taken into account, etc. The simulations are still subject to numerous uncertainties. One of the models commonly used to interpolate biological responses to climate change is the BIOME model based on biogeography/biogeochemistry coupling (Prentice *et al.* 1992). The BIOME4 generation simulates the potential distribution of biomes on a global scale by combining PFTs with bioclimatic, ecophysiological (water availability, CO_2 availability, etc.) and tolerance/requirement indices (cold, drought tolerance, etc.) (Kaplan *et al.* 2003). Based on these parameters, coupled with general atmospheric and oceanic circulation models, BIOME4 estimates the competition between PFTs as a function of primary productivity (which is derived from a seasonal leaf area index) and finally classifies the biomes according to dominant and sub-dominant PFTs by assigning environmental limits to them. Thus, BIOME4 simulates the distribution of *potential* vegetation by associating phenological, hydrological and biogeochemical properties. The diagnostic model has been used in particular to reconstruct vegetation-climate feedbacks in the past, from one of the most documented syntheses, that of Hoogakker *et al.* which presents a paleohistorical reconstruction of the 120 ka, 84 ka, 64 ka, 54 ka, 21 ka and 6 ka BP biomes from pre-industrial data (Hoogakker *et al.* 2016). These retropredictive models are relatively robust, but present two major methodological shortcomings in the context of prospective simulations. Indeed, the relationships and correlations resulting from current models propose global maps of potential biomes, which do not (or only to a limited extent) take into account the complexity and non-linearity of disturbances at regional and local scales (i.e. fire regimes), particularly those due to direct or indirect human activities. Moreover, these rather deterministic models (the relationships established are of a mathematical, statistical or logical but not probabilistic type), still exclude the historical dimensions of the occupation and use of the biomes.

Among the proposed multifactorial models, that of Gonzales *et al.* (2010) proposes a simulation of biomes based on a Dynamic Global Vegetation Model (DGVM), established from field observations, associated with emission scenarios of the IPCC and global circulation climate models; it relies on vulnerability indicators, thus integrating the role of disturbances into the model, to ultimately assess the sensitivity of biomes to current climate change (see Figure 13.4). According to this study, one tenth to half of the terrestrial biomes are potentially vulnerable to climate change. Temperate mixed forests, boreal coniferous forest, tundra and alpine biomes show the greatest vulnerability, often due to potential changes in forest fire regimes, while tropical broadleaf forests and desert biomes appear to be the least vulnerable (Hufnagel and Garamvölgyu 2014).

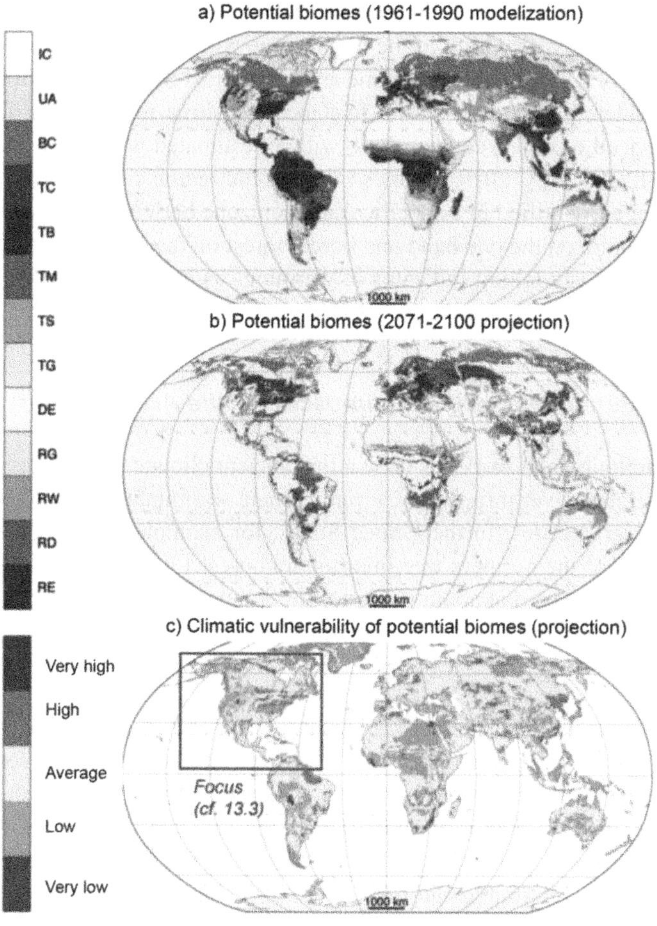

Figure 13.4. *Biome modeling. For a color version of this figure, see www.iste.co.uk/mercier/climate.zip*

COMMENT ON FIGURE 13.4.– *(a) Potential biomes modeled from climate observations over the period 1961–1990. (b) Potential biomes modeled for a climate projection 2071–2100. (c) Climate vulnerability of potential biomes. Biomes in (a) and (b) from the poles to the equator: ice (IC), tundra and alpine grassland (UA), boreal conifer forest (BC), temperate conifer forest (TC), temperate broadleaf forest (TB), temperate mixed forest (TM), temperate shrubland (TS), temperate grassland (TG), desert (DE), tropical grassland (RG), tropical woodland (RW), tropical deciduous broadleaf forest (RD), tropical evergreen broadleaf forest (RE) (source: modified from Gonzales* et al. *2010).*

Due to stronger local warming than the global average over the last century, rates of change are particularly high in the Arctic zone with an increase in woody species accompanied by a change in fire regime and permafrost degradation (Gonzales *et al.* 2010). Considered to be one of the five most sensitive regions – in terms of the forms and speed of observed changes – with an estimated global warming of 1.5 to 2°C by the IPCC (IPCC 2019), studies in the Arctic region can illustrate the types of current changes identified in a specific transition zone between differentiated growth forms, low tundra on the one hand and boreal forest on the other, where the presence of woody plants is a telling indicator of warming, with a strong vulnerability index depending on the model (see Figure 13.4).

13.3.2. *Observed and projected impacts for the Arctic region*

Paleoenvironmental work has shown that ancient climate change has affected the Earth's vegetation over millennia. Current climate warming tends to alter the ranges of biomes over decades. In the United States, for example, the average latitudinal upwelling rate of the northern tree line (tundra/boreal forest ecotone) between 1960 and 2009 was measured at about 0.4 km/year, compared to 0.002 km/year estimated for the end of the Last Glacial Maximum (LGM) about 20,000 years ago (Grimm *et al.* 2013).

From Arctic North America to Scandinavia and Eurasia, the circumpolar tundra/taiga ecotone forms a band about 1,000 km wide and extends over 13,000 km (see Figure 13.4). The boreal forest is limited to the north by cold weather that tends to eliminate trees in favor of grasses and mosses, the domain of tundra. The transition from boreal forest to tundra is marked by a relatively clear change in landscape that is identifiable and measurable at regional scales from the analysis of satellite imagery, for example. Monitoring this transition zone provides an indication of vegetation responses to regional climate change. In recent decades, boreal forest has been gradually gaining ground towards higher latitudes (see Figure 13.5). Global warming in high latitudes (+2 to 4°C observed) is stimulating tree growth. Boreal forests benefit from warmer and longer summers, which results in earlier and more intense snowmelt and allows trees to benefit from water more efficiently. The early summer thaw also gives shrubs the opportunity to take root and thus push the tundra northward, something that diachronic monitoring using high-resolution satellite imagery has made it possible to quantify more globally in recent years (Ranson *et al.* 2011). This boreal greening is evidence of a relatively rapid reactive process, accompanied at the regional scale by feedbacks. Indeed, it is recognized that the northward progression of trees following summer warming leads to local increases in ground temperature, progressively favoring forest colonization. The boreal forest has a "buffer effect" that partially conserves the sun's rays, and therefore heat, while the snow promotes a "mirror effect" that accentuates the cold (the sun's rays being reflected back to the atmosphere before warming the ground surfaces). Once the forest is in place, the ground freezes less easily and the snow depth is less, allowing shrubs to gradually colonize the environment.

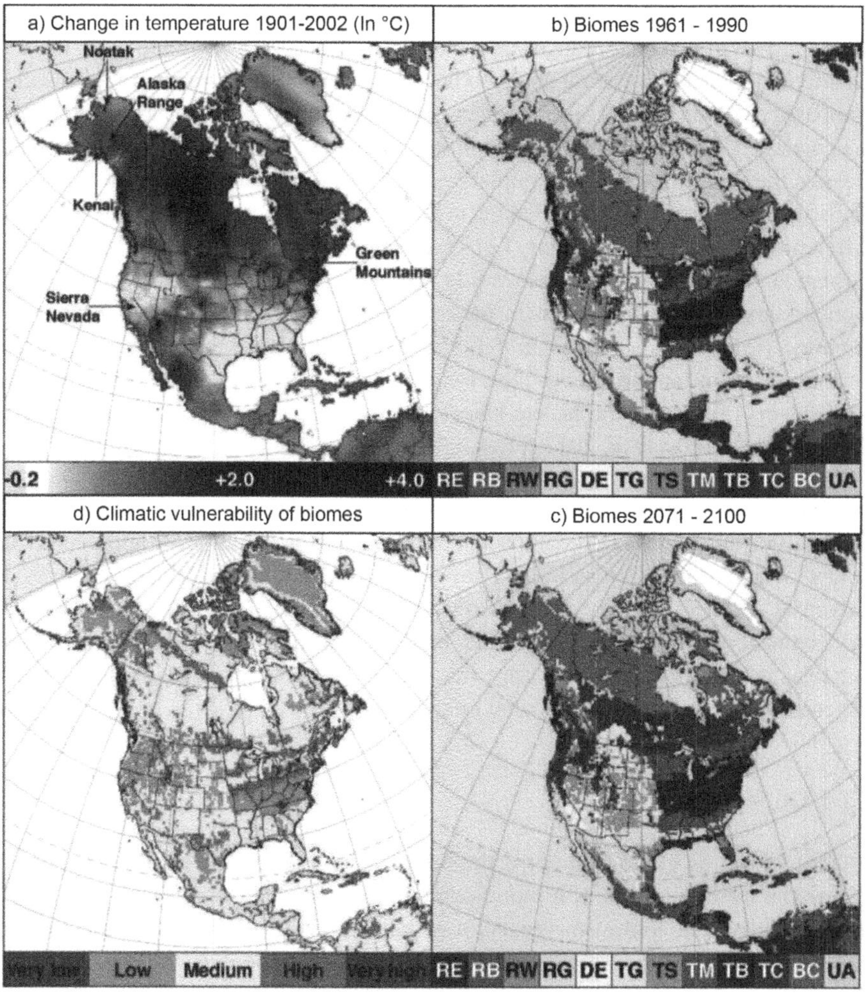

Figure 13.5. *Changes in assigned biomes to climate change in the United States. For a color version of this figure, see www.iste.co.uk/mercier/climate.zip*

COMMENT ON FIGURE 13.5.– *(a) Linear temperature trend between 1901 and 2002. (b) Potential biomes according to the observed climate between 1961 and 1990. (c) Potential biomes for climate projections 2071–2100. (d) Biome vulnerability to climate change. Biomes: see Figure 13.4 (source: Grimm* et al. *2013).*

While the Arctic zone can be used as an example because of its spatial extent (relatively zonal, continuous and continental scale), it is not necessarily representative of potential changes on a global scale because it is unique in its thermal sensitivity to current warming and also because of its lower degree of anthropization compared to other biomes in the temperate and equatorial zones. The complexity of the combined sets of factors involved in the current evolution of the biosphere partly isolates the Arctic zone as a "simpler" case. Despite this, the sensitivity of Arctic landscapes to global climate change is locally heterogeneous given the spatially differentiated regime (and history) of cryogenic disturbances, which facilitates the formation of landscape patterns. While the trend towards tundra overgrowth by pioneer shrubs (alder and birch in particular) is predominantly observed in North America (Canada, Alaska) as in Siberia, the relative changes measured are not uniform from one site to another, the wet-prone ecotones of the topographic troughs (floodplains, alluvial terraces, etc.) have a tendency to change more rapidly than drier ecotones (Frost *et al.* 2013; Frost and Epstein 2014).

Given the complex interactions between ecological and bio-geographic mechanisms in response to the impacts of climate change, multi-scale approaches are a key avenue for the search for knowledge on current and future changes in terrestrial biomes. The availability of increasingly rich field databases and the development of increasingly accurate models that consider many interdependent variables make it possible to envisage future improvements that could reduce some of the current uncertainties due to spatial inferences.

13.4. Conclusion

In ecology, as in biogeography, the biome is considered a useful tool because it allows the materialization of a biotic community at small spatial scales. The biome approach gives rise to the emergence of global and regional patterns of organization and functioning of living organisms, without neglecting the dynamic aspects. Thus, the profile of a biome is roughly identifiable on the global macroclimatic scale, but it is regionally identified by a weighted grid of mesological factors (soil, water, disturbances) and processes with interactive mechanisms (interspecific competition, etc.) or even retroactive mechanisms. Biophysical and biogeochemical feedback of vegetation on climate are indeed identified and tend to show that biomes modify climate change. It therefore seems sensible to strictly measure the impacts (direct, indirect and induced) of global change on the biomes. Current meta-analyses nevertheless make it possible to propose a synthesis of potential responses in the short (species behavior, physiological adaptations, etc.), medium (population dynamics, biotic deregulation, selection pressures, etc.) and long (speciation, mutation, etc.) terms.

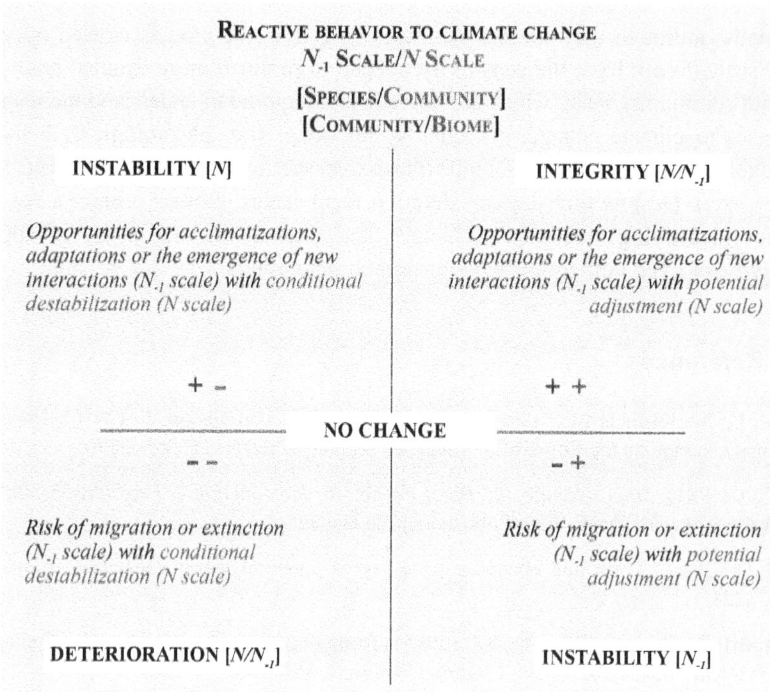

Figure 13.6. *Changes in biomes attributed to climate change (source: Gramond 2019). For a color version of this figure, see www.iste.co.uk/mercier/climate.zip*

A schematic typology of the reactive behaviors of terrestrial biomes to global change can serve as a framework for interpreting the observations made and, in the long term, can guide actions (and non-actions) with regard to biodiversity management (see Figure 13.6). This grid theoretically identifies the four life positions of a species in relation to its community (but also of a community in relation to the reference biome) from integrity, where there is optimal adjustment between species and habitat to alteration or, on the contrary, the migration or even extinction of a species can lead to deregulation of the environment (case of engineering species). Between these two positions, two forms of instability may be expressed, the first when the species acclimatizes but the environment destabilizes (modification of resource pressures, for example), the second when a species disappears but the community adjusts quickly (case of species that are not very competitive, tolerant and not very active in plant dynamics). Finally, there may be cases of no change, that is, of environments that are resistant (at a given scale and time frame) to current climate change.

Finally, although they meet a synthesis objective, the models carried out at the biome scale do not have the sensitivity needed to perform more detailed analyses at the plant community scale, which are nevertheless required to understand the responses of species to climate change, particularly functional (i.e. physiological) behavioral fluctuations, which are more difficult to map than structural modifications (size, shape, density, etc.). Despite their lack of detail at local scales, biomes remain a key plant unit for identifying small-scale changes and thus identifying areas of apparent sensitivity, or even vulnerability, to current climate trends.

13.5. References

Bellard, C., Bertelsmeier, C., Leadley, P., Thuillier, W., Courchamp, F. (2016). Impacts of climate change on the future of biodiversity. *Ecology Letters*, 15, 365–377.

Box, E.O. (1981). *Macroclimate and Plant Forms: An Introduction to Predictive Modeling in Phytogeography*. Dr W. Junk Publishers, The Hague.

Box, E.O. (2019). Form and character diversity of potential world vegetation. *Flora*, 254, 203–221.

Buitenwerf, R., Rose, L., Higgins, S.I. (2015). Three decades of multi-dimensional change in global leaf phenology. *Nature Climate Change*, 5, 364–368.

Clements, F.E. (1917). The development and structure of biotic communities. *Journal of Ecology*, 5, 120–121.

Doukouchaev, V.V. (1899). *One Theory of Natural Zones. Horizontal and Vertical Soil Zones*. Mayor's Office Press, St. Petersburg.

Feeley, K.J., Reem, E.M., Machovina, B. (2012). The response of tropical forest species to global climate change: Acclimate, adapt, migrate, or go extinct? *Frontiers of Biogeography*, 4(2), 69–84.

Foley, J.A., Costa, M.H., Delire, C., Ramankutty, N., Snyder, P. (2003). Green surprise? How terrestrial ecosystems could affect earth's climate. *Frontiers in Ecology and the Environment*, 1(1), 38–44.

Frost, G.V. and Epstein, H.E. (2014). Tall shrub and tree expansion in Siberian tundra ecotones since the 1960s. *Global Change Biology*, 20, 1264–1277.

Frost, G.V., Epstein, H.E., Walker, D.A., Matyshak, G., Ermokhina, K. (2013). Patterned-ground facilitates shrub expansion in Low Arctic tundra. *Environmental Research Letters*, 8, 1–9.

Garamvölgyu, A. and Hufnagel, L. (2013). Impacts of climate change on vegetation distribution no. 1: Climate change induced vegetation shift in the palearctic region. *Applied Ecology and Environmental Research*, 11(1), 19–122.

Gonzales, P., Neilson, R.P., Lenihan, J.M., Drapek, R.J. (2010). Global patterns in the vulnerability of ecosystems to vegetation shifts due to climate change. *Global Ecology and Biogeography*, 19, 755–768.

Grimm, N.B., Stuart Chaplin III, F., Bierwagen, B., Gonzalez, P., Groffman, P.M., Luo, Y., Melton, F., Nadelhoffer, K., Pairis, A., Raymond, P.A. *et al.*, Schimel, J., Williamson, C.E. (2013). The impacts of climate change on ecosystem structure and function. *The Ecological Society of America*, 11(9), 474–482.

Heinrich, D. and Hergt, M. (1993). *Atlas de l'ecologie*. Hachette, Paris.

Hoegh-Guldberg, O., Jacob, D., Taylor, M. (eds) (2019). Impacts of 1.5°C global warming on natural and human systems. In Global warming of 1.5°C. Special report, Masson-Delmotte V. *et al.* (eds) [Online]. Available at: https://www.ipcc.ch/sr15/.

Holdrigde, L.R. (1947). Determination of world plant formations from simple climatic data. *Science*, 105, 367–368.

Hoogakker, B.A.A., Smith, R.S., Singarayer, J.S., Marchant, R., Prentice, I.C., Allen, J.R.M., Anderson, R.S., Bhagwat, S.A., Behling, H., Borisova, O. *et al.* (1947). Terrestrial biosphere changes over the last 120 Kyr. *Climate of the Past*, 12, 51–73.

Hufnagel, L. and Garamvölgyu, A. (2014). Impacts of climate change on vegetation distribution no. 2: Climate change induced vegetation shift in the new world. *Applied Ecology and Environmental Research*, 12(2), 355–422.

IPBES (2019). The Global Assessment report on Biodiversity and Ecosystem Services of the Intergovernmental Science-Policy Platform on Biodiversity and Ecosystem Services. *IPBES Global Assessment* [Online]. Available at: https://ipbes.net/global-assessment.

Kaplan, J.O., Bigelow, N.H., Prentice, I.C., Harrison, S.P., Bartlain, P.J., Christensen, T.R., Cramer, W., Matveyeva, N.V., McGuire, A.D., Murray, D.F. Razzhivin, V.Y., Smith, B., Walker, D.A., Anderson, P.M., Andreev, A.A., Brubaker, L.B., Edwards, M.E., Lozhkin, A.V. (2003). Climate change and arctic ecosystems 2: Modeling, paleodata-model comparisons, and future projections. *Journal of Geophysical Research*, 108(19), 8171.

Köppen, W. (1900). Versuch einer Klassifikation der Klimate, vorzugweise nach ihren Beziehungen zur Pflanzenwelt. *Geographische Zeitschrift*, 6, 593–611/657–679.

Lavorel, S., Lebreton, D., Le Maho, Y. (2017). Les mécanismes d'adaptation de la biodiversité aux changements climatiques et leurs limites. Report, Académie des Sciences, Paris.

Merriam, C.H. (1894). Laws of temperature control of the geographic distribution of terrestrial animals and plants. *National Geographic Magazine*, 6, 229–238.

Parmesan, C. and Yohe, G. (2003). A globally coherent fingerprint of climate change impacts across natural systems. *Nature*, 421, 37–42.

Prentice, I.C., Cramer, W., Harrison, S.P., Leemans, R., Monserud, R.A., Solomon, A.M. (1992). A global biome model based on plant physiology and dominance, soil properties and climate. *Journal of Biogeography*, 19, 117–134.

Ranson, K.J., Montesano, P.M., Nelson, R. (2011). Object-based mapping of the circumpolar taiga-tundra ecotone with MODIS tree cover. *Remote Sensing of Environment*, 115, 3670–3680.

Reyes-Fox, M., Steltzer, H., Trlica, M.J., McMaster, G.S., Andales, A.A., LeCain, D.R., Morgan, J.A. (2014). Elevated CO_2 further lengthens growing season under warming conditions. *Nature*, 510, 259–268.

Scheffers, B.R., De Meester, L., Bridge, T.C.L., Hoffmann, A.A., Pandoolfi, J.M., Corlett, R.T., Butchart, S.H.M., Pearce-Kelly, P., Kovacs, K.M., Dudgeon, D., Pacifici, M., Rondinini C., Foden, W.B., Martin, T.G., Mora, C., Bickford, D., Watson, J.E.M. (2016). The broad footprint of climate change from genes to biome to people. *Science*, 354, 6313.

Schimper, A.F.W. (1898). *Pflanzengeographie auf Physiologischer Grundlafe*. Gustav Fischer.

Walter, H. (1954). Klimax und zonale vegetation. *Angewandte Pflanzensoziologie*, 1, 144–150.

Walter, H. and Box, H., (1976). Global classification of natural terrestrial ecosystems. *Vegetatio*, 32(2), 75–91.

Whittaker, R.H. (1970). *Communities and Ecosystems*. MacMillan, New York.

Yates, D.N., Kittel, T.G., Cannon, R.F. (2000). Comparing the correlative Holdrigde model to mechanistic biogeographical models for assessing vegetation distribution response to climatic change. *Climatic Change*, 44, 59–87.

14

Spatial Impacts of Climate Change on Birds

Laurent GODET
CNRS, Nantes, France

14.1. Introduction

Birds' ranges are largely limited by environmental conditions (Hoffmann and Parsons 1997; Bridle and Vines 2006). These conditions can be biotic as well as abiotic (Orton 1920; Setchell 1920; Root 1988a). Geographic boundaries of distribution ranges towards the equator are known to be largely determined by biotic parameters with an increase in interspecific competition as latitude decreases (Brommer and Møller 2010; Svenning *et al.* 2014), often combined with an increase in predation and parasitism as heat and humidity conditions increase (Dobzhansky 1950; Mac Arthur 1972; Brown and Lomolino 1998; Hofer *et al.* 1999; Gross and Price 2000). For example, McKinnon *et al.* (2010) showed that shorebirds have an interest in nesting at very high latitudes, where nest predation is much lower, rather than in similar environments at lower latitudes. On the other hand, the limits of the distribution area towards the poles and at high altitudes are rather determined by abiotic factors and first and foremost by climatic conditions, especially temperature (e.g. Gaston 2003; Newton 2003; Brommer and Møller 2010), but also by the available favorable areas, especially with regard to altitude (see Figure 14.1).

Spatial Impacts of Climate Change,
coordinated by Denis MERCIER.

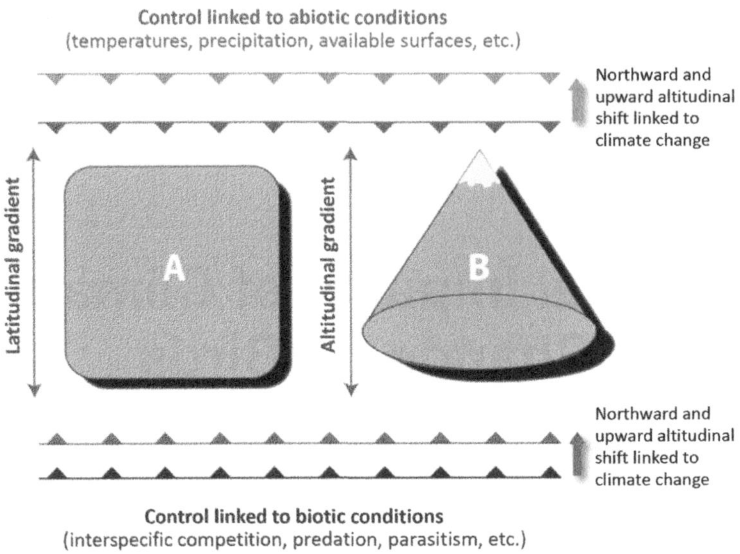

Figure 14.1. *The latitudinal (A) and altitudinal (B) limits of the birds' ranges are mainly controlled by abiotic factors (in red) at high latitudes (A) and altitudes (B), whereas the limits at low latitudes (A) and altitudes (B) are primarily controlled by biotic factors (in violet) (source: design and production, Laurent Godet). For a color version of this figure, see www.iste.co.uk/mercier/climate.zip*

There are many examples that illustrate these biogeographic determinants. For example, Root (1988b) showed the correspondence between the spatial distribution of isotherms and the ranges of birds in North America more than 30 years ago (Root 1988b): 62 species of birds in this region had their northern limits of distribution associated with particular January isotherms. Also in North America, the range of the Eurasian Magpie (*Pica pica*) reaches its southern most limit at less than 40°C, unlike that of the Yellow-billed Magpie (*Pica nuttalli*), a species which is more heat tolerant (Hayworth and Weathers 1984). In France, the second atlas of breeding birds (Yeatman-Berthelot and Jarry 1995) also showed the correspondence of isotherms with the northern limit of distribution of certain species (18°C in July for the Woodchat Shrike (*Lanius senator*), 3°C in January for the Dartford Warbler (*Sylvia undata*), 8°C in March for the Red-legged Partridge (*Alectoris rufa*) or their southern limit, 27°C in August for the Grey Partridge (*Perdix perdix*). In addition to temperature, precipitation patterns are also a determining factor (MacArthur 1972; Newton 2003) either directly by affecting survival (the northern limit of distribution of the Western Capercaillie (*Tetrao urogallus*) in Great Britain is for example constrained by the abundance of rainfall in the north, which affects the survival of chicks (Moss 1986) or indirectly by playing on the abundance of the trophic resource (Newton 2003).

In this context, it is understood that climate change can determine changes in the spatial distribution of birds. Although there are few paleontological data on birds (light, hollow bird bones are poorly preserved), various studies provide evidence of past changes in bird distribution related to climate change (e.g. Buehler *et al.* 1994; Sun *et al.* 2000; Wagner and Melle 2001), notably through genetic markers. As as example, Baker *et al.* (1994) address the phylogeography of Red Knot (*Calidris canutus*) in relation to climate change, which can be linked to episodes of glaciation and deglaciation (Piersma 1994). Birds, on the other hand, are good models for studying the impact of contemporary climate change on biodiversity. Indeed, they are a taxonomic group with a fairly large number of species (around 10,000), most of which are fairly mobile and easy to identify, whose ecology is fairly well known, and which interest enough enthusiasts to be mobilized for large-scale monitoring (Wormworth and Sekercioglu 2011).

It can be said somewhat schematically that birds have four solutions to current climate change (Peterson *et al.* 2001): adaptation (i.e. evolutionary change), sufficient phenotypical plasticity, movement, or extinction. Although the four solutions are closely related, the focus of this chapter is on the "movements" of birds, that is, changes in their spatial distribution, which may be linked both to processes of colonization of new areas and/or extinction at local or regional scales. We will first present how these changes in distribution occur (section 14.2), and then highlight the fact that not all species respond in the same way to climate change (section 14.3). These changes lead to changes in bird communities that find themselves out of step with the communities of other species with which they interact and raise new conservation issues (section 14.4).

14.2. Contemporary distributional changes

Although it is not the only factor, climatic conditions determine the distribution ranges of species, and it is therefore logical to note that the climate warming of recent decades favors a redistribution of certain species towards high latitudes (Beever *et al.* 2003; Walther 2005a; Cannone *et al.* 2007; Pauli *et al.* 2007; Holzinger *et al.* 2008) and high altitudes (Parmesan *et al.* 1999; Parmesan *et al.* 2000; Walther 2005b; Lemoine *et al.* 2007).

14.2.1. *Latitudinal shifts*

The first major studies, which can be described as "pioneering", to demonstrate northward range shifts of bird species were derived from temporal analyses of breeding bird atlases at national or sub-national scales. Thomas and Lennon (1999) were among the first to show that in 20 years British birds extended their northern limit of distribution northwards by about 0.95 km per year. In Finland, Brommer (2004) showed that in 12 years the country's 116 southern species also extended their range northwards by 1.57 km per year, whereas he did not observe this for the country's northern species. In North America, Hitch and Leberg (2007) also noted a northern expansion of southern species over 26 years of about 2.30 km per year, but no southern expansion of northern species. Also in North America, Zuckerberg *et al.* (2009) showed a northward progression in 20 years of 0.18 km per year of the 129 species studied in New York state, but also that the southern limit of northern species extends northward by 0.57 km per year. More recently, Auer and King (2014) estimated the average latitudinal return of 40 North American passerine species to 1.84 km per year from 1996 to 2011. Since then, studies calculating northward returns of passerine species have increased at different national, international and continental scales, largely due to data from participatory science monitoring, which has even allowed comparisons between countries at different latitudes and for taxonomic groups other than birds (e.g. Devictor *et al.* 2012).

Studies showing latitudinal shifts of birds in wintering grounds are less common; however, and, apart from studies of wintering birds in the United States showing northward movement of their wintering grounds over nearly 30 years (La Sorte and Thompson 2007), they have mainly concerned shorebirds wintering in Europe (see our literature review on the subject: Godet and Luczack 2012). In Great Britain, wintering shorebirds have tended to redistribute from the southwest to the northeast of the island, which corresponds schematically to an approximation of their tundra nesting areas and thus a shortening of their migration route (Rehfisch *et al.* 2004; Austin and Rehfisch 2005). At the scale of northwestern Europe, this same shift in the wintering grounds of shorebirds has been well documented for seven species (Maclean *et al.* 2008) and in 20 years has even reached 1.5 km per year for the Redshank (*Tringa totanus*) and 1.9 km per year for the Eurasian Oystercatcher (*Haematopus ostralegus*), 3.75 to 4.45 km for the Dunlin (*Calidris alpina*), Red Knot (*Calidris canutus*) and Bar-tailed Godwit (*Limosa lapponica*) and over 5.5 km for the Grey Plover (*Pluvialis squatarola*) or the Common Curlew (*Numenius arquata*). On the scale of Eurasia, Zöckler has also shown a shift in the range of the

Ruff (*Philomachus pugnax*) towards the northeast, which he attributes to global warming (Zöckler 2002).

In general, high-latitude bird species will be particularly affected by climate change because changes in precipitation patterns and global warming are more rapid at the poles (on average two to three times greater in the Arctic; see IPCC 2018), but also because the progression of these species towards the poles is (or is about to be) physically "blocked", either in the Arctic Ocean to the north or in the Southern Ocean to the south.

14.2.2. *Altitudinal shifts*

The altitudinal distribution of birds has also been well documented. For example, Auer and King estimate that 40 species of North American passerine birds will ascend 3.6 m per year from 1996 to 2011 (Auer and King 2014). As with high-latitude birds, biodiversity at high altitudes is strongly affected by climate change (Brunetti *et al.* 2009; Dirnböck *et al.* 2011), which is faster than the rest of the world (Böhm *et al.* 2001). This vulnerability of mountain biodiversity is all the greater as the rates of endemism (Essl *et al.* 2009) and the number of threatened species (Viterbi *et al.* 2013) are high. The distribution areas of mountain bird species are generally reduced, in particular due to relict distributions in the form of refuges following deglaciation episodes, which makes them vulnerable as the reduced size of the distribution area of terrestrial species is linked to their risk of extinction (Harris and Pimm 2007). An altitudinal rise in the range of birds is almost inevitably accompanied by a narrowing of their range, since the available surface area of a mountain decreases with altitude.

The mountain habitats available for birds are in a state of upheaval, as already evidenced by large changes in mountain plant communities (Pauli *et al.* 2007) and the sometimes-significant rise in the tree line (Harsch *et al.* 2009). Modeling studies therefore logically predict an increase in bird species associated with closed bush and forest environments in the Alps (environments that tend to progress and rise in altitude), to the detriment of species associated with open high-altitude grassland environments (Chamberlain *et al.* 2013). Similarly, in the Giant Mountains (Krkonoše) in the Czech Republic, in 20 years, Reif and Flousek showed a faster altitudinal ascent of birds in open environments than that of forest birds (Reif and Flousek 2012).

But it is probably in the intertropical zone that mountain birds are most threatened (Sekercioglu *et al.* 2008; Forero-Medina *et al.* 2011; Wormworth and Sekercioglu 2011; Freeman 2014; see Figure 14.2). Large cloud forests are the tropical environments where birds are likely to be among the most affected by climate change. Cascading effects of climate change are expected to be observed here, since many epiphytic species, which are the basis of food and building materials for birds' nests, have a thermal niche and very low resistance to global warming (Sekercioglu *et al.* 2012). In addition, warming of lower altitude areas will tend to progressively isolate bird populations from higher altitudes, which are often sedentary species (Sekercioglu *et al.* 2012).

Figure 14.2. *The Royal Sunbird (*Cinnyris regius*, bottom), endemic to the Albertine Rift, and its biotope (top), here the Nyungwe montane rainforest in the south of Rwanda, are under serious threat from climate change (source: © Laurent Godet, October 2018, Nyungwe National Park, Rwanda). For a color version of this figure, see www.iste.co.uk/mercier/climate.zip*

COMMENT ON FIGURE 14.2.– *Polar birds and environments are generally thought to be the main components of biodiversity threatened by global warming, but tropical mountain birds and environments, which are often endemic, are among the most threatened by global warming. This Nectariniidae is one of the tropical species whose range is expected to decline dramatically as a result of global warming (BirdLife International 2008) (source: © Laurent Godet, October 2018, Nyungwe National Park, Rwanda).*

14.2.3. *Spatial manifestations of range changes*

Schematically, several major types of distribution changes related to climate change can be distinguished, involving changes in the perimeter of distribution areas or a redistribution of abundance of individuals within these areas (Breshears *et al.* 2008; Kelly and Goulden 2008; Godet 2015; see Figure 14.3). These different patterns may combine as they occur for the same population or species.

– *Slippage*: the size of the distribution area is maintained but shifts in latitude or altitude. The study by Zuckerberg and colleagues (see section 14.2.1) argues in favor of this, as it shows both latitudinal shift from the northern and southern limits of the species (Zuckerberg *et al.* 2009).

– *Contraction*: the range moves up in latitude or altitude but is constrained to its latitudinal or altitudinal limits. The best examples are probably the bird species associated with open environments at high altitudes, whose low limits are pushed back in altitude by a progression of the forest front, whereas their high altitudinal limits are constrained by the top of the relief (see Angert *et al.* 2011 for this last point). Similarly, bird ranges may be compressed as they progress towards a sea or ocean.

– *Extirpation*: the rise in latitude or altitude is no longer possible beyond the current range and the population or species disappears. This disappearance can be considered the extreme case of contraction.

– *Stretching:* the range extends polewards or upwards, but its low latitude or altitude limit is stable. Maclean *et al.* (2008) tend to show such a pattern for waders wintering in Europe, as their ranges expand northward while their abundance does not decline in the south. However, their interpretation should be qualified in the sense that the authors only use data from north-western Europe, an area that does not include the true southern limits of the species studied. Birds associated with forest cover in mountains are also good examples of species with expanding ranges, as they follow the altitudinal progression of the forest front. Species that colonize new sites at higher elevations or latitudes and whose populations are increasing in size are also likely to experience range stretching rather than just shifting, since population size increases with range size (Brown 1984). Finally, since some North American birds have been found to move northward at a faster rate than the center of their range (La Sorte and Thompson 2007), this is a stretch (which may possibly precede a longer-term shift).

– *Tilting*: nuclei of high bird abundance shift towards the poles or upwards while the perimeter of the distribution area remains stable. This can be considered as the stage that precedes a true range shift, with colonizations of individuals at high latitudes or altitudes and extinctions at low latitudes or altitudes.

– *The general decrease or increase in population uniformly throughout the distribution area*: the population is not sufficiently plastic and/or does not adapt sufficiently, leading to a collapse of its numbers or, on the other hand, it shows good plasticity or adaptation and experiences an increase in numbers.

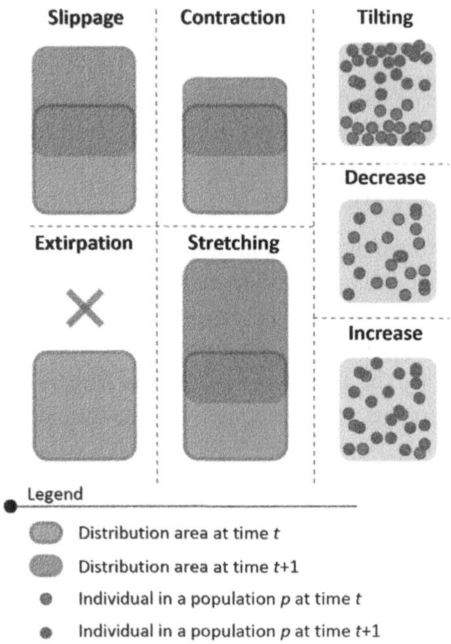

Figure 14.3. *Different patterns of change in species distribution as a consequence of climate change (source: design and production: Laurent Godet). For a color version of this figure, see www.iste.co.uk/mercier/climate.zip*

However, observing such changes empirically remains difficult as it requires monitoring data at the scale of the entire species' range and even a little beyond this range, to be able to record potential colonization of new areas. Otherwise, most data on birds are available at national scales and understanding changes in distribution requires working on species that have their spatial distribution limits within these countries. Indeed, studies conducted in countries centered on the core range of most species are not very informative (Brommer and Møller 2010).

It should be noted that these changes in distribution patterns can be determined by climate change in complex combination with land use and land cover patterns (Williams and Newbold 2019). While the effect of climate change is considered by some authors to be the main threat to terrestrial vertebrates, its combined effect with changes in land use is expected to lead to the loss of nearly 38% of the species in this group by 2070 (Newbold 2018).

Finally, the question of whether these changes in range boundaries and changes in abundance within ranges are due to movements of the same individuals over time (reflecting phenotypic plasticity) or are due to generational changes (reflecting adaptation) remains open for many species. This issue could be the subject of a whole chapter that is beyond the scope of this book (e.g, see Bradshaw and Holzapfel 2006 on this subject).

14.3. Different responses for different species

Several life-history traits are associated with varying degrees of vulnerability of birds to climate change: dispersal capacity (Sekercioglu *et al.* 2008), low temperature and rainfall tolerance (Deutsch *et al.* 2008), dependence on snow or ice cover (Jenouvrier *et al.* 2009), specialist habitat selection (Hilbert et *al.* 2004), high interspecific dependence, low genetic diversity or low phenological response to climate change (Both and Visser 2001). Comprehensive syntheses on the relationship between life-history traits in birds and susceptibility to distribution change in the face of climate change have recently been provided (Gregory *et al.* 2009; Chen *et al.* 2011; Estrada *et al.* 2016; Maclean and Beissinger 2017). We are interested here in three families of life-history traits known to be related to their ability to change or not change distribution in the context of climate change: their dispersal capacities; their reproductive capacities; and their degree of specialization.

14.3.1. *Dispersion capabilities*

It is expected that the bird species with the greatest dispersal capacity will also be the most likely to change their range by colonizing previously unoccupied sites more easily. The species most likely to disperse would be large, highly migratory species, with a high capacity for movement, or those whose young are not very philopatric (high dispersion between birth sites and sites of settlement during their first breeding season).

Several studies have tested (among other traits) the influence of the size (often body size) of bird species on their ability to alter their range (Austin and Rehfisch 2005; Brommer 2008; Tingley *et al.* 2009; Angert *et al.* 2011) but few have shown truly significant and unambiguous effects. Nevertheless, Brommer shows that, using the example of Finnish birds, large birds have the most northerly range shift

(Brommer 2008). Conversely, for waders wintering in Great Britain, the dependence between distribution shift and temperature is more pronounced for smaller species (Austin and Rehfisch 2005).

Similarly, the links between migration strategies and distributional change remain to be clarified despite the numerous studies exploring them (La Sorte and Thompson 2007; Brommer 2008; Tingley *et al.* 2009; Zuckerberg *et al.* 2009; Brommer and Møller 2010; Angert *et al.* 2011; Auer and King 2014). Sekercioglu *et al.* (2012) consider that sedentary birds are much more likely to become extinct than migratory birds due to a change in distribution linked to climate change. Hockey *et al.* (2011) show that birds in South Africa that have undergone distribution changes related to climate change are mostly the most "mobile" species (in the sense of fully migratory species – as opposed to partially migratory or sedentary species). Similarly, Brommer and Moller (2010) highlight that in Finland and the UK, the most mobile southern bird species are also more likely to be migratory species. Although this remains to be demonstrated, it can also be expected that species that have adopted migration strategies requiring a large number of migratory stopovers have knowledge of a large number of sites and would be more capable of changing their range. In contrast, bird species that learn to migrate in family groups from generation to generation and often over long distances, such as geese (Elder and Elder 1949) or swans, using traditional wintering, breeding and staging sites, are probably unlikely to change their range easily. Similarly, long-lived bird species that are very loyal to their breeding or wintering sites, such as many species of shorebirds (e.g. the Grey Plover, *Pluvialis squatarola*, see Townshend 1985) will also be less likely to change their distribution.

14.3.2. *Reproductive capacity*

Species with high reproductive capacity are expected to be more able to colonize new sites and thus change their distribution (Angert *et al.* 2011; Buckley 2012).

Auer and King show that large egg-laying North American passerines did indeed shift to higher altitudes than others, but less in latitude, which the authors are still struggling to explain in their study (Auer and King 2014). At most, they draw a parallel between small egg-laying species and a change in distribution with a study (Sol *et al.* 2012) showing that the bird species most likely to be "invasive" (i.e. to successfully colonize new sites) were species with slow cycles because they would be better adapted to abrupt environmental changes.

14.3.3. *Generalist nature*

Specialist species are less flexible than generalists in the face of environmental change (Travis 2003) and are more prone to extinction (Sekercioglu *et al.* 2004). On the contrary, a species is expected to be all the more capable of colonizing new sites at a given latitude or altitude the more generalist it is (in terms of habitat and/or diet selection), as it is expected to benefit more easily from new habitats or food resources it encounters (Buckley 2012).

Like Davey *et al.* (2013), who show that birds in Sweden that are generalist in their habitat extend their range more, Hockey *et al.* (2011) also show that birds in South Africa that have shown distribution changes related to climate change are mostly generalist in their habitat.

On the other hand, studies linking dietary specialization to the ability to change distribution offer rather contradictory results. A first group of studies establishes a positive link between the generalist nature of birds' diets and their ability to change distribution. Among North American passerines, generalist diet appears to be the life-history trait most closely (and positively) related to the ability to change distribution (Angert *et al.* 2011). In the intertropical domain, omnivorous (but also frugivorous) mountain birds of New Guinea also move higher in altitude (Freeman 2014). However, other studies show, on the contrary, that it is rather in species with specialized diets that changes in distribution are most marked. By studying the southernmost bird species in Finland and the United Kingdom, it appears that species specialized in their diet (mainly insectivores and some herbivores) have experienced significant changes in distribution towards the north, unlike birds of prey and omnivores, whose ranges have remained more stable (Brommer 2008; Brommer and Møller 2010). Similarly, Auer and King show that the more specialized the diet of a bird species, the higher the latitude of its range (Auer and King 2014). The authors explain this by the fact that specialist birds are forced to follow their prey, which also undergo rapid changes in distribution. For example, terrestrial and aquatic invertebrates are known to experience faster range shifts than birds and mammals (Hickling *et al.* 2006), so insectivorous birds tend to have range shifts that change more rapidly than others to "follow" their prey.

14.4. Conservation implications

14.4.1. *Ecological consequences*

The consequences of changes in bird distribution related to climate change can be of different magnitudes.

One of the first consequences of range changes is the population decline of species that will not be likely to find suitable habitat at higher latitudes or altitudes. This has been seen previously for polar species or species living at higher altitudes, but it is also true for coastal bird species that will experience major disruptions in the availability of their habitats as a result of sea level rise (Galbraith *et al.* 2002; Hugues 2004).

The speed of change in the distribution of a bird species may not be synchronous with that of its prey. Birds and butterflies, for example, have experienced very different latitudinal ranges in Europe: 37 km for birds compared to 114 km for butterflies in about 20 years (Devictor *et al.* 2012). The breeding period of some species may be out of synchrony with the period of prey availability, as shown in Dutch populations of the Pied Flycatcher (*Ficedula hypoleuca*) where those that are least synchronous with prey emergence peaks are subject to decline (Hickling *et al.* 2006). Bradshaw and Holzapfel (2006) have for example shown the desynchronization between the hatching of chicks of the Great Tit (*Parus major*) and the emergence of caterpillars that serve as food for the chicks. Caterpillars tend to emerge earlier and earlier, even before the chicks hatch. Some individuals have been able to adjust their egg-laying dates to the new caterpillar emergence phenology, thus maintaining their reproductive success, in contrast to non-adapting individuals whose reproductive success has dropped.

A change in the distribution of a bird species can also lead to its confrontation with a predator to which it is not adapted, or with other species with which it will compete (Miller-Rushing *et al.* 2010). For example, in Costa Rica, the altitudinal ascent of the Keel-billed Toucan (*Ramphastos sulfuratus*) has led to the colonization of the range of the Resplendent Quetzal (*Pharomachrus mocinno*, see Pounds *et al.* 1999), whose eggs and chicks it may plunder (Miller-Rushing *et al.* 2010).

Finally, since not all species see their distribution area modified with the same intensity and speed, this leads to profound community changes at the local scale (see Godet *et al.* 2011) that can significantly alter the functioning of ecosystems.

14.4.2. *Conservation measures*

The solutions to combat the negative impact of climate change on birds are multifaceted. We will not go into detail here on the first of these solutions, which obviously consists of limiting global warming itself by reducing greenhouse gas emissions. This is beyond the scope of this chapter.

– Improving our knowledge is a necessity for making good decisions on how to combat the effects of climate change on birds. Despite the impressive accumulation of scientific articles dealing with the impact of climate change on birds (as of November 12, 2019, 8,089 articles published since 1983 and listed in the Web of Science are linked to the keywords "climate change" and "bird", including more than 500 articles published annually since 2011), This literature is very recent (2 articles listed in the Web of Science in the 1980s, and less than 100 in the 1990s, based on these keywords), many have still contradictory results as we have seen (for example, about the link between certain life history traits and changing distribution) and few propose conservative solutions. However, the lack of knowledge should not mask the evidence put forward by the scientific literature: in the face of climate change, the distribution areas of birds are being disrupted, not all species react in the same way, and many are threatened by these changes.

– Considering the sensitivity of species to climate change has therefore become necessary to assess the vulnerability of a species and, ultimately, its protection. Currently, the classification of species on the International Union for Conservation of Nature (IUCN) Red List is based on criteria of population size and trend, geographical distribution, probability of extinction in the wild (based on the population viability analysis, itself based on life history characteristics, habitat requirements, threats and management options specific to each species (UICN 2012). Miller-Rushing *et al.* (2010) proposed that climate threat should be included in these criteria, which is still not the case. This has also been highlighted by other authors who point out that 35% of the world's bird species have life history traits that make them sensitive to climate change (Vié *et al.* 2009).

– Before proposing proactive measures to protect birds threatened by climate change, it is first urgent to conserve them within their current range and against the best-known threats to them in addition to climate change (habitat fragmentation, invasions, over-exploitation, etc.), as it is often cumulative pressures that endanger biodiversity. Anticipation is also absolutely necessary in the sense that it is important to conserve those areas that allow for the greatest resistance or resilience of bird communities in the face of climate change. Gaüzère *et al.* (2017) have shown that landscape diversity, topography, and naturalness of landscapes are important factors in reducing the "climate debt" of birds (understood as the accumulation of delay in the response of birds to rising temperatures; the "climate debt" can be expressed in km per year, where "a debt of 3 km per year" means that birds are "3 km behind the northward migration of isotherms"). These areas must therefore be conserved as a priority.

– Protected areas, in their current state, are a relevant tool in the sense that they are areas within which the impact of climate change on birds is mitigated, as shown by Gaüzère *et al.* (2016) both at the scale of bird communities (which adjust better to climate change) and at the scale of individual species (the more a species benefits from a significant part of its population within a protected area network, the less vulnerable it is to climate change). However, the effectiveness of protected areas, fixed in time and space while species have shifting ranges, in countering the effects of climate change has also been widely questioned. Reviews by Halpin (1997), Hannah (2008) and us (Godet 2015) summarize measures to adjust the spatial distribution of protected areas to climate change. The creation of redundant protected areas in the same biogeographic region, protecting similar habitats and species are expected to reduce the risk of potential impacts on one of the protected areas. The extension in latitude or altitude of protected area boundaries should also anticipate changes in species distribution. A very popular approach is also the development or maintenance of ecological corridors allowing for pathways between protected areas. The spatial (re)organization of the entire network of protected areas seems more utopian. The simplest way is to think upstream about the creation of a network of protected areas according to the problem of climate change, as proposed in the marine domain by Salm and Coles (2001) or West and Salm (2003). The creation of new protected areas in the areas of future distribution of species, towards the poles (Scott and Suffling 2000) or in altitude (Halpin 1997), is also proposed by some authors. The proposal of protected areas moving in time and space (see Pressey *et al.* 2007 or Soto 2001 in the marine realm) seems somewhat illusory, although flexible perimeters for authorized hunting and fishing are already in place (Hannah 2008).

– More interventionist measures are also proposed. Assisted migrations of species may be proposed, that is, the movement of individuals outside their original range to sites that are or will be favorable in the future in a context of climate change. These measures may be proposed for species whose natural colonization capacities are particularly low for several reasons: low dispersal capacities (see Sekercioglu 2007) and/or restricted availability of habitats during latitudinal or altitudinal shift of the range, and/or populations that are too small or with too low demographic trends to establish new populations. These measures should, in our view, be handled with extreme caution. It is not only rather difficult to ensure that the habitats at the place of introduction are adequate, but also, and more importantly, that the effects of the introduction will not be detrimental to the ecosystems (Mueller and Hellmann 2000). It should not be forgotten that the introduction of species is itself one of the major causes of the current collapse of biodiversity.

14.5. Conclusion

The only current representatives of dinosaurs, birds have experienced many phases of global warming during their 150-million-year evolutionary history.

However, the archaeozoological remains of this group are few in number and therefore provide little evidence of their ability to cope with these past episodes. On the other hand, it is indisputable that contemporary climate change has already induced changes in the spatial distribution of birds in just a few decades. Their distribution areas are slipping, contracting or stretching towards the poles and in altitude, sometimes leading to population collapses already noted or to come, especially for polar or tropical and temperate mountain species. These changes in distribution also mark a spatial and temporal delay in relation to latitudinal and altitudinal temperature rises, causing birds to suffer a "climate debt" that they accumulate over time. Bird species are more or less likely to change their distribution in response to contemporary climate change depending on their life history traits. Those species that are the least mobile and the most specialist in terms of habitat selection are probably the least able to make such changes and are therefore the most threatened. All the consequences of these changes in distribution are still difficult to anticipate precisely, but asynchronous changes in distribution between birds and their prey or predators and local-scale changes in bird communities are already occurring and are likely to significantly alter the functioning of ecosystems. Conservation measures of a purely technical nature, such as species-assisted migration, should not lead us to believe that technoscience will be able to respond to the crisis situation facing biodiversity. The first solution to combat the pressure of climate change on birds is to act on the anthropogenic sources of global warming, first and foremost the fight against greenhouse gas emissions. It is also urgent to protect bird communities in their current and future distribution areas and first of all in the areas that are most resilient to climate change, including protected areas and areas offering great naturalness and landscape diversity. Climate change is indeed combined with massive habitat fragmentation and overexploitation of resources by humans. Isolating the effects of climate change from these other global changes is still delicate. The conservation of the last remaining areas of nature that are little or not anthropogenic is therefore a matter not only of urgency but also of common sense in the current Anthropocene context.

14.6. References

Angert, A.L., Crozier, L.G., Rissler, L.J., Gilman, S.E., Tewksbury, J.J., Chunco, A.J. (2011). Do species' traits predict recent shifts at expanding range edges? *Ecology Letters*, 14, 677–689.

Auer, S.K. and King, D.I. (2014). Ecological and life-history traits explain recent boundary shifts in elevation and latitude of western North American songbirds. *Global Ecology and Biogeography*, 23, 867–875.

Austin, G.E. and Rehfisch, M.M. (2005). Shifting nonbreeding distributions of migratory fauna in relation to climate change. *Global Change Biology*, 11, 31–38.

Baker, A.J., Piersma, T., Rosenmeier, L. (1994). Unraveling the intraspecific phylogeography of Knots *Calidris canutus*: Progress report on the search for genetic markers. *Journal of Ornithology*, 135, 599–608.

Beever, E.A., Brussard, P.F., Berger, J. (2003). Patterns of apparent extirpation among isolated populations of pikas (*Ochotona princeps*) in the Great Basin. *Journal of Mammalogy*, 84, 37–54.

BirdLife International (2008). *State of the World's Birds: Indicators for Our Changing World.* BirdLife International, Cambridge.

Böhm, R., Auer, I., Brunetti, M., Maugeri, M., Nanni, T., Schöner, W. (2001). Regional temperature variability in the European Alps; 1769–1998 from homogenized instrumental time series. *International Journal of Climatology*, 21, 1779–1801.

Both, C. and Visser, M.E. (2001). Adjustment to climate change is constrained by arrival date in a long-distance migrant bird. *Nature*, 411, 296–298.

Bradshaw, W. and Holzapfel, C. (2006). Evolutionary response to rapid climate change. *Science*, 312, 1477–1478.

Breshears, D.D., Huxman, T.E., Adams, H.D., Zou, C.B., Davison, J.E. (2008). Vegetation synchronously leans upslope as climate warms. *Proceedings of the National Academy of Sciences of the USA*, 105, 11591–11592.

Bridle, J.R. and Vines, T.H. (2006). Limits to evolution at range margins: When and why does adaptation fail? *Trends in Ecology and Evolution*, 22, 140–147.

Brommer, J.E. (2004). The range margins of northern birds shift polewards. *Annales Zoologici Fennici*, 41, 391–397.

Brommer, J.E. (2008). Extent of recent polewards range margin shifts in Finnish birds depends on their body mass and feeding ecology. *Ornis Fennica*, 85, 109–117.

Brommer, J.E. and Møller, A.P. (2010). Range margins, climate change, and ecology. In *Effects of Climate Change on Birds*, Møller, A.P., Fiedler, W., Berthold, P. (eds). Oxford University Press, Oxford.

Brown, J.H. (1984). On the relationship between abundance and distribution of species. *American Naturalist*, 124, 255– 279.

Brown, J.H. and Lomolino, M.V. (1998). *Biogeography*, 2nd edition. Sinauer Associates, Inc., Sunderland.

Brunetti, M., Lentini, G., Maugeri, M., Nanni, T., Auer, I., Böhm, R., Schöener, W. (2009). Climate variability and change in the Greater Alpine Region over the last two centuries based on multi-variable analysis. *International Journal of Climatology*, 29, 2197–2225.

Buckley, L.B. (2012). Functional and phylogenetic approaches to forecasting species' responses to climate change. *Annual Review of Ecology, Evolution, and Systematics*, 43, 205–226.

Buehler, D.M., Baker, A., Piersma, T. (1994). Reconstructing palaeoflyways of the late Pleistocene and early Holocene Red Knot. *Calidris canutus. Ardea*, 94, 485–498.

Cannone, N., Sgorbati, S., Guglielmin, M. (2007). Unexpected impacts of climate change on alpine vegetation. *Frontiers in Ecology and the Environment*, 5, 360–364.

Chamberlain, D.E., Negro, M., Caprio, E., Rolando, A. (2013). Assessing the sensitivity of alpine birds to potential future change in habitat and climate to inform management strategies. *Biological Conservation*, 167, 127–135.

Chen, I.C., Hill, J.K., Ohlemüller, R., Roy, D.B., Thomas, C.D. (2011). Rapid range shifts of species associated with high levels of climate warming. *Science*, 333, 1024–1026.

Davey, C.M., Devictor, V., Jonzén, N., Lindström, A., Smith, H.H. (2013). Impact of climate change on communities: Revealing species' contribution. *Journal of Animal Ecology*, 82, 1365–2656.

Deutsch, C.A., Tewksbury, J.J., Huey, R.B., Sheldon, K.S., Ghalambor, C.K., Haak, D.C., Martin, P.R. (2008). Impacts of climate warming on terrestrial ectotherms across latitude. *Proceedings of the National Academy of Sciences of the USA*, 105, 6668–6672.

Devictor, V., van Swaay, C., Brereton, T., Brotons, L., Chamberlain, D., Heliölä, J., Herrando, S., Julliard, R., Kuussaari, M., Lindström, A., Reif, J., Roy, D.B., Schweiger, O., Settele, J., Stefanescu, C., Van Strien, A., Van Turnhout, C., Vermouzek, Z., WallisDeVries, M., Wynhoff, I., Jiguet, F. (2012). Differences in the climatic debts of birds and butterflies at a continental scale. *Nature Climate Change*, 2, 121–124.

Dirnböck, T., Essl, F., Babitsch, W. (2011). Disproportional risk for habitat loss of high altitude endemic species under climate change. *Global Change Biology*, 17, 990–996.

Dobzhansky, T. (1950). Evolution in the tropics. *American Scientist*, 38, 209–221.

Elder, W.H. and Elder, N.L. (1949). Role of the family in the formation of goose flocks. *Wilson Bulletin*, 61, 133–140.

Essl, F., Staudinger, M., Stöhr, O., Schratt-Ehrendorfer, L., Rabitsch, W., Niklfeld, H. (2009). Distribution patterns, range size and niche breadth of Austrian endemic plants. *Biological Conservation*, 142, 2547–2558.

Estrada, A., Morales-Castilla, I., Caplat, P., Early, R. (2016). Usefulness of species traits in predicting range shifts. *Trends in Ecology and Evolution*, 31, 190–203.

Forero-Medina, G., Terborgh, J., Socolar, S.J., Pimm, S.L. (2011). Elevational ranges of birds on a tropical montane gradient lag behind warming temperatures. *PLoS ONE*, 6.

Freeman, B.G. and Class Freeman, A.M. (2014). Rapid upslope shifts in New Guinean birds illustrate strong distributional responses of tropical montane species to global warming. *Proceedings of the National Academy of Sciences of the USA*, 111, 4490–4494.

Galbraith, H., Jones, R., Park, R., Clough, J., Herrod-Julius, S., Harrington, B., Page, G. (2002). Global climate change and sea level rise: Potential losses of intertidal habitat for shorebirds. *Waterbirds*, 25, 173–183.

Gaston, K.J. (2003). *The Structure and Dynamics of Geographic Ranges*. Oxford University Press, Oxford.

Gaüzère, P., Jiguet, F., Devictor, V. (2016). Can protected areas mitigate the impacts of climate change on bird's species and communities? *Diversity and Distributions*, 22, 625–637.

Gaüzère, P., Princé, K., Devictor, V. (2017). Where do they go? The effects of topography and habitat diversity on reducing climatic debt in birds. *Global Change Biology*, 23, 2218–2229.

Godet, L. (2015). La conservation de la biodiversité dans un contexte de changements globaux. In *Pour une geographie de la conservation, biodiversites, natures et societes*, Mathevet, R., Godet, L. (eds). L'Harmattan, Paris.

Godet, L. and Luczack, C. (2012). Changement climatique et limicoles en hiver. In *Manuel d'étude et de gestion des oiseaux et de leurs habitats en zones côtières*, Triplet, P. (ed.). Estuarium/Forum des Marais Atlantiques, Rochefort.

Godet, L., Jaffré, M., Devictor, V. (2011). Waders in winter: Long-term changes of bird assemblages facing climate change. *Biology Letters*, 7, 714–717.

Gregory, R.D., Willis, S.G., Jiguet, F., Vorisek, P., Klvanova, A., van Strien, A., Huntley, B., Collingham, Y.C., Couvet, D., Green, R.E. (2009). An indicator of the impact of climatic change on European bird populationsl. *PLoS ONE*, 4.

Gross, S.J. and Price, T.D. (2000). Determinants of the northern and southern range limits of a warbler. *Journal of Biogeography*, 27, 869–878.

Halpin, P.N. (1997). Global climate change and natural-area protection: Management responses and research directions. *Ecological Applications*, 7, 828–843.

Hannah, L. (2008). Protected areas and climate change. *Year in Ecology and Conservation Biology*, 1134, 201–212.

Harris, G. and Pimm, S.L. (2007). Range size and extinction risk in forest birds. *Conservation Biology*, 22, 163–171.

Harsch, M.A., Hulme, P.E., McGlone, M.S., Duncan, R.P. (2009). Are treelines advancing? A global meta-analysis of treeline response to climate warming. *Ecology Letters*, 12, 1040–1049.

Hayworth, A.M. and Weathers, W. (1984). Temperature regulation and climatic adaptation in Black-billed and Yellow-billed Magpies. *Condor*, 86, 19–26.

Hickling, R., Roy, D.B., Hill, J.K., Fox, R., Thomas, C.D. (2006). The distribution of a wide range of taxonomic groups are expanding polewards. *Global Change Biology*, 12, 450–455.

Hilbert, D.W., Bradford, M., Parker, T., Westcott, D.A. (2004). Golden Bowerbird (*Prionodura newtonia*) habitat in past, present and future climates: Predicted extinction of a vertebrate in tropical highlands due to global warming. *Biological Conservation*, 116, 367–377.

Hitch, A.T. and Leberg, P.L. (2007). Breeding distributions of North American bird species moving north as a result of climate change. *Global Change Biology*, 12, 450–455.

Hockey, P.A.R., Sirami, C., Ridely, A.R., Midgley, G.F., Babiker, H.A. (2011). Interrogating recent range changes in South African birds: Confounding signals from land use and climate change present a challenge for attribution. *Diversity and Distributions*, 17, 254–261.

Hofer, U., Bersier, L.F., Borcard, D. (1999). Spatial organization of a herpetofauna on an elevational gradient revealed by null model tests. *Ecology*, 80, 976–988.

Hoffmann, A.A. and Parsons, P.A. (1997) *Extreme Environmental Change and Evolution*. Cambridge University Press, Cambridge.

Holzinger, B., Hülber, K., Camenisch, M., Grabherr, G. (2008). Changes in plant species richness over the last century in the eastern Swiss Alps: Elevational gradient, bedrock effects and migration rates. *Plant Ecology*, 195, 179–196.

Hughes, R.G. (2004). Climate change and loss of saltmarshes: Consequences for birds. *Ibis*, 146, 21–28.

IPCC, Intergovernmental Panel on Climate Change (2018). Global warming of 1.5°C. Summary for policymakers [Online]. Available at: https://report.ipcc.ch/sr15/pdf/sr15_spm_final.pdf.

Jenouvrier, S., Caswell, H., Barbraud, C., Holland, M., Stroeve, J., Weimerskirch, H. (2009). Demographic models and IPCC climate projections predict the decline of an Emperor Penguin population. *Proceedings of the National Academy of Sciences of the USA*, 106, 1844–1847.

Kelly, A.E. and Goulden, M.L. (2008). Rapid shifts in plant distribution with recent climate change. *Proceedings of the National Academy of Sciences of the USA*, 105, 11823–11826.

La Sorte, F.A. and Thompson, F.R. (2007). Poleward shifts in winter ranges of North American birds. *Ecology*, 88, 1803–1812.

Lemoine, N., Schaefer, H.-C., Böhning-Gaese, K. (2007). Species richness of migratory birds is influenced by global climate change. *Global Ecology and Biogeography*, 16, 55–64.

Maclean, I.M.D., Austin, G.E., Rehfish, M.M., Blew, J., Crowe, O., Delany, S., Devos, K., Deceuninck, B., Günther, K., Laursen, K., Van Roomen, M., Wahl, J. (2008). Climate change causes rapid changes in the distribution and site abundance of birds in winter. *Global Change Biology*, 14, 2489–2500.

Maclean, S.A. and Beissinger, S.R. (2017). Species' traits as predictors of range shifts under contemporary climate change: A review and meta-analysis. *Global Change Biology*, 23, 4094–4105.

MacArthur, R.H. (1972). *Geographical Ecology. Patterns in the Distribution of Species*. Harper and Row, New York.

McKinnon, L., Smith, P.A., Nol, E., Martin, J.L., Doyle, F.I., Abraham, K.F., Gilchrist, H.G., Morrison, R.I.G., Bêty, J. (2010). Lower predation risk for migratory birds at high latitudes. *Science*, 327, 326–327.

Miller-Rushing, A.J., Primack, R.B., Sekercioglu, C.J. (2010). Conservation consequences of climate change for birds. In *Effects of Climate Change on Birds*, Moller, A.P., Fielder, W., Berthold, P. (eds). Oxford University Press, Oxford.

Moss, R. (1986). Rain, breeding success and distribution of Capercaillie *Tetrao urogallus* and Black Grouse *Tetrao tetrix* in Scotland. *Ibis*, 128, 63–72.

Mueller, J.M. and Hellmann, J.J. (2008). An assessment of invasion risk from assisted migration. *Conservation Biology*, 22, 562–567.

Newbold, T. (2018). Future effects of climate and land-use change on terrestrial vertebrate community diversity under different scenarios. *Proceedings of the Royal Society B*, 285, 2018–0792.

Newton, I. (2003). *The Speciation and Biogeography of Birds*. Academic Press, Cambridge.

Orton, J.H. (1920). Sea-temperature, breeding, and distribution in marine animals. *Journal of the Marine Biology Association*, 12, 339–366.

Parmesan, C., Ryrholm, N., Stefanescu, C., Hill, J.K., Thomas, C.D., Descimon, H., Huntley, B., Kaila, L., Kullberg, J., Tammaru, T., Tennent, W.J., Thomas, J.A., Warren, M. (1999). Poleward shifts in geographical ranges of butterfly species associated with regional warming. *Nature*, 399, 579–583.

Parmesan, C., Root, T.L., Willig, M.R. (2000). Impacts of extreme weather and climate on terrestrial biota. *Bulletin of the American Meteorological Society*, 81, 443–450.

Pauli, H., Gottfried, M., Reiter, K., Klettner, C., Grabherr, G. (2007). Signals of range expansions and contractions of vascular plants in the high Alps: Observations (1994–2004) at the GLORIA master site Schrankogel, Tyrol, Austria. *Global Change Biology*, 13, 147–156.

Peterson, A.T., Sánchez-Cordero, V., Soberón, J., Bartley, J., Buddemeier, R.W., Navarro-Sigüenza, A.G. (2001). Effects of global climate change on geographic distributions of Mexican Cracidae. *Ecological Modeling*, 144, 21–30.

Piersma, T. (1994). *Close to the Edge: Energetic Bottlenecks and the Evolution of Migratory Pathways in Knot*. Het Open Boek, Den Burg.

Pounds, J.A., Fogden, M.P.L. Campbell, J.H. (1999). Biological response to climate change on a tropical mountain. *Nature*, 398, 611–615.

Pressey, R.L., Cabeza, M., Watts, M.E., Cowling, R.M., Wilson, K.A. (2007). Conservation planning in a changing world. *Trends in Ecology and Evolution*, 22, 583–592.

Rehfisch, M.M., Austin, G.E., Freeman, S.N., Armitage, M.J.S., Burton, N.H.K. (2004). The possible impact of climate change on the future distributions of waders on Britain's non-estuarine coast. *Ibis*, 146, S70–S81.

Reif, J. and Flousek, J. (2012). The role of species' ecological traits in climatically driven altitudinal range shifts of central European birds. *Oikos*, 121, 1053–1060.

Root, T.L. (1988a). Environmental factors associated with avian distributional boundaries. *Journal of Biogeography*, 15, 489–505.

Root, T.L. (1988b). Energy constraints on avian distributions and abundances. *Ecology*, 69, 330–339.

Salm, R.V. and Coles, S.L. (2001). *Coral Bleaching and Marine Protected Areas*. The Nature Conservancy, Comté d'Arlington.

Scott, D. and Suffling, R. (2000). *Climate Change and Canada's National Park System*. Environment Canada, Ottawa.

Sekercioglu, C.H. (2007). Conservation ecology: Area trumps mobility in fragment bird extinctions. *Current Biology*, 17, R283–R286.

Sekercioglu, C.H., Daily, G.C., Ehrlich, P.R. (2004). Ecosystem consequences of bird decline. *Proceedings of the National Academy of Sciences of the USA*, 101, 18042–18047.

Sekercioglu, C.H., Schneider, S.H., Fay, J.P., Loarie, S.R. (2008). Climate change, elevational range shifts, and bird extinctions. *Conservation Biology*, 22, 140–150.

Sekercioglu, C.H., Primack, R.B., Wormworth, J. (2012). The effects of climate change on birds. *Biological Conservation*, 148, 1–18.

Setchell, W.A. (1920). The temperature interval in the geographical distribution of marine algae. *Science*, 52, 187–190.

Sol, D., Maspons, J., Vall-Llosera, M., Bartomeus, I., Garcia-Pena, G.E., Pinol, J., Freckleton, R. (2012). Unraveling the life history of successful invaders. *Science*, 337, 580–583.

Soto, C.G. (2001). The potential impacts of global climate change on marine protected areas. *Reviews in Fish Biology and Fisheries*, 11, 181–195.

Sun, L., Xie, Z., Zhao, J. (2000). A 3,000-year record of penguin populations. *Nature*, 407, 858.

Svenning, J.C., Gravel, D., Holt, R.D., Schur, F.M., Thuiller, W., Münkemüller, T., Schiffers, K.H., Dullinger, S., Edwards Jr., T.C., Hickler, T., Higgins, S.I., Nabel, J.E.M.S., Pagel, J., Normand, S. (2014). The influence of interspecific interactions on species range expansion rates. *Ecography*, 37, 1198–1209.

Thomas, C.D. and Lennon, J.J. (1999). Birds extend their ranges northwards. *Nature*, 399, 213.

Tingley, M.W., Monahan, W.B., Bessinger, S.R., Moritz, C. (2009). Birds track their Grinnellian niche through a century of climate change. *Proceedings of the National Academy of Sciences of the USA*, 106, 19637–19643.

Townshend, D.J. (1985). Decisions for a lifetime: Establishment of spatial defence and movement patterns by juvenile Grey Plovers (*Pluvialis squatarola*). *Journal of Animal Ecology*, 54, 267–274.

Travis, J.M.J. (2003). Climate change and habitat destruction: A deadly anthropogenic cocktail. *Proceedings of the Royal Society of London Series B – Biological Sciences*, 270, 467–473.

UICN (2012). *Catégories et criteres de la liste rouge de l'UICN : version 3.1*, 2nd edition. CSE, Gland, Switzerland.

Vié, J.-C., Hilton-Taylor, C., Stuart, S.N.E. (2009). *Wildlife in a Changing World: An Analysis of the 2008 IUCN Red List of Threatened Species*. IUCN, Gland, Switzerland.

Viterbi, R., Cerrato, C., Bassano, B., Bionda, R., von Hardenberg, A., Provenzale, A., Bogliani, G. (2013). Patterns of biodiversity in the northwestern Italian Alps: A multitaxa approach. *Community Ecology*, 14, 18–30.

Wagner, B. and Melles, M. (2001). A Holocene seabird record from Raffles Sø sediments, East Greenland, in response to climatic and oceanic changes. *Boreas*, 30, 228–239.

Walther, G.-R., Beißner, S., Burga, C.A. (2005a). Trends in the upward shift of alpine plants. *Journal of Vegetation Science*, 16, 541–548.

Walther, G.-R., Berger, S., Sykes, M.T. (2005b). An ecological "footprint" of climate change. *Proceedings of the Royal Society B*, 272, 1427–1432.

West, J.M. and Salm, R. (2003). Resistance and resilience to coral bleaching: Implications for coral reef conservation and management. *Conservation Biology*, 17, 956–967.

Williams, J.J. and Newbold, T. (2019). Local climatic changes affect biodiversity responses to land use: A review. *Diversity and Distributions*, 26, 76–92.

Wormworth, J. and Sekercioglu, C.H. (2011). *Winged Sentinels: Birds and Climate Change*. Cambridge University Press, Cambridge.

Yeatman-Berthelot, D. and Jarry, G. (1995). *Nouvel atlas des oiseaux nicheurs de France*, Seof, Paris.

Zöckler, C. (2002). Declining Ruff *Philomachus pugnax* populations: A response to global warming? *Wader Study Group Bulletin*, 97, 19–29.

Zuckerberg, B., Wood, A., Porter, W.F. (2009). Poleward shifts in breeding bird distributions in New York State. *Global Change Biology*, 15, 1866–1883.

List of Authors

Damien ARVOR
CNRS
Rennes
France

Éric CANOBBIO
University of Paris 8 –
Vincennes – Saint-Denis
Paris
France

Alain CARIOU
Sorbonne University
Paris
France

Étienne COSSART
Jean Moulin University Lyon 3
France

Axel CREACH
Sorbonne University
Paris
France

Gilles DROGUE
University of Lorraine
Metz
France

Vincent DUBREUIL
University of Rennes 2
France

Beatriz FUNATSU
CNRS
Nantes
France

Emmanuèle GAUTIER
University of Paris 1
Panthéon-Sorbonne
France

Laurent GODET
CNRS
Nantes
France

Delphine GRAMOND
Sorbonne University
Paris
France

Renan LE ROUX
CIRAD
Montpellier
France

Neli DE MELLO-THÉRY
University of São Paulo
Brazil

Denis MERCIER
Sorbonne University
Paris
France

Vincent NÉDÉLEC
University of Rennes 2
France

Hervé QUÉNOL
CNRS
Rennes
France

Florian RAYMOND
University of Burgundy
Dijon
France

Albin ULLMANN
University of Burgundy
Dijon
France

Index

D

E

F

G

H

I

J, K, L

M

N

O, P

R

S

T

U, V

W, Y

Printed and bound by CPI Group (UK) Ltd, Croydon, CR0 4YY

13/09/2026

14961043-0007